METHODS OF COMPUTATION

METHODS OF COMPUTATION

The Linear Space Approach to Numerical Analysis

Jens A. Jensen

John H. Rowland
The University of Wyoming

Scott, Foresman and Company Glenview, Illinois
Dallas, Tex. Oakland, N.J. Palo Alto, Cal.
Tucker, Ga. Brighton, England

Library of Congress Catalog Card Number: 74-79034.

AMS 1970 Subject Classification 6501.

ISBN: 0-673-05394-6.

Printed in the United States of America.

Preface

This book provides a broad survey of the computational aspects of mathematics and is intended for use in an upper-division course enrolling students of engineering, mathematics, and the sciences. The text is written to present some of the flavor of modern functional analysis without losing sight of the practical aspects of computation. For the most part, the presentation is oriented toward techniques rather than individual problems. For example, iteration techniques are studied in general and then applied to a variety of problems such as non-linear equations, differential equations, integral equations, and matrix problems. I hope that this orientation will help to avoid the impression that numerical analysis is nothing but a disconnected bag of tricks.

Methods of Computation presupposes only that students have had an elementary course in calculus. If they are familiar with basic linear algebra and matrix operations, then the instructor may want to skip or move rapidly through the more elementary sections of Chapters 2 and 3. The section on Fortran programming can easily be skipped if it is not appropriate for a given class. Certain more advanced sections, marked by an asterisk preceding the section title, can also be omitted without loss of continuity.

At the University of Wyoming this text has been used for two one-semester courses on numerical methods. The first course covers Chapters 1–4 and emphasizes the computational aspects of algebra: machine arithmetic, matrix problems, and the solution of non-linear equations. The second course takes up Chapters 5–7 and stresses the computational aspects of analysis: interpolation and its applications to differentiation, integration, and the solution of differential equations. An introduction to approximation theory (Chapter 7) concludes the second semester's study. These courses can be taught independently at the expense of some duplication in Chapters 1 and 2, or by de-emphasizing the linear space aspects of Chapters 5–7.

Many examples and some of the exercises presented here involve rather simple calculations that can be done by hand. In some cases this approach does not fully illustrate the power of a technique to deal with difficult problems. However, I believe that it is better not to have students bog down in computational complexity in their first encounter with a numerical procedure. I would recommend that the student write a number of computer programs to test selected techniques on a variety of problems. (On a time sharing system my students write about six required programs per semester. They are permitted to use the computer for any of their homework.) The exercises involve some theoretical questions along with straightforward practice in computation. Answers are supplied for selected problems only, in order to encourage students to develop self-reliance by devising their own checks on the reasonableness of their solutions.

The inspiration and first draft for this book were provided by Jens Jensen, my colleague at the University of Wyoming. Unfortunately, he was not able to complete the project before his untimely death. I hope that the final product will be a worthy memorial to the courage, good humor, and devotion displayed by Jens during his lifetime.

I am indebted to David Anderson, Stanley Brown, Peter Perkins, Verne Varineau, and Yako Varol for helpful conversations concerning portions of the book. Peter Lax and Malcolm Harrison suggested a number of improvements, and Paul Berg read the entire manuscript and offered many useful comments. Eleanor Rowland cheerfully typed the manuscript and class notes (many at the eleventh hour), while Lareda Dolan provided secretarial assistance. To these people and to the many students who helped shape the book, I offer hearty thanks.

J. H. R.

Contents

CHAPTER 1

Fundamentals

1. INTRODUCTION

This book is concerned with practical methods of computation. It includes methods for the computation of integrals, derivatives, functional values, and solutions to algebraic, differential, and integral equations. These topics are frequently discussed within the framework of a linear space, which serves to unify many ideas that might at first glance appear unrelated.

The science of computation is a very old one. The early Greeks knew how to approximate the area enclosed by certain plane figures and were perhaps the first mathematicians concerned with numerical integration. Newton (1642–1727) developed a technique that is widely used today for solving polynomial and transcendental equations. In this book we will study procedures associated with the names of Maclaurin (1698–1746), Lagrange (1736–1813), and Gauss (1777–1855). Even though the subject is quite old, there is vigorous interest in numerical analysis today. This renewed interest was inspired primarily by the advent of the electronic computer in the 1940s.

Although we will be concerned with practical methods of computation, it is nonetheless important to know something of the theoretical aspects of mathematics. It frequently happens that a limit theorem suggests an approximation procedure, while the proof of the theorem contains the ideas needed for making an error estimate. This is not to say that all theoretical mathematics is crucial for computation—certain existence theorems, for example, give no clue for finding the solution to a problem. Furthermore, a thorough knowledge of theoretical mathematics does not necessarily imply a knowledge of the many pitfalls involved in obtaining numerical solutions to problems. For practical computation one needs to know something of both the theoretical and practical aspects of numerical analysis.

Frequently we want to know which is the best method of computation for a given type of problem. This usually presents a difficult decision because a method which works well in one situation might work very poorly in another. Following are some of the criteria used in comparing different computational procedures.

Accuracy. Usually a numerical method involves several approximations. A method is sought in which the approximations produce a reasonably small error.

Cost. If the accuracy is reasonable, the preferred method is the one which is least expensive in terms of time or money. The cost of a computation can be broken down into costs for analysis, programming, and calculating. *Analysis* is the process of learning possible techniques and deciding which one should be used. *Programming* is the process of writing out and checking detailed instructions which tell a person or machine how to carry out the calculations. The term *calculation* refers to the performance of the arithmetic and logical steps required to obtain the solution to the problem.

Miscellaneous. Other criteria such as simplicity, sensitivity to round-off errors, and generality are also used to compare computational methods. Of course, these factors may have a bearing on questions of cost and accuracy.

2. SOURCES OF ERRORS

The applied scientist is forced to make many approximations when working with mathematical models. Each approximation introduces an error which has some effect on the final result. If an error introduced at one stage of a computational process tends to grow rapidly in later stages, then the process is said to be *unstable.* The question of stability is quite important, especially for computations involving a large number of steps. In Chapters 3 and 6 we will study the propagation of error and techniques for avoiding instability. For the moment, let us list the principle sources of errors.

1. *Model Errors.* A mathematical model rarely describes a physical, economic, social, or biological system exactly. For example, air resistance is frequently neglected in the equation of motion for a falling object.

2. *Measurement Errors.* The parameter values for a mathematical model may come from measurements of time, length, money, etc. Normally these measurements are accurate only to within a finite number of decimal places.

3. *Round-Off Errors.* It is seldom practical to retain all of the digits for intermediate numbers occurring in a calculation. For example, try computing $(1.005)^{10}$ by direct multiplication without incurring any round-off error.

4. *Truncation Errors.* The term truncation error originally referred to errors caused by stopping (truncating) an infinite process after a finite number of steps. For example, the error produced by approximating

$$\sum_{j=0}^{\infty} \frac{1}{2^j j!} \quad \text{by} \quad \sum_{j=0}^{5} \frac{1}{2^j j!}$$

is called a truncation error. We will use this term to include any error that does not fall into one of the other categories.

5. *Mistakes.* A very frequent source of error is a mistake on the part of the person making the computation. Desk calculators and electronic computers can also make mistakes; but a person should not be too hasty in accusing the machine—unless he likes to eat his hat.

Model errors are important; but they are not really computational errors and therefore will not be discussed here. Truncation errors will be studied with the individual approximation methods. As a general rule for catching mistakes, one should be a bit suspicious of one's computations and step back and ask if the results are reasonable. Specific checks which will help to uncover mistakes will be discussed for some individual problems. Let us now turn our attention to the analysis of measurement and round-off errors.

3. ANALYSIS OF MEASUREMENT ERRORS

A simple but important principle in the analysis of errors is to give each error a name (symbol) and see how the error affects the end result. Errors are described in terms of error bounds and error estimates. The term *error bound* means an upper bound on the absolute value of the error. An *error estimate* is an approximation to the magnitude of the error—that is, a number which one has reason to believe is close to the absolute value of the error.

EXAMPLE 3.1. The radius r of a circle is measured to be 25.3 feet, correct to within a tenth of a foot. To find a bound on the error in the area calculated from this measured value, let $r_0 = r + \epsilon$, where r_0 is the true value, r the measured value, and ϵ the error, $|\epsilon| \leqslant 0.1$. Now the error E in the calculated area is given by $E = \pi r_0^2 - \pi r^2$. A bound on the error is given by the inequality

$$|E| = \pi |r_0 - r| \cdot |r_0 + r| = \pi |\epsilon| (r_0 + r) \leqslant 0.1\pi(2r + 0.1) \leqslant 16.0.$$

A convenient tool for obtaining error bounds and estimates is the differential approximation. Here we use the fact that if f is a differentiable function, then for small values of ϵ,

$$f(x+\epsilon)-f(x) \doteq \epsilon f'(x).$$

(The symbols $a \doteq b$ mean that a is close to b or that a is to be approximated by b.)

EXAMPLE 3.2. In the right triangle shown below, $c = 10$ feet and φ is measured with an error of ϵ.

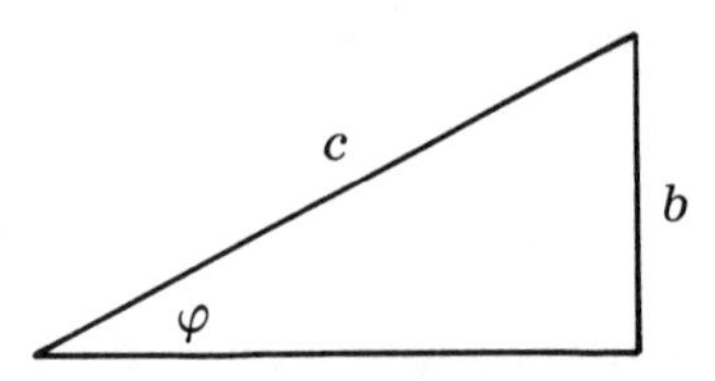

To find an error estimate in the side b as computed from the measured value of φ, we let $\varphi_0 = \varphi + \epsilon$ as in Example 3.1. Then

$$E = c\sin(\varphi+\epsilon) - c\sin(\varphi) \doteq c\epsilon\cos(\varphi) = 10\epsilon\cos(\varphi).$$

Here we have assumed that φ and ϵ are given in radian measure. (Why?)

PROBLEMS

3.1. The sides of a rectangle are measured to be 10.2 ± 0.05 and 20.0 ± 0.05. Find a bound on the error in the area calculated from these measurements.

3.2. In the right triangle shown in Example 3.2, b and φ are measured to within ϵ_1 feet and ϵ_2 radians, respectively.
   (a) Using the differential approximation for a function of two variables, find an error estimate for the value of c calculated from the measured values of b and φ.
   (b) For values of φ near 0 would it be better to measure b and calculate c or vice versa?

3.3. How would one go about finding a bound on the error produced by the differential approximation in Example 3.2? (This problem will be considered in detail in Chapter 5.)

4. MACHINE ARITHMETIC AND ROUND-OFF ERRORS

Before examining round-off errors, let us discuss some common ways of handling the round-off process in machine computations. First of all, a number can be truncated, which means discarding all digits in its decimal expansion after a prescribed position. For example, 2.083 is the number obtained by truncating 2.08375 to three decimal places. (For our purposes we will refer to the error caused by truncating a number as a round-off error rather than a truncation error as described in Section 2 above.) To round a number means the same as to truncate, except that the last retained digit is increased by one unit if the discarded part has a value greater than half the value of a unit in the last retained digit. For example, in rounding to three decimals, 2.08375 would be rounded to 2.084, while 2.08332 would be rounded to 2.083. One could round 2.0835 to either 2.083 or 2.084. A common practice in this case is to round so that the last retained digit is even. Thus 2.0835 would be rounded to 2.084, and 2.0825 would be rounded to 2.082. The rounding process has an advantage over truncating in that the errors have less tendency to accumulate. (See Problem 4.6.)

Let us assume that we are going to round numbers rather than truncate. There are two systems that are commonly used for representing numbers and performing arithmetic operations in machine computation. In the first method, referred to as *fixed-point arithmetic*, numbers are represented in the usual decimal system. Each number used as an input parameter and the result of each computation is rounded to a fixed number of decimal places. The accountant commonly uses this system when dealing with money to the nearest cent.

EXAMPLE 4.1. The sides of a rectangle are 5.022 ft and 10.0135 ft. The approximation to the area A calculated by fixed-point arithmetic to two decimal places is

$$A \doteq (5.02)(10.01) = 50.2502 \doteq 50.25.$$

In the second system, referred to as *floating-point arithmetic*, numbers are represented in a manner similar to the so-called scientific notation. That is, x is written in the form $x = r_0 \cdot 10^k$, where $0.1 \leqslant r_0 < 1$ and k is an integer. (We will call this the standard form of the number.) Then r_0 is rounded to a prescribed number of decimal places to obtain $x \doteq r \cdot 10^k$. The number x is then said to be rounded to n significant digits, where n is the number of decimal places carried in r. In floating-point arithmetic each input number and the result of each computation are rounded to a fixed number of significant digits.

EXAMPLE 4.2. Using floating-point arithmetic, the area A in Example 4.1 would be approximated to four significant digits as follows:

$$A \doteq (0.5022) \cdot 10^1 (0.1001) \cdot 10^2 = (0.05027022) \cdot 10^3$$
$$\doteq 0.5027 \cdot 10^2 = 50.27.$$

It should be mentioned that many electronic computers employ (internally) a binary or hexadecimal system for representing numbers.

In the binary system one writes $x = a_n a_{n-1} \cdots a_1 a_0 \, . \, b_1 b_2 \cdots$, where the a_i's and b_i's are 0's and 1's. It is understood that this is shorthand for the equation

$$x = a_n 2^n + a_{n-1} 2^{n-1} + \cdots + a_0 2^0 + b_1 2^{-1} + b_2 2^{-2} + \cdots.$$

For example, if $x = 12$ in decimal notation, then $x = 2^3 + 2^2$, and thus $x = 1100$ in the binary notation. This would be represented in a four-digit binary floating-point system as $x = (0.1100) \cdot 2^4$. The hexadecimal system uses a base of 16.

The theorem below gives information about the round-off error in floating-point operations. Since the initial round-off error upon entering a number into a machine behaves just like a measurement error, we will assume that the numbers under discussion have already been rounded to n significant digits.

THEOREM 4.1 [16]* *Let $x \odot y$ represent the floating-point product of x and y with n significant digits in the decimal system. Then*

$$x \odot y = xy(1 + \epsilon),$$

where $|\epsilon| \leqslant 5 \cdot 10^{-n}$.

Proof. We assume that x and y are positive since the sign does not affect the magnitude of the error. Let $x = r \cdot 10^k$ and $y = s \cdot 10^m$, where $0.1 \leqslant r < 1$ and $0.1 \leqslant s < 1$. Now

$$x \odot y = (rs + \eta) \cdot 10^{k+m} = xy + \eta \cdot 10^{k+m},$$

where η is the error incurred in rounding rs to n significant digits. If $rs \geqslant 0.1$, then rs is rounded to n decimal places; while if $0.01 \leqslant rs < 0.1$, then rs is rounded to $n + 1$ decimal places. Thus

$$|\eta| \leqslant 5 \cdot 10^{-(n+1)} \qquad \text{if } 0.1 \leqslant rs,$$

and

$$|\eta| \leqslant 5 \cdot 10^{-(n+2)} \qquad \text{if } 0.01 \leqslant rs < 0.1.$$

Let $\epsilon = \eta/(rs)$. Now if $0.1 \leqslant rs$, we have

$$|\epsilon| \leqslant \frac{5 \cdot 10^{-(n+1)}}{0.1} = 5 \cdot 10^{-n}.$$

Similar reasoning shows that the above inequality also holds when $0.01 \leqslant rs < 0.1$. The proof is completed by noting that

$$x \odot y = xy(1 + \epsilon).$$

The *relative error* is defined as the magnitude of the error divided by the magnitude of the true value. Thus Theorem 4.1 states that the

*Numbers in brackets refer to the bibliography at the end of the chapter.

relative error in floating-point multiplication is at most $5 \cdot 10^{-n}$. The same bound also applies to the relative error in addition, subtraction, and division. Similar results apply to binary or hexadecimal arithmetic [16].

A special warning should be issued for subtraction. If two numbers are nearly equal, then a previous rounding error could produce a large relative error in the difference. For example, suppose $a = 1.0114$, $b = 1.012$, and a is rounded to 1.011 before the subtraction. Then $b - a = 0.0006$, but the rounded result is 0.001. This gives a relative error of 67%. The difficulty illustrated here is referred to as *loss of significance.* In some cases it is possible to perform algebraic manipulations which avoid the loss of significance (see Problem 4.5 and Chapter 1 Froberg [6] for examples). In other situations, such as numerical differentiation, loss of significance is an inherent difficulty. Ashenhurst and Metropolis [1] have proposed a floating-point system, known as *significance arithmetic*, which would give some warning of significance loss.

When a fixed-point operation produces more than the allowable number of digits to the left of the decimal point, we say that an *overflow* has occurred. Assuming no overflow, there is no error produced by fixed-point addition or subtraction after the input rounding error. The error in fixed-point multiplication or division is bounded by $5 \cdot 10^{-(n+1)}$, where n is the number of digits carried to the right of the decimal point.

PROBLEMS

4.1. Let $[\cdot]$ and $[+]$ represent fixed-point multiplication and addition. Which of the following algebraic rules hold?
(a) $x \, [\cdot] \, y = y \, [\cdot] \, x$
(b) $(x \, [\cdot] \, y) \, [\cdot] \, z = x \, [\cdot] \, (y \, [\cdot] \, z)$
(c) $x \, [\cdot] \, (y \, [+] \, z) = (x \, [\cdot] \, y) \, [+] \, (x \, [\cdot] \, z)$

4.2. Is there a preferred order for approximating xyz by fixed-point multiplication? Explain.

4.3. Prove the analog of Theorem 4.1 for floating-point addition. What happens if the binary system is used?

4.4. Find an upper bound for the relative error in n factor floating-point multiplication. See [16].

4.5. Let $\odot, \oplus, \ominus$ represent floating-point multiplication, addition, and subtraction. Approximate $(1.0149)^2 - 1$ using floating-point operations with three significant digits by the following two methods. Compare the relative errors.
(a) $(1.00 \oplus 0.0149) \odot (1.00 \oplus 0.0149) \ominus 1.00$
(b) $(2.00) \odot (0.0149) \oplus (0.0149) \odot (0.0149)$
This illustrates the so-called loss of significance which can occur when subtracting two numbers that are almost equal.

4.6. (Optional problem for students of probability.) Suppose that 100 numbers are rounded to the nearest integer and then added. An upper bound on the error in the sum caused by round-off is 50.
   (a) What is the approximate probability that the magnitude of the error will not exceed 5?
   (b) How would things change if the numbers were truncated rather than rounded?

4.7. Prove that for each positive number x there is a unique number r_0 and integer k such that $0.1 \leqslant r_0 < 1$ and $x = r_0 \cdot 10^k$.

5. ALGORITHMS AND FLOW CHARTS

An *algorithm* is a complete set of instructions describing a step-by-step procedure for approximating the solution to a given type of problem. A convenient way to describe an algorithm is in terms of a *flow chart*. The flow chart is a diagram which shows the steps required by the algorithm, the order in which the steps should be carried out, and the logical decisions which affect the order of the steps.

EXAMPLE 5.1. The chart below represents an algorithm for computing the maximum of two real numbers A and B.

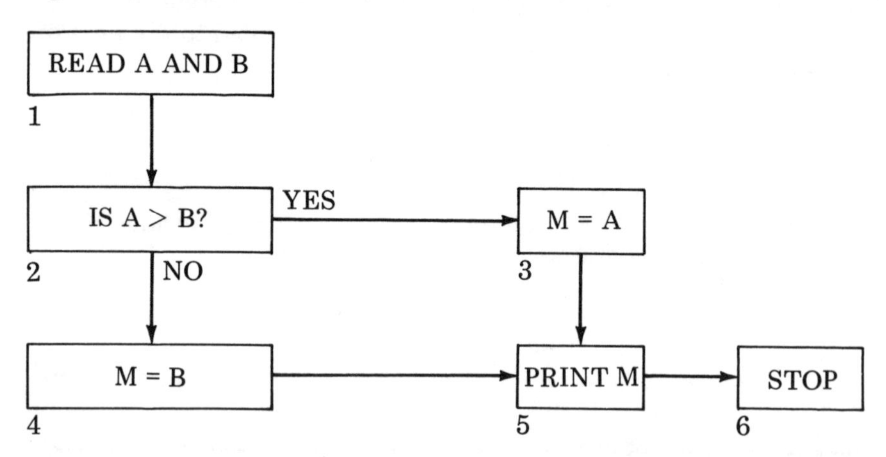

FIGURE 5.1. Flow Chart for Computing the Maximum of Two Numbers.

The first step is to read the numbers A and B (the term "read" refers to the process of getting the numbers into an electronic computer). The second step is to answer the question, Is A > B? If the answer is yes, proceed to step 3 and set M = A. Otherwise go to step 4 and set M = B. Then print the number M, which is the maximum of A and B.

EXAMPLE 5.2. The chart below illustrates an algorithm for approximating the factorial of an integer.

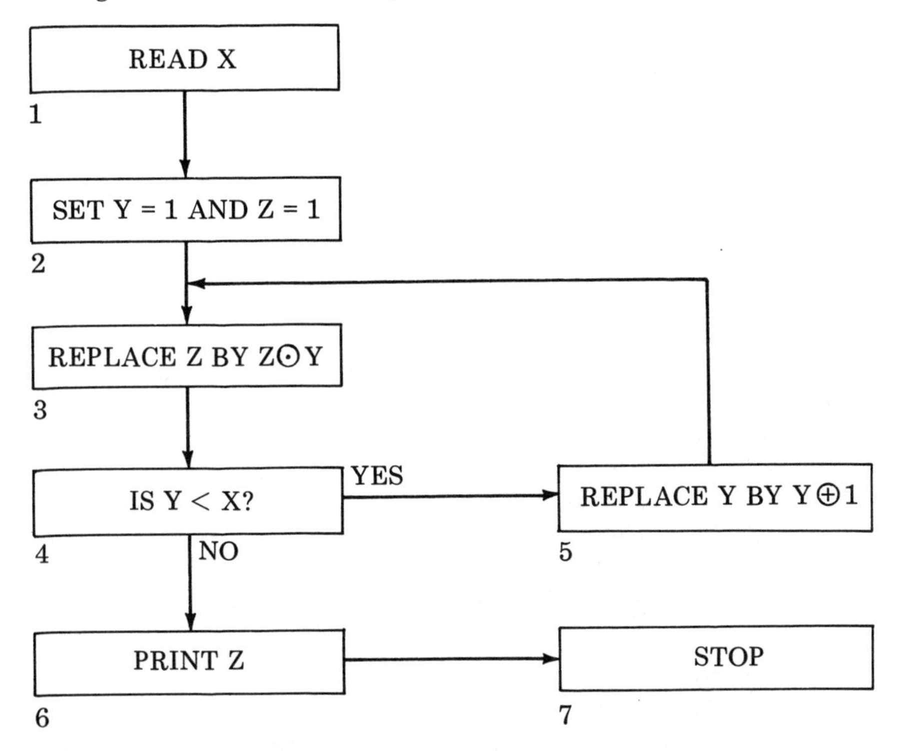

FIGURE 5.2. Flow Chart for Computing Factorials.

A circle around an operation represents a floating-point operation. It will be convenient here to think of X, Y, and Z as the labels on three bins, each capable of storing only one number. The first statement then means that the number whose factorial is to be computed is put in the bin labeled X. Then the number 1 is stored in the bins labeled Y and Z. In the third step, the floating-point product of the numbers stored in Y and Z is computed, and the result replaces the previous value in Z. The fourth step is to answer the question, Is the number stored in Y less than the number stored in X? If the answer is yes, we increment the number stored in Y by one and repeat the third and fourth steps. When the content of Y is equal to the content of X we stop cycling since the number stored in Z will then represent the desired factorial.

It is not necessary to distinguish between the bin labeled X and its content; the content of the bin is simply referred to as X. (Keep in mind that the value of X might change many times in the course of a calculation.)

The flow chart is a helpful device in planning a set of computations for an electronic computer or desk calculator, much as a wiring diagram is helpful in designing electric circuits. Many mistakes can be caught at this stage which will save time and money later on. Those interested in more details on flow charts may consult Brooks and Iverson [2], who propose a more compact form of flow chart, Leeds and Weinberg [10], or Davis [3].

PROBLEMS

5.1. Draw a flow chart for computing $(1 + r)^n$ for integer values of n and real values of r.

5.2. Draw a flow chart for determining the maximum of n real numbers.

5.3. An electric company charges 3¢/kwh for the first 100 kwh, 2¢/kwh for the next 50 kwh, and 1¢/kwh thereafter. Draw a flow chart for calculating the bills of this company.

5.4. In Example 5.2, suppose that for some reason (Can you give a plausible reason?) the number 1 was approximated by 0.999 in the fifth step. Find the approximation to 3! produced by the algorithm in this case. (In Section 6 we will see how to avoid this problem in a Fortran program.)

6. COMPUTERS AND PROGRAMS

For our purposes a computer is a machine which has the following features:

1. It has a memory, that is, a set of bins where it can store numbers.
2. It can read, write, add, subtract, multiply, divide, determine algebraic signs, branch (described below), and transfer data.
3. It can follow a sequence of properly coded instructions.

Let us consider a hypothetical computer which we will call the JAM 359. The JAM has 5000 storage locations, each of which can store one ten-digit decimal number along with two algebraic signs. (Actually, most computers store numbers in binary form, but the general idea is the same.) These locations are numbered from 0001 to 5000. The JAM can read data from a magnetic tape or from punched cards. It can write data onto magnetic tape or onto paper by means of a high-speed printer.

The instructions for the JAM consist of ten-digit numbers. The first two digits form a number called the *operation code*; the next four digits are called the *P address*; and the last four digits are called the *Q address*. For example, in the instruction 1010302567 the operation code is 10, the P address is 1030, and the Q address is 2567.

A sequence of instructions makes up what is called a *program*. The instructions for a program are usually stored in consecutive memory locations with each instruction occupying one location. During execution of the program, the computer ordinarily executes the first instruction and then goes to the next location for the second instruction. After executing the second instruction it moves on to the next location for the third instruction, and so on. An exception to this pattern occurs when the instruction is a branch or stop. A *branch* directs the computer to jump to some other location for its next instruction if certain conditions are satisfied. A *stop* instruction tells the computer to quit working on this program.

Let us now turn to the operations for the JAM computer.

Fixed-Point Operations. The operation codes for fixed-point addition, subtraction, multiplication, and division are 11, 12, 13, and 14, respectively. Let x and y be the numbers stored in the respective P and Q storage locations, and let M be the operation code. Then a ten-digit number of the form MPQ is an instruction for computing:

$$\begin{array}{ll} x\ [+]\ y & \text{if M} = 11, \\ x\ [-]\ y & \text{if M} = 12, \\ x\ [\cdot]\ y & \text{if M} = 13, \\ x\ [\div]\ y & \text{if M} = 14. \end{array}$$

The result of an arithmetic operation on the JAM is always stored in location 5000.

Floating-Point Operations. The operation code for a floating-point operation on the JAM is obtained by adding 10 to the corresponding fixed-point operation code. The conventions regarding the P and Q addresses are the same as those for fixed-point operations.

Branching Operations. An instruction of the form MPQ tells the JAM computer to go to the location Q (branch to Q) for the next instruction in the program if one of the following conditions is satisfied:

(1) M = 15 and the content of P is positive.
(2) M = 16 and the content of P is positive or zero.
(3) M = 17 and the content of P is negative.
(4) M = 18 and the content of P is negative or zero.
(5) M = 19 (regardless of the content of P).

Otherwise the computer obtains its next instruction from the location which immediately follows the branch instruction. For example, the instruction 15 1000 2000 located in memory cell 0300 would cause JAM to branch to 2000 if the content of 1000 is positive, or go to location 0301 for its next instruction if the content of 1000 is negative or zero. The branch operation enables a computer to make decisions.

Transfer Instruction. The instruction 20 PQ tells the computer to copy the content of location P into location Q (without erasing the content of P).

Read and Write Instructions. An instruction of the form 31 PQ means to read a ten-digit number and two signs and store it in location Q. The number is to be read from tape if P = 0001 and from a punched card if P = 0002. An instruction of the form 32 PQ means to write the content of location Q on tape if P = 0001 and on paper if P = 0002.

Stop. An instruction of the form 40 PQ causes the computer to stop.

A summary of the JAM operation codes and the time required for each operation is given below.

Operation Code	*Operation*	*Time in Microseconds*
11	fixed-point addition	10
12	fixed-point subtraction	10
13	fixed-point multiplication	20
14	fixed-point division	20
15	branch positive	2
16	branch positive or zero	2
17	branch negative	2
18	branch negative or zero	2
19	branch unconditionally	1
20	transfer	4
21	floating-point addition	20
22	floating-point subtraction	20
23	floating-point multiplication	40
24	floating-point division	40
31	read	50
32	write	50
40	stop	1

TABLE 6.1. Operation Codes and Times for the JAM 359.

Most computers have many more instructions than our hypothetical JAM 359. The description here is intended to give the reader only a general idea of machine instructions, and the discussion has been oversimplified to some extent for the sake of brevity. Those interested in the details of such instructions should consult manufacturers' manuals or one of the many books on computers and programming such as Davis [3] or Germain [7].

EXAMPLE 6.1. A program for computing factorials according to the algorithm given in Example 5.2 is shown in the table below. The symbols X, Y, and Z correspond to locations 1001, 1002, and 1003, respectively. The instructions for this program will be stored in locations 0001 to 0013. Some type of loading routine must be executed in order to load the instructions into these locations and branch to location 0001. The computer then executes the instruction in 0001 and continues to execute instructions in successive locations (unless a branch instruction commands otherwise) until it reaches a stop instruction.

Location	*Instruction*	*Explanation*
0001	31 0002 1001	Read a number from a card and store it in 1001 (READ X).
0002	31 0002 1004	Read the constant 1 into 1004.
0003	20 1004 1002	Transfer the content of 1004 into 1002 ($Y = 1$).
0004	20 1004 1003	Transfer the content of 1004 into 1003 ($Z = 1$).
0005	23 1002 1003	Multiply the content of 1002 and 1003 ($Y \odot Z$).
0006	20 5000 1003	Transfer above product to 1003 (Replace Z by $Y \odot Z$).
0007	22 1002 1001	Subtract content of 1001 from that of 1002 ($Y \ominus X$).
0008	16 5000 0012	Branch to 0012 if content of 5000 is positive or zero (Is $Y < X$?)
0009	21 1002 1004	Add content of 1002 and 1004 ($Y \oplus 1$).
0010	20 5000 1002	Transfer content of 5000 to 1002 (Replace Y by $Y \oplus 1$).
0011	19 0000 0005	Branch to 0005.
0012	32 0002 1003	Print content of 1003 (PRINT Z).
0013	40 0000 0000	STOP.

TABLE 6.2. Machine Language Program for Factorials.

This type of program is called a *machine language program* because we are giving the instructions to the machine in its native tongue. In the next section we will see that it is usually more convenient to communicate with the machine in a language closer to our own native tongue.

PROBLEMS

Write a machine language program for the JAM 359 corresponding to the flow chart given in:

6.1. Example 5.1.

6.2. Problem 5.1.

6.3. Problem 5.3.

6.4. Estimate the time required for the program in Problem 6.1. (See Table 6.1). How much will it cost at $200/hr?

6.5. Estimate the time and cost at $200/hr to compute 20! by the program in Example 6.1.

7. FORTRAN PROGRAMS

There are several disadvantages to writing programs in machine language. First of all, operation codes vary from computer to computer, so that a program written for one computer usually has to be completely rewritten if it is to be used on another. Furthermore, the programmer must learn a new vocabulary every time he switches machines. In a large program it is difficult to remember where the variables are stored; thus the programmer spends a lot of time looking up the addresses of the variables. There are a number of languages which alleviate these difficulties, among which the most common are Fortran [12], Basic [4] [13], Algol [15], and Cobol [11]. Following is a description of some of the fundamental features of the Fortran language.

The first step in Fortran programming is to write a program in the Fortran language. The program is then translated from the Fortran language into machine language by means of a *compiler program.* Each machine must of course have its own compiler program; but the compiler is independent of the particular program being processed. In this way a Fortran program can be run on different computers with only minor modifications to take care of the idiosyncrasies of the individual compiler programs.

There are several versions of Fortran. The instructions which are described below will all work with the version known as Fortran IV, although most of them will also work with the other versions. We are not attempting to give here all the rules and regulations of Fortran. More complete discussions can be found in McCracken [12], Kuo [9], and in manuals put out by computer manufacturers.

Allocation of Space. For ordinary variables space is automatically allocated in Fortran by mentioning the variable in an instruction.

For example, the instruction Z = A + B serves as an instruction to assign storage locations to the variables A, B, and Z, even though it is primarily an instruction to add the contents of the locations assigned to A and B and store the result in the location assigned to Z.

The variable names may consist of combinations of up to six letters and numbers, the first of which must be a letter. The first letter determines whether or not the values of the variable are to be handled by fixed-point or floating-point arithmetic. If the first letter is I, J, K, L, M, or N, the values are stored in the computer as integers and are handled by fixed-point arithmetic with no digits after the decimal point. Otherwise, the values of the variable are stored in the computer in the standard form $r \cdot 10^k$, with $0.1 \leqslant r < 1$, and handled with floating-point arithmetic. Normally, about eight significant digits are carried. These conventions can be overruled by methods described in Chapter 3 of McCracken [12].

EXAMPLE 7.1.

Fixed-Point Variables	*Floating-Point Variables*
J	A
JA	SPEED
J2	VEL
MAX	VMAX

There is another type of variable called a *dimensioned* or *subscripted variable.* Suppose we want to work with three variables called x_1, x_2, and x_3. Fortran allows neither lower case variables nor subscripts, so we could name these variables X(1), X(2), and X(3). Space for these variables would be reserved by the statement

```
DIMENSION X(3)
```

which must occur before the first use of one of these variables. It is also possible to use double indices on dimensioned variables.

EXAMPLE 7.2. The statement

```
DIMENSION X(2), Z(2,2)
```

reserves space for the variables X(1), X(2), Z(1,1), Z(1,2), Z(2,1), and Z(2,2).

One could also use X1, X2, . . . , X10 to indicate the variables $x_1, x_2, \ldots, x_{10}$. The advantage of using the dimensioned variable is shown by considering the problem of reading in initial values of x_1, $x_2, \ldots, x_{10}$. With the dimensioned variable this can be handled by a read statement of the form

```
READ 1, (X(N), N = 1, 10)
```

whereas the use of X1, X2, . . . , X10 would require the statement

READ 1, X1, X2, X3, X4, X5, X6, X7, X8, X9, X10.

(Try, for example, using $x_1, x_2, \ldots, x_{100}$, keeping in mind that Fortran does not allow the use of ". . .".)

Input and Output Statements. The command READ tells the computer to accept input data from a card reader (with some compilers it means instead to accept input data from a particular magnetic tape unit). The general form of the instruction is:

READ N, A, B, C

where N is the number of a format statement, and A, B, and C are variables whose values are to be obtained from punched cards. The format statement, which will be explained below, tells the computer which columns of the punched card should be used for the values of A, B, and C. The number of variables in the list following the format number is limited only by the storage capacity of the machine.

EXAMPLE 7.3. The statement

READ 2, A, B, C, D

means to read four numbers from one or more punched cards and store the numbers in the locations associated with A, B, C, and D. The number 2 signifies that the computer should look at format statement 2 in order to decide how many cards and what columns should be read.

A variation of the READ statement is

READ 3, Z, (X(N), N = 1,100).

This statement tells the computer to read 101 numbers according to format statement 3 and to store the numbers in the locations associated with Z, X(1), X(2), . . . , X(100). Similar statements can be used for two-dimensional arrays.

The command PRINT is strictly analogous to the READ command. It instructs the computer to put data on paper by means of a high-speed printer.

Format Statements. The format statement specifies the location of variables on a punched card, tape, or paper. First, let us look at the so-called I format which is used for integer variables. The specification I followed by a number *n* indicates that the corresponding variable occupies *n* columns. For example, the statements

```
      READ 5, J, K, L
    5 FORMAT (I2, I3, I5)
```

indicate that the variable J will be found in columns 1 and 2 of a punched card; K will be in columns 3, 4, and 5; and L will be in columns 6 through 10. On data cards a blank is interpreted as a zero, so that a 1 in column 1 followed by a blank in column 2 would be interpreted as J = 10, while a blank in column 1 followed by a 1 in column 2 would be interpreted as J = 1.

One way of handling a floating-point variable is to use what is called an F format. A specification of the form F*n*.*m* indicates that the variable is punched in n columns of the card with m digits to the right of an understood decimal point. For example, the statements

```
      READ 3, X, Y
    3 FORMAT (F5.2, F8.3)
```

indicate that X will be punched in the first 5 columns of the card, and the decimal point is understood to be between the digits in the third and fourth columns. The value for Y is punched in columns 6 to 13 with a decimal point understood to be just after column 10. If a decimal point is actually punched in the card, it overrules the position called for in the format statement.

Another format used for floating-point variables is called the E format. This is used when the programmer wants to write a number in the form $r \cdot 10^k$. In a program or on a data card the number $r \cdot 10^k$ is abbreviated *r*E*k*. For example, $10.57 \cdot 10^3$ is abbreviated as 10.57E3, while $1.36 \cdot 10^{-5}$ is abbreviated as 1.36E–5. A format specification of the form E*n*.*m* indicates that the variable in question occupies n columns and has an understood decimal point m places to the left of the letter E which signifies the start of the exponent. Again, a decimal point punched in the card overrules that called for in the format statement.

EXAMPLE 7.4. Consider the following statements:

```
      READ 7, X, Y, J
    7 FORMAT (F5.2, E7.3, I3)
```

Table 7.1 gives several ways in which the first 15 columns of a card might be punched and the corresponding values of X, Y, and J which would be stored in the computer memory.

The format used with the PRINT is just like that for the READ, except that it should contain an initial specification for controlling the spacing on the printer. This is accomplished by the specification 1H*n*, which tells the computer to single-space if n is a blank, double-space if $n = 0$, go to the top of the next page if $n = 1$, and give no

Values Punched on Card			*Values Stored in Computer*		
Col. 1-5 (X)	Col. 6–12 (Y)	Col. 13–15 (J)	X	Y	J
98341	23745E4	bb3*	983.41	$23.745 \cdot 10^4$	3
6.23b	2.37bE1	b3b	6.23	$2.37 \cdot 10^1$	30
.6bbb	−2.3E−1	3bb	.6	$-2.3 \cdot 10^{-1}$	300
bb732	2137E5b	bbb	7.32	$2.137 \cdot 10^{50}$	0

TABLE 7.1. Illustration of Format Conventions

space if n is the symbol +. The H format can also be used for printing out messages by specifying nH followed by the message, where n is the number of characters in the message. A message can be combined with the printer spacing control by writing mH followed by the spacing control and message, where m is the total number of characters in both the spacing control and the message.

EXAMPLE 7.5. Consider the following statements:

```
   PRINT 60, J, A, B
60 FORMAT (1H ,I5,2F10.3)
   PRINT 70, K
70 FORMAT (19H1LOOP A IS COMPLETE, I5)
```

The first statement tells the computer to single-space and then print J in the first five columns of the paper, A in the next ten columns with three digits after the decimal point, and B in the next ten columns with three digits after the decimal point. The symbols 2F10.3 are short for F10.3, F10.3.

The second print statement instructs the computer to move to the top of the next page, then print the message LOOP A IS COMPLETE followed by the value of K in the five columns after the message.

Arithmetic Instructions. Arithmetic instructions in Fortran are very similar to ordinary algebraic expressions. The symbols used for arithmetic operations are:

+ addition
— subtraction
* multiplication
/ division
** exponentiation

*The b stands for *blank*, that is, no punch in that column.

An arithmetic instruction is given by writing a statement of the general form

$$A = F(A, B, C, \ldots)$$

where F(A, B, C, . . .) is an expression involving a number of variables (possibly including A). This instruction tells the computer to combine the contents of the locations assigned to these variables according to the arithmetic expression and to store the result in the location assigned to A. For example, the statement

$$Z = X + Y$$

instructs the computer to add the contents of the X and Y locations and to store the result in the Z location. A statement of the form

$$A = A + B$$

should be thought of in terms of the equation

$$\text{New content of A} = \text{old content of A} + \text{content of B.}$$

Note that this statement does not imply that the value of B is zero, as it would in algebra.

With the exception of integer-valued exponents, the variables on the right-hand side of an arithmetic statement must all be of the same type—for example, all floating-point variables or all integer variables. Thus the statement

$$Z = A + N$$

would not be accepted by the compiler because A is a floating-point variable and N is an integer variable.* Constants must have a decimal point in a floating-point expression, and they must not have a decimal point in a fixed-point expression. Division of integer variables will yield the value of the quotient truncated to an integer. If the right-hand side of an instruction is a floating-point expression and the left-hand side is an integer variable, then the final result of the arithmetic operations is truncated to an integer. If the right-hand side involves integer variables and the left-hand side is a floating-point variable, then the final result is converted to floating-point form. In this last step there is the possibility of a small round-off error with a binary machine. For this reason it is better to use integer variables for such things as counting the number of times a program has gone through a loop (see Problem 5.4).

EXAMPLE 7.6. The following table gives some algebraic equations along with correct and incorrect Fortran instructions corresponding to those equations.

*Some recent compilers do permit mixed-mode expressions [12].

Algebraic Equation	*Correct Instruction*	*Incorrect Instruction*
$z = ab$	Z = A * B	Z = AB
$y = ax + b$	Y = A * X + B	Y = AX + B
$y = x^6$	Y = X ** 6	
$y = x + 1$	Y = X + 1	Y = X + 1
$m = n + 1$	M = N + 1	M = N + 1

TABLE 7.2. Sample Arithmetic Instructions.

Values of Common Functions. Fortran systems provide for the evaluation of certain commonly occurring functions by means of arithmetic instructions. The exact list of available functions and corresponding abbreviations depends on the particular Fortran system, but these would normally include the square root, exponential, natural logarithm, and some trigonometric functions. For example, the expression $e^{2x} \cos y + \sqrt{x} \sin y$ would be evaluated by an instruction such as

```
Z = EXP(2.*X)*COS(Y) + SQRT(X)*SIN(Y).
```

Branch Instructions. We will consider two types of branch instructions. The first instruction has the general form

GO TO n

where n is a statement number. This instructs the computer to go to statement number n for its next instruction. The second instruction has the general form

IF (X) j, k, m

where j, k, and m are statement numbers. This tells the computer to go to statement number j for its next instruction if the value of X is negative; to go to statement number k if the value of X is zero; and to go to statement number m if the value of X is positive. It is permissible to use an arithmetic expression in place of X in the IF statement.

EXAMPLE 7.7. A Fortran program for computing the maximum of two numbers according to the flow chart in Figure 5.1 is given below. We will use ZM in place of M in order to have a floating-point variable for the maximum.

```
   READ 1,A,B
1  FORMAT (2F10.0)
   IF (A—B)3,3,2
2  ZM=A
   GO TO 4
```

```
    3  ZM=B
    4  PRINT 5,ZM
    5  FORMAT (1H ,F10.4)
       STOP
```

The Loop Instruction. Loops occur quite frequently, and it is convenient to have an easy way to set them up. A statement of the form

DO *m* J=1,N

where *m* is a statement number, sets up a loop. This statement instructs the computer to set J=1, then execute all instructions between the DO instruction and statement number *m* (including statement *m*); set J=2 and execute again all the statements between the DO and statement *m*, . . .; set J=N and execute all these statements; and finally execute the first statement after statement *m*. The variable N can be replaced by a constant in the DO instruction.

EXAMPLE 7.8. The following is a program for computing N!. The algorithm given in Figure 5.2 has been slightly modified for this program.

```
       READ 1,N
    1  FORMAT (I10)
       Z=1
       DO 2 J=1,N
       Y=J
    2  Z=Z*Y
       PRINT 3,N,Z
    3  FORMAT (1H ,I10,E15.7)
       STOP
```

Punching the Program on Cards. The Fortran program is put on punched cards with each instruction starting on a new card. The instruction is punched in columns 7–72 and may be continued by punching anything but a blank or zero in column 6 of the next card. Except for an H format, blanks are ignored in columns 7–72 of a statement card. Columns 73–80 are ignored in statement cards (but not on data cards). A statement number, which is not required unless the statement is referenced by another part of the program, is punched in columns 1–5. C punched in column 1 tells the compiler to ignore the whole card—the card is just a comment reminding the programmer what the program is doing at that point. A card with END in columns 7–9 must be placed at the end of the list of Fortran instructions. The data cards corresponding to the various READ instructions are placed after the END card in the order in which they are to be read.

Most computer installations require that certain cards known as *header cards* precede the Fortran instructions. Typically these will include a card giving the programmer's name and account number followed by a card telling the machine that the program is written in Fortran. There may also be certain cards (for example, a card punched with the word DATA) required between the END card and the first data card.

PROBLEMS

Write Fortran programs to carry out the following tasks:

7.1. Compute $(1 + r)^n$ for a given real number r and integer n (see Problem 5.1).

7.2. Compute the electric bills described in Problem 5.3.

7.3. Approximate $e^{1/2}$ by using five terms in the Maclaurin series for the exponential function. Compare your result with that obtained from the instruction Z = EXP(0.5).

7.4. Consider the dimension statement:

DIMENSION X(2), Y(4), Z(2).

Computer manuals sometimes advise not to use a 0 or 5 for an index on Y. Write a program to determine what X or Z locations are assigned to Y(0) and Y(5). (One might expect Y(0) and X(2) to share the same location.)

REFERENCES

1. R. L. Ashenhurst and N. Metropolis, "Unnormalized floating-point arithmetic," *Journal of the Association for Computing Machinery*, 6 (1959), 415–428.
2. F. P. Brooks, Jr., and Kenneth E. Iverson, *Automatic Data Processing*, Wiley, 1963.
3. G. B. Davis, *An Introduction to Electronic Computers*, McGraw-Hill, 1965.
4. R. C. Dorf, *Introduction to Computers and Computer Science*, Boyd and Fraser, 1972.
5. A. Forsythe, T. A. Keenan, E. I. Organick, and W. Stenberg, *Computer Science, A First Course*, Wiley, 1969.
6. C-E. Froberg, *Introduction to Numerical Analysis*, Addison-Wesley, 1969.
7. C. B. Germain, *Programming the IBM 360*, Prentice-Hall, 1967.
8. J. Jeenel, *Programming for Digital Computers*, McGraw-Hill, 1959.
9. S. S. Kuo, *Numerical Methods and Computers*, Addison-Wesley, 1965.

10. H. D. Leeds and G. M. Weinberg, *Computer Programming Fundamentals*, McGraw-Hill, 1961.
11. D. D. McCracken, *A Guide to COBOL Programming*, Wiley, 1965.
12. D. D. McCracken, *A Guide to FORTRAN IV Programming*, Wiley, 1965.
13. P. W. Murrill and C. W. Smith, *BASIC Programming*, Intext, 1971.
14. A. Ralston, *Introduction to Programming and Computer Science*, McGraw-Hill, 1971.
15. H. Rutishauser, *Description of ALGOL 60*, Springer-Verlag, 1967.
16. J. H. Wilkinson, "Error analysis of floating-point computation," *Numerische Mathematik*, 2 (1960), 319–340.
17. H. S. Wilf, *Programming for a Digital Computer in the Fortran Language*, Addison-Wesley, 1969.

CHAPTER 2

Linear Spaces

1. INTRODUCTION

Many problems involving approximations can be conveniently formulated in terms of an abstract mathematical structure known as a *linear space*. The linear space concept provides a uniform terminology and facilitates recognition of certain analogies that might otherwise be obscure. In this section we will state the axioms and discuss some of the terminology associated with linear spaces.

Let R be the set of real numbers and let X be a set of objects (in many of our applications X will be a set of functions). Let there be defined an additive operation + which associates a unique element $x + y \in X$ with each pair (x,y) of elements from X. Furthermore, suppose there is a multiplicative operation which associates a unique element $\alpha x \in X$ with each pair (α,x) when $\alpha \in R$ and $x \in X$. We will then call X a real linear (vector) space if and only if the following conditions hold whenever x, y, and $z \in X$, α and $\beta \in R$.

(i) $x + y = y + x$
(ii) $x + (y + z) = (x + y) + z$
(iii) There exists an element $\emptyset \in X$ such that $x + \emptyset = x$ for all x. (This element is necessarily unique—see Problem 1.7.)
(iv) Each element x has an additive inverse $-x$ such that $x + (-x) = \emptyset$. (The notation $x - y$ will be short for $x + (-y)$. The additive inverse is also unique—see Problem 1.7.)
(v) $\alpha(\beta x) = (\alpha\beta)x$
(vi) $\alpha(x + y) = \alpha x + \alpha y$
(vii) $(\alpha + \beta)x = \alpha x + \beta x$
(viii) $1(x) = x$

In algebraic terms axioms (i)–(iv) say that X forms an Abelian group under addition. A complex linear space is defined similarly by substituting the set of complex numbers K for the set R. The elements of a linear space are frequently referred to as *vectors*.

EXAMPLE 1.1. Let $C[a,b]$ represent the collection of all continuous real-valued functions defined on the closed interval $[a,b]$, where a and

b are fixed real numbers with $a < b$. For f and $g \in C[a,b]$ and $\alpha \in R$, we define $f + g$ and αf in the usual way by the equations

$$(f + g)(t) = f(t) + g(t),$$

$$(\alpha f)(t) = \alpha f(t)$$

for all $t \in [a,b]$. Note that $f + g$ and $\alpha f \in C[a,b]$ because finite sums and multiples of continuous functions are again continuous. It is easy to show that axioms (i)–(viii) hold by using the corresponding properties of real numbers. Consider, for example, axiom (vii). For each $t \in [a,b]$ we have

$$\begin{aligned}[(\alpha + \beta)f](t) &= (\alpha + \beta)f(t) \\ &= \alpha f(t) + \beta f(t) \\ &= (\alpha f + \beta f)(t).\end{aligned}$$

This follows from the distributive law for real numbers. Since $(\alpha + \beta)f$ and $\alpha f + \beta f$ have the same value for each $t \in [a,b]$, we must have $(\alpha + \beta)f = \alpha f + \beta f$. (In discussions concerning $C[a,b]$, two functions are considered to be equal if their values coincide on $[a,b]$—regardless of what might happen outside this interval.)

EXAMPLE 1.2. An ordered n-tuple is a finite sequence of objects $(x_1, x_2, \ldots, x_n)$. If $x = (x_1, x_2, \ldots, x_n)$ is an n-tuple, then x_1 is called the first coordinate of x; x_2 is the second coordinate, and so on. Euclidean n-space, denoted by E_n, is the collection of all ordered n-tuples of real numbers. Two members of E_n are considered to be equal iff* the corresponding coordinates are equal. If $x = (x_1, x_2, \ldots, x_n)$, $y = (y_1, y_2, \ldots, y_n) \in E_n$ and $\alpha \in R$, then addition and multiplication are defined by

$$x + y = (x_1 + y_1, x_2 + y_2, \ldots, x_n + y_n),$$

$$\alpha x = (\alpha x_1, \alpha x_2, \ldots, \alpha x_n).$$

With these definitions it is easy to verify that E_n is a linear space. For most practical purposes the elements of E_n are the same as the vectors used in discussions of mechanics, electricity and magnetism, etc.

EXAMPLE 1.3. Complex Euclidean n-space, denoted by EK_n, is the set of all ordered n-tuples of complex numbers. Addition and multiplication are defined as in Example 1.2; of course α can now be any complex number. It is also easy to show that EK_n is a linear space. It should be noted that the term "Euclidean" used in the above context would normally imply that an inner product is defined on the space. This will be discussed in Section 7.

*The abbreviation *iff* means *if and only if*.

The usual convention is that if $x \in E_n$ or EK_n, then $x_1, x_2, \ldots, x_n$ represent the coordinates of x. However, we frequently consider a number of vectors, in which case it is convenient to subscript the vectors. In this situation we would use double subscripts to represent the coordinates. For example, if x_1 and $x_2 \in E_n$, then we would write

$$x_1 = (x_{11}, x_{12}, \ldots, x_{1n}),$$

$$x_2 = (x_{21}, x_{22}, \ldots, x_{2n}).$$

Later on we will write vectors in column form. For vectors x_1 and x_2 in column form we would interchange the order of the double subscripts and write

$$x_1 = \begin{bmatrix} x_{11} \\ x_{21} \\ \cdot \\ \cdot \\ \cdot \\ x_{n1} \end{bmatrix}, \quad x_2 = \begin{bmatrix} x_{12} \\ x_{22} \\ \cdot \\ \cdot \\ \cdot \\ x_{n2} \end{bmatrix}.$$

Some of the basic linear space terminology is explained in the paragraphs below. In this discussion X will denote a linear space and F the number system under consideration (either R or K).

DEFINITION 1.1. *The vector y is said to be a linear combination of the vectors $x_1, x_2, \ldots, x_n$ iff there exist numbers $\alpha_1, \alpha_2, \ldots, \alpha_n$ such that*

$$y = \alpha_1 x_1 + \alpha_2 x_2 + \ldots + \alpha_n x_n.$$

EXAMPLE 1.4. In E_3, let $x_1 = (2,0,0)$, $x_2 = (0,4,0)$, $w = (6,8,0)$, $y = (1,3,0)$, and $z = (1,2,6)$. Then w and y are linear combinations of x_1 and x_2 since

$$w = 3x_1 + 2x_2,$$

$$y = \tfrac{1}{2} x_1 + \tfrac{3}{4} x_2.$$

On the other hand, z is not a linear combination of x_1 and x_2 because it is impossible to obtain a non-zero third coordinate by adding multiples of x_1 and x_2.

DEFINITION 1.2. *Let A be a subset of X. The set of all finite linear combinations of members of A is called the span of A. In the language of set theory,*

$$\text{span } A = \{z\colon z = \alpha_1 x_1 + \ldots + \alpha_n x_n \textit{ for some combination } x_1, \ldots, x_n \in A \textit{ and } \alpha_1, \ldots, \alpha_n \in F\}.$$

EXAMPLE 1.5. Let x_1 and x_2 be the same as in Example 1.4. Then span $\{x_1, x_2\}$ is the set of all vectors in E_3 which have a third coordinate of zero.

In some cases a subset of a linear space is itself a linear space. These subsets are called *linear subspaces.*

DEFINITION 1.3. *Let A be a non-empty subset of X. Then A is said to be a linear subspace of X iff*

$$\alpha x + \beta y \in A$$

whenever $x,y \in A$ *and* $\alpha,\beta \in F$.

Problem 1.11 will show that a linear subspace is again a linear space. Note, in particular, that every subspace contains the zero element.

The subset $A \subset X$ is said to be *closed under addition* iff the sum of two elements of A is always a member of A. It is said to be *closed under scalar multiplication* iff $\alpha x \in A$ whenever $x \in A$ and $\alpha \in F$. It is a straightforward matter to show that a non-empty set A is a linear subspace iff it is closed under addition and scalar multiplication (Problem 1.12).

EXAMPLE 1.6. Let $\mathcal{P}_n$ be the set of all polynomials defined on $[0,1]$ whose degree does not exceed n. Then $\mathcal{P}_n$ is a linear subspace of $C[0,1]$.

The notion of dimension is very important in the theory of linear equations. First, let us distinguish between finite and infinite dimensional spaces.

DEFINITION 1.4. *The linear space X is said to be finite dimensional if it has a finite subset A such that span A = X. Otherwise the space is said to be infinite dimensional.*

The dimension of a finite dimensional space can be defined as follows:

DEFINITION 1.5. *The dimension of X, denoted by dim X, is the smallest number of vectors which span X. The dimension of the trivial linear space* $\{\emptyset\}$ *is defined as zero.*

The same definition can also be applied to infinite dimensional spaces. However, in this text we will not assign a specific dimension to these spaces because that would involve concepts that lie beyond the scope of this course.

The set of real numbers R is a good example of a one-dimensional linear space, and the Euclidean space E_2 is a two-dimensional space (see Problem 1.13). We will show in Section 4 that Euclidean n-space has dimension n; at this point it should be clear that the dimension of E_n is at most n since the vectors $e_1 = (1, 0, \ldots, 0)$, $e_2 = (0, 1, 0, \ldots, 0)$, $\ldots, e_n = (0, 0, \ldots, 1)$ span the space.

PROBLEMS

1.1. Sketch the graph of two functions f and g belonging to the linear space $C[a,b]$. Then sketch the graph of:
(a) $f + g$
(b) $f - g$
(c) $\emptyset$
(d) $3f$

1.2. Let $x = (4,7,5,9)$ and $y = (2,1,3,6)$ be members of E_4. Compute:
(a) $x + y$
(b) $x - y$
(c) $8x$
(d) $5x + 10y$

1.3. Prove that axioms (i), (ii), (iv), and (vi) hold for E_n.

1.4. Is E_n a subset of EK_n? Explain.

1.5. In E_5, solve the following equations for x:
(a) $x + (1,2,3,6,2) = \emptyset$
(b) $x + (2,0,3,5,1) = (1,7,2,3,5)$
(c) $x + (1,1,1,1,1) = 2x + 5(0,1,0,0,0)$

1.6. Write a computer program for computing the sum of two vectors in E_n.

1.7. Let X be a space which satisfies axioms (i)–(iii). Show that there can be at most one zero element; that is, an element λ such that $x + \lambda = x$ for all $x \in X$. Also show that in a space X which satisfies axioms (i)–(iv), the additive inverse is unique for each $x \in X$.

1.8. Let X be a linear space. Show that for each $x \in X$ and $\alpha \in R$:
(a) $0x = \emptyset$
(b) $(-1)x = -x$
(c) $\alpha\emptyset = \emptyset$

1.9. At first glance, it may appear that the proof given to show that axiom (vii) holds in Example 1.1 is not valid because we assume the fact which we are trying to prove. Explain why this is not the case.

1.10. Let $\varphi_0(x) = 1, \varphi_1(x) = x, \varphi_2(x) = x^2, \varphi_3(x) = x^3$, and $f(x) = 5x^2 + 2x + 6$.
   (a) Is f a linear combination of φ_0, φ_1, and φ_2? Explain.
   (b) Is φ_3 a linear combination of φ_0, φ_1, and φ_2? Explain.
   (c) Describe the span of $\{\varphi_0, \varphi_1, \varphi_2, \varphi_3\}$.

1.11. Explain why a linear subspace is a linear space.

1.12. Let A be a non-empty subset of a linear space X. Prove the following:
   (a) A is a linear subspace iff A is closed under addition and scalar multiplication.
   (b) A is a linear subspace iff A = span A.
   (c) The intersection of all linear subspaces which contain A is equal to span A.

1.13. Show that the set of real numbers R has dimension 1 while the space E_2 has dimension 2.

2. LINEAR TRANSFORMATIONS

Let us first discuss the notion of a transformation. A transformation T from a set X into a set Y is a function which associates an element $T(x) \in Y$ with each element $x \in X$. Technically, the word "transformation" is synonymous with the word "function." In actual practice there is a tendency to use the word "function" when X and Y are sets of numbers, and to use the word "transformation" when X or Y is a set of functions or vectors. For example, the equation

$$T(f) = \int_0^1 f(x)dx$$

defines a transformation which associates the number $\int_0^1 f(x)dx$ with each continuous real-valued function defined on the interval $[0,1]$. Similarly, the equation

$$T(f) = f'$$

defines a transformation T which associates the function f' with each differentiable real-valued function f.

In the study of calculus it is shown that the derivative of the sum of two functions is the sum of the derivatives. Also, the derivative of a constant multiple of a function is the constant multiple of the derivative. Similar properties hold for integrals. A transformation of this type is called a *linear transformation.*

DEFINITION 2.1. *Let X and Y be linear spaces, F the number system under consideration (either R or K), and let T be a transformation mapping X into Y. Then T is said to be a linear transformation iff*

$$T(\alpha x + \beta z) = \alpha T(x) + \beta T(z) \tag{2.1}$$

whenever $x,z \in X$ and $\alpha,\beta \in F$.

If we choose $\alpha = \beta = 1$ in (2.1), we obtain the equation

$$T(x + z) = T(x) + T(z). \tag{2.2}$$

If we choose $\beta = 0$, then (2.1) implies that

$$T(\alpha x) = \alpha T(x). \tag{2.3}$$

A transformation that satisfies (2.2) for all x and $z \in X$ is said to be *additive*, while one that satisfies (2.3) for all $x \in X$ and $\alpha \in F$ is said to be *homogeneous*. It is easy to see that (2.2) and (2.3) imply (2.1); thus a transformation T is linear if and only if it is both additive and homogeneous.

It is very important to observe the difference between the left and right sides of an equation such as (2.2). The expression on the left is evaluated by first adding x and z, and then sending the sum through the transformation T. The expression on the right is evaluated by first sending x and z through T, and then adding the results. Example 2.3 shows that these two operations do not always produce the same result.

EXAMPLE 2.1. Let $C'[0,1]$ be the set of real-valued functions which have continuous derivatives on $[0,1]$. For each $f \in C'[0,1]$, let

$$T(f) = f'.$$

Then T is a linear transformation from $C'[0,1]$ into $C[0,1]$ since

$$T(\alpha f + \beta g) = (\alpha f + \beta g)' = \alpha f' + \beta g' = \alpha T(f) + \beta T(g).$$

A special name is used for transformations which map into the number system F.

DEFINITION 2.2. *A transformation T from a linear space X into the number system F is called a functional. If T is also linear, then T is called a linear functional or linear form.*

EXAMPLE 2.2. For each $f \in C[0,1]$, let $T(f)$ be defined by

$$T(f) = \int_0^1 f(t)dt.$$

Then T is a linear functional on $C[0,1]$.

EXAMPLE 2.3. For each $f \in C[0,1]$, let

$$T(f) = 1 + \int_0^1 f(t)dt.$$

Now

$$\begin{aligned} T(f + g) &= 1 + \int_0^1 [f(t) + g(t)]\,dt \\ &= 1 + \int_0^1 f(t)dt + \int_0^1 g(t)dt. \end{aligned}$$

On the other hand,

$$\begin{aligned} T(f) + T(g) &= 1 + \int_0^1 f(t)dt + 1 + \int_0^1 g(t)dt \\ &= 2 + \int_0^1 f(t)dt + \int_0^1 g(t)dt. \end{aligned}$$

Thus (2.2) does not hold and T is a non-linear functional.

Linear functionals are important in the theory of interpolation and numerical integration. The set of all linear functionals defined on X is called the *algebraic conjugate (dual) space* and is denoted by X^*. If S and $T \in X^*$ and α is a scalar, then addition and scalar multiplication are normally defined for X^* by the equations

$$(S + T)(x) = S(x) + T(x),$$

$$(\alpha S)(x) = \alpha[S(x)]$$

for each $x \in X$. With these definitions X^* is itself a linear space. An important fact concerning X^* is that it has the same dimension as the original space X.

We will need some more terminology concerning transformations. Let T be a transformation from X into Y.

DEFINITION 2.3. *The range of T, denoted by $\mathcal{R}(T)$, is the set of all points in Y which are images of some point in X under the transformation T. In set notation:*

$$\mathcal{R}(T) = \{y \in Y: \; y = T(x) \text{ for some } x \in X\}.$$

DEFINITION 2.4. *The nullspace of T, denoted by $\mathcal{N}(T)$, is the set of vectors in X which are mapped into the zero element of Y by T. In symbols:*

$$\mathcal{N}(T) = \{x \in X: \; T(x) = \emptyset\}.$$

EXAMPLE 2.4. Let $X = Y = E_3$ and $T(x) = (x_1 - x_2, x_1 - x_2, x_3)$ for each point $x = (x_1, x_2, x_3) \in E_3$. In this case

$$\mathcal{N}(T) = \{x \in E_3: \; x_1 = x_2 \text{ and } x_3 = 0\}.$$

Geometrically $\mathcal{N}(T)$ is the line going through the origin with slope 1 in the x_1, x_2 plane. The range of T is given by

$$\mathcal{R}(T) = \{y \in E_3: \; y_1 = y_2\}.$$

This represents a vertical plane making an angle of 45° with the y_1 axis.

If T is linear, then $\mathcal{R}(T)$ and $\mathcal{N}(T)$ are both linear subspaces (see Problem 2.8). It will be shown in Section 5 that if X is finite dimensional, then

$$\dim \mathcal{R}(T) + \dim \mathcal{N}(T) = \dim X.$$

This fact is of great importance in the theory of linear equations.

PROBLEMS

2.1. Consider the transformation T mapping R into R which is defined by

$$T(x) = x^2$$

for every $x \in R$.

(i) Compute the following:

(a) $T(0)$
(b) $T(2)$
(c) $T(5)$
(d) $T(7)$
(e) $T(z)$
(f) $T(x + z)$
(g) $T(x) + T(z)$

(ii) Is $T(5) = T(0) + T(5)$?

(iii) Is $T(7) = T(5) + T(2)$?

(iv) Is T linear? Explain.

2.2. The equation $T(x) = 5x + (0,0,1)$ defines a transformation from E_3 into E_3.

(i) Compute the following:

(a) $T(2,1,4)$
(b) $T(6,1,3)$
(c) $T(z)$
(d) $T(x + z)$
(e) $T(x) + T(z)$
(f) $T(\alpha x)$ if $\alpha \in R$
(g) $\alpha T(x)$

(ii) Is T linear? Explain.

2.3. Which of the following equations define linear transformations? Explain.

(a) $T(x) = x_1 + x_2$ if $x = (x_1,x_2) \in E_2$.
(b) $T(x) = 3x$ for every $x \in E_2$.
(c) $T(x) = x_1 + x_2 + 3$ if $x = (x_1,x_2) \in E_2$.
(d) $T(f) = \int_a^b f(t) \cos 3t dt$ for each $f \in C[a,b]$.
(e) $T(x) = x_1^2 + x_2^2 + \ldots + x_6^2$ for each $x \in E_6$, where $x = (x_1, x_2, \ldots, x_6)$.
(f) $f(x) = 8x$ for each $x \in R$.
(g) $f(x) = 8x + 1$ for each $x \in R$.

The function defined in part (g) is not linear according to Definition 2.1 even though its graph is a straight line. In this text the word "linear" is always to be taken in the sense of Definition 2.1.

2.4. Which of the transformations in Problem 2.3 are functionals?

2.5. Show that $T(\emptyset) = \emptyset$ if T is linear. (More precisely, we should write $T(\emptyset_x) = \emptyset_y$, where $\emptyset_x$ and $\emptyset_y$ are the zero elements of X and Y, respectively.)

2.6. Show that the composite of two linear transformations is linear.

2.7. Show that the inverse of a one-to-one linear transformation is linear.

2.8. Let T be a linear transformation. Show that the range and nullspace of T are linear subspaces.

2.9. Describe the range and nullspace of the transformations given in Problem 2.3 (a) and (b).

3. EIGENVALUES

The concept of an *eigenvalue* is very important in applied mathematics. In many physical problems eigenvalues are related to the natural frequencies of an electrical or mechanical system. These natural frequencies are important in the analysis of vibrations, resonance, and stability. Eigenvalues are also used in statistics and in the theory of extrema for functions of many variables.

DEFINITION 3.1. *Let T be a transformation mapping the linear space X into itself. The number λ is said to be an eigenvalue for T iff*

$$T(x) = \lambda x \tag{3.1}$$

for some non-zero vector $x \in X$. A vector x which satisfies (3.1) *is said to be an eigenvector (eigenfunction) corresponding to λ.*

Note that if x is an eigenvector, then $T(x)$ is a multiple of x. Geometrically, this means that the direction of x is unchanged by the transformation T if $\lambda > 0$, or is reversed by T if $\lambda < 0$.

EXAMPLE 3.1. Let $C^{(\infty)}$ be the set of real-valued functions which have continuous derivatives of all orders on the real line R. For each $y \in C^{(\infty)}$, let

$$D(y) = y'.$$

Then D is a linear transformation from $C^{(\infty)}$ into $C^{(\infty)}$. Let $E_\lambda(x) = e^{\lambda x}$. Now

$$D(E_\lambda) = \lambda E_\lambda$$

for every real number λ. Thus every number $\lambda \in R$ is an eigenvalue of D, and E_λ is a corresponding eigenvector.

A very important property of eigenvectors is their simple behavior under operators which are functions of T. To illustrate this we will consider a polynomial operator of the form

$$p(T) = \sum_{j=0}^{n} a_j T^j,$$

where $T^0(x) = x$, $T^1(x) = T(x)$, $T^2(x) = T(T(x)), \ldots, T^j(x) = T(T^{j-1}(x))$ for $j > 2$. Note that if x is an eigenvector of T corresponding to λ, then

$$T^2(x) = T(T(x)) = T(\lambda x) = \lambda T(x) = \lambda^2 x.$$

It is easy to show by induction that

$$T^j(x) = \lambda^j x, \qquad j = 0,1,2, \ldots . \tag{3.2}$$

From (3.2), it follows that

$$p(T)(x) = \sum_{j=0}^{n} a_j T^j(x) = \left(\sum_{j=0}^{n} a_j \lambda^j \right) x. \tag{3.3}$$

Now, when λ is a number, we will interpret the symbol $p(\lambda)$ in the usual polynomial sense, that is:

$$p(\lambda) = \sum_{j=0}^{n} a_j \lambda^j,$$

where λ^j is the jth power of λ. Using this convention, we can rewrite (3.3) simply as

$$p(T)(x) = p(\lambda)x \tag{3.4}$$

whenever x is an eigenvector of T corresponding to the eigenvalue λ. The next example shows how this equation is used in the theory of differential equations.

EXAMPLE 3.2. Consider the following linear homogeneous nth order differential equation with constant coefficients:

$$a_n y^{(n)} + a_{n-1} y^{(n-1)} + \ldots + a_0 y = \emptyset. \tag{3.5}$$

Let $D(y) = y'$ as in Example 3.1. Now (3.5) can be written in the form

$$p(D)(y) = \emptyset,$$

where $p(D) = \Sigma_{j=0}^{n} a_j D^j$ is a polynomial operator formed from D. With E_λ as before (Example 3.1), we can apply equation (3.4) to obtain

$$p(D)(E_\lambda) = p(\lambda)E_\lambda.$$

If λ_0 is a zero of the polynomial p, then

$$p(D)(E_{\lambda_0}) = \emptyset,$$

that is, E_{λ_0} is a solution to (3.5). Thus solutions to (3.5) can be found by computing the zeros of p. (Imaginary zeros of p do not present any essential difficulties—see Problem 3.5.)

PROBLEMS

3.1. Find the eigenvalues of T if $T(x) = 3x$.

3.2. For each $f \in C[0,1]$, let $T(f)$ be the function g defined by $g(t) = e^t f(t)$ for each $t \in [0,1]$. Find the eigenvalues of T.

3.3. Show that if x is an eigenvector corresponding to λ, then αx is likewise an eigenvector corresponding to λ, where $\alpha \in F$ is not zero.

3.4. Let $T(x) = (x_1 + 2x_2, 2x_1 + x_2)$ for each $x \in EK_2$.
(a) Find the eigenvalues of T.
(b) Find an eigenvector for each eigenvalue. Hint: The system of equations

$$ax_1 + bx_2 = 0,$$
$$cx_1 + dx_2 = 0$$

has a non-trivial solution iff $ad = bc$.

3.5. Let $CK^{(\infty)}$ be the set of complex-valued functions having continuous derivatives of all orders on R. (If $f = f_1 + if_2$, where f_1 and f_2 are real-valued, then f' is defined to be $f_1' + if_2'$.) Let

$$e^{x+iy} = e^x(\cos y + i \sin y)$$

whenever x and y are real. Using the notation of Examples 3.1 and 3.2, show that:
(a) $D(E_z) = zE_z$ for each $z \in K$.
(b) E_z is a solution to (3.5) if z is a zero of p.

3.6. Let K^∞ be the set of all sequences

$$x = (x_1, x_2, \ldots, x_n, \ldots),$$

where $x_1, x_2, \ldots, x_n, \ldots$ are complex numbers. For each $x \in K^\infty$, let $\Delta(x)$ be the sequence defined by

$$\Delta(x) = (x_2 - x_1, x_3 - x_2, \ldots, x_{n+1} - x_n, \ldots).$$

(a) Show that Δ is a linear operator.
(b) Show that every number $z \in K$ is an eigenvalue of Δ.
(c) Find an eigenvector for each $z \in K$.
(d) Find two non-trivial solutions to the equation

$$\Delta^2(x) - 3\Delta(x) + x = \emptyset. \tag{3.6}$$

The operator Δ is called the *forward difference operator* and an equation such as (3.6) is called a *difference equation.*

3.7. Suppose that $\int_{-\infty}^{\infty} |K(u)| du < \infty$. For each complex-valued function f which is continuous and bounded on $(-\infty, \infty)$, let $T(f)$ be the function g defined by

$$g(t) = \int_{-\infty}^{\infty} f(u)K(t-u)du.$$

For a fixed number ω, let $E_{i\omega}(t) = e^{i\omega t}$. Show that $E_{i\omega}$ is an eigenfunction (eigenvector) for T. One of the reasons that Fourier series and Fourier transforms are used in models of linear electrical and mechanical systems is that $E_{i\omega}$ is an eigenfunction of a transformation of the above type. Hint: The eigenvalue for $E_{i\omega}$ is given by

$$\lambda = \int_{-\infty}^{\infty} e^{-i\omega z} K(z) dz.$$

4. LINEAR INDEPENDENCE

The concept of linear independence is extremely important in many areas of mathematics. This concept is especially important in the study of differential and integral equations, difference equations, and linear systems of equations (see Section 5). Let X be a linear space and F the number system under consideration.

DEFINITION 4.1. *The elements $x_1, x_2, \ldots, x_n \in X$ are said to be linearly independent iff the equation*

$$(4.1) \qquad \alpha_1 x_1 + \alpha_2 x_2 + \ldots + \alpha_n x_n = \emptyset,$$

with $\alpha_1, \alpha_2, \ldots, \alpha_n \in F$, implies that $\alpha_1 = \alpha_2 = \ldots = \alpha_n = 0$.

Recall that an expression like that on the left side of equation (4.1) is called a *linear combination* of $x_1, x_2, \ldots, x_n$. An equivalent definition of linear independence is to say that none of the elements $x_1, x_2, \ldots, x_n$ can be written as a linear combination of the other elements.

EXAMPLE 4.1. Let $f(x) = e^x$ and $g(x) = e^{2x}$ for every $x \in [0,1]$. Then f and g are linearly independent elements of $C[0,1]$. To prove this suppose that $\alpha_1 f + \alpha_2 g = \emptyset$. This means that

$$\alpha_1 f(x) + \alpha_2 g(x) = \emptyset(x) = 0$$

for every $x \in [0,1]$. In particular,

$$\alpha_1 f(0) + \alpha_2 g(0) = 0,$$

$$\alpha_1 f(1) + \alpha_2 g(1) = 0.$$

This implies that $\alpha_2 = -\alpha_1$ and hence that

$$\alpha_1 e - \alpha_1 e^2 = 0.$$

But then $\alpha_1 = 0$ since $e - e^2 \neq 0$. It follows that $\alpha_2 = -\alpha_1 = 0$ and thus f and g are linearly independent.

DEFINITION 4.2. *A subset $A \subset X$ is said to be linearly independent iff every finite subset of A is linearly independent.*

EXAMPLE 4.2. Let $A = \{f\colon$ for some non-negative integer n, $f(x) = x^n$ for all $x \in [0,1]\}$. Let $\{f_1, \ldots, f_k\}$ be a finite subset of A with $f_j(x) = x^{n_j}$. Then

$$\alpha_1 f_1 + \ldots + \alpha_k f_k = \emptyset$$

implies that

$$\alpha_1 x^{n_1} + \ldots + \alpha_k x^{n_k} = 0$$

for all $x \in [0,1]$. But this polynomial can only have a finite number of zeros unless $\alpha_1 = \ldots = \alpha_k = 0$. Thus $f_1, \ldots, f_k$ are linearly independent. It follows that A is linearly independent.

DEFINITION 4.3. *The subset $A \subset X$ is said to be a basis for X iff A is linearly independent and every member of X can be expressed as a finite linear combination of elements from A.*

EXAMPLE 4.3. Let X be the set of all polynomials defined on $[0,1]$. One can readily show that X is a linear space and that the set A defined in Example 4.2 is a basis for X.

If A is a basis which can be indexed, say $A = \{a_1, a_2, \ldots,\}$, then each $x \in X$ has a unique representation of the form $x = \Sigma \alpha_j a_j$ where $\alpha_1, \alpha_2, \ldots$ are numbers (Problem 4.4). The number α_j is then referred to as the jth coordinate of x with respect to the basis A.

Now suppose X is a linear space with finite dimension n. It turns out that every basis for X has n elements. In order to establish this result we will need the following lemma.

LEMMA 4.1. *Let $x_1, x_2, \ldots, x_m$ and $y_1, y_2, \ldots, y_k$ be elements of a linear space. If $k > m$ and*

$$\{y_1, y_2, \ldots, y_k\} \subset \text{span}\,\{x_1, x_2, \ldots, x_m\},$$

then $y_1, y_2, \ldots, y_k$ are linearly dependent.

Proof. (Faddeev and Fadeeva [6].) We will prove this lemma by using induction on m. If $m = 1$, then $y_1, y_2, \ldots, y_k$ are all multiples of x_1 and are thus linearly dependent. Suppose the lemma holds for m. If $k > m + 1$, and

$$\{y_1, y_2, \ldots, y_k\} \subset \text{span}\,\{x_1, x_2, \ldots, x_{m+1}\}, \tag{4.2}$$

then we infer from (4.2) that there exist constants c_{ij} such that

$$\begin{aligned} y_1 &= c_{11}x_1 + c_{12}x_2 + \ldots + c_{1,m+1}x_{m+1}, \\ y_2 &= c_{21}x_1 + c_{22}x_2 + \ldots + c_{2,m+1}x_{m+1}, \\ &\;\;\vdots \\ y_k &= c_{k1}x_1 + c_{k2}x_2 + \ldots + c_{k,m+1}x_{m+1}. \end{aligned} \tag{4.3}$$

If $y_1 = \emptyset$, then clearly $y_1, y_2, \ldots, y_k$ are dependent. Otherwise there is an index ν such that $c_{1\nu} \neq 0$. Now let

$$z_i = y_i - \frac{c_{i\nu}}{c_{1\nu}}\, y_1, \qquad i = 2, 3, \ldots, k. \tag{4.4}$$

Using (4.3) and (4.4) we obtain the equations

$$z_i = \sum_{j=1}^{m+1} c_{ij}x_j - \sum_{j=1}^{m+1} \frac{c_{i\nu}}{c_{1\nu}} c_{1j}x_j, \qquad i = 2, 3, \ldots, k. \tag{4.5}$$

But the coefficient of x_ν on the right side of (4.5) is zero; hence

$$\{z_2, z_3, \ldots, z_k\} \subset \text{span}\,\{x_1, x_2, \ldots, x_{\nu-1}, x_{\nu+1}, \ldots, x_{m+1}\}.$$

By the inductive hypothesis $z_2, z_3, \ldots, z_k$ are dependent. It follows readily that $y_1, y_2, \ldots, y_k$ are dependent and the proof is complete.

THEOREM 4.2. *Let X be a linear space with finite dimension n. Then every basis for X has exactly n elements.*

Proof. From the definition of dimension (Definition 1.5), there exist vectors $x_1, x_2, \ldots, x_n$ such that $X \subset \text{span}\,\{x_1, x_2, \ldots, x_n\}$. Let $\{y_1, y_2, \ldots, y_k\}$ be a basis for X. Now $k \geqslant n$ because X cannot be spanned by fewer than n vectors. Also

$$\{y_1, y_2, \ldots, y_k\} \subset \text{span}\,\{x_1, x_2, \ldots, x_n\}.$$

If k were greater than n, then by Lemma 4.1, $y_1, y_2, \ldots, y_k$ would be dependent. But a basis must be independent. Thus $k = n$ and the proof is complete.

Several corollaries follow from Lemma 4.1 and Theorem 4.2.

COROLLARY 4.3. *Let X be a linear space with finite dimension n. Then*

(i) *Every subspace of X (including X) has a basis of at most n elements.*
(ii) *Any linearly independent subset of X containing n elements is a basis for X.*
(iii) *If A is a linearly independent subset of X, then there is a basis B of X such that $A \subset B$.*
(iv) *Any subset of X containing more than n elements is linearly dependent.*

COROLLARY 4.4. *The Euclidean spaces E_n and EK_n have dimension n.*

The proofs of these corollaries are left as exercises (Problems 4.6 and 4.9).

PROBLEMS

4.1. Give an example of three linearly independent elements from E_4. Also produce a set of three vectors from E_4 which are linearly dependent (i.e., not independent).

4.2. In $C[0,1]$, let $f(x) = \sin x$ and $g(x) = \cos x$. Show that f and g are linearly independent.

4.3. Show that the vectors $e_1 = (1,0,0)$, $e_2 = (0,1,0)$, and $e_3 = (0,0,1)$ form a basis for E_3.

4.4. Suppose that A is a basis for X and $x \in X$. Show that the representation of x as a finite linear combination of elements of A is (essentially) unique.

4.5. Show that if X has more than one member, then a basis for X must be a proper subset of X. Also show that X must be an infinite set.

4.6. Show that E_n and EK_n have dimension n.

4.7. Show that a linear transformation is completely determined by its values on a basis.

4.8. Show that the set A in Example 4.2 is not a basis for $C[0,1]$.

4.9. Prove Corollary 4.3.

4.10. Let Y be a finite dimensional linear space and suppose that Z is a subspace of Y. Show that if $\dim Z = \dim Y$, then $Z = Y$.

5. THE THEORY OF LINEAR EQUATIONS

Let X and Y be linear spaces, and let T be a linear transformation from X into Y. In this section we will discuss solutions to equations of the form

(5.1) $\qquad T(x) = y.$

Such equations arise in the study of differential and integral equations, difference equations, and linear systems of equations.

The basic questions which we will consider are:

1. *Existence.* Do any solutions exist?
2. *Uniqueness.* Is there more than one solution?
3. *Characterization.* What is the general form of the solutions?

For convenience these problems will be treated somewhat independently, even though interest in the second and third questions is highest when the answer to the first question is affirmative.

The equation

(5.2) $\qquad T(x) = \emptyset$

is referred to as the *homogeneous equation* associated with (5.1.). This equation is relatively easy to handle, and its solutions play a fundamental role in the study of the more general equation (5.1). Recall that the nullspace of $T, \mathcal{N}(T)$, is the set of all solutions to (5.2).

THEOREM 5.1. *If T is linear, then the following statements hold for the homogeneous equation* (5.2):

(i) *At least one solution exists.*
(ii) *The solution is unique if and only if*

$$\dim \mathcal{N}(T) = 0.$$

(iii) *Let $\{x_1, x_2, \ldots, x_k\}$ be a basis for $\mathcal{N}(T)$. Then x is a solution if and only if*

$$x = c_1x_1 + c_2x_2 + \ldots + c_kx_k$$

for some combination of numbers $c_1, c_2, \ldots, c_k$.

Proof. By Problem 2.5, $T(\emptyset) = \emptyset$; thus at least one solution exists. If a solution other than $\emptyset$ exists, then $\dim \mathcal{N}(T) \geqslant 1$. Conversely, if the solution is unique, then by Definition 1.5, $\dim \mathcal{N}(T) = \dim \{\emptyset\} = 0$. Part (iii) follows readily from the definition of a basis and the fact that $\mathcal{N}(T)$ is a subspace (Problem 2.8).

From Theorem 5.1 one can see that solutions to the homogeneous equation can be completely characterized in terms of a basis for $\mathcal{N}(T)$. We will now proceed to show that such a basis can also be used to characterize the solutions of the non-homogeneous equation (5.1).

Recall that the range of $T, \mathcal{R}(T)$, is the set of all vectors $y \in Y$ for which a solution to (5.1) exists.

THEOREM 5.2. *Let $y \in \mathcal{R}(T)$ and suppose that $T(x_0) = y$. Then the following statements hold for equation* (5.1):

(i) *The solution x_0 is unique if and only if* $\dim \mathcal{N}(T) = 0$.
(ii) *If $\{x_1, x_2, \ldots, x_k\}$ is a basis for $\mathcal{N}(T)$, then x is a solution if and only if*

$$x = x_0 + c_1x_1 + c_2x_2 + \ldots + c_kx_k \tag{5.3}$$

for some combination of numbers $c_1, c_2, \ldots, c_k$.

Proof. In view of the linearity of T, we see that $T(x) = y = T(x_0)$ if and only if $T(x - x_0) = \emptyset$—that is, $x - x_0$ is a solution to the homogeneous equation. The result then follows from Theorem 5.1.

The solution x_0 in (5.3) is frequently referred to as a *particular solution*, while $c_1x_1 + \ldots + c_kx_k$ is called the *complementary solution.* The general method for finding all solutions to (5.1) consists of two steps:

1. Obtain a particular solution x_0.
2. Obtain the complementary solution by finding k linearly independent solutions to the homogeneous equation, where $k = \dim \mathfrak{N}(T)$.

Specific methods for carrying out these steps depend on the nature of T and y. In Example 5.1 we will outline some of the procedures which are used for difference equations.

So far nothing has been said about the existence question. The following theorem will enable us to establish an existence theorem for finite dimensional spaces.

THEOREM 5.3. *Let T be a linear transformation defined on a finite dimensional space X. Then*

$$\dim \mathfrak{R}(T) + \dim \mathfrak{N}(T) = \dim X.$$

Proof. Let $k = \dim \mathfrak{N}(T)$, $n = \dim X$, and let $\{x_1, x_2, \ldots, x_k\}$ be a basis for $\mathfrak{N}(T)$. According to Corollary 4.3, the basis for $\mathfrak{N}(T)$ can be extended to a basis for X, say $x_1, x_2, \ldots, x_k, x_{k+1}, \ldots, x_n$. Now let $y_i = T(x_i)$, $i = 1, 2, \ldots, n$. We will show that $y_{k+1}, y_{k+2}, \ldots, y_n$ is a basis for $\mathfrak{R}(T)$.

First, we must show that $y_{k+1}, \ldots, y_n$ are linearly independent. Suppose there exist numbers $\alpha_{k+1}, \ldots, \alpha_n$ such that

$$\alpha_{k+1} y_{k+1} + \ldots + \alpha_n y_n = \emptyset.$$

Then

$$\begin{aligned} T(\alpha_{k+1} x_{k+1} + \ldots + \alpha_n x_n) &= \alpha_{k+1} T(x_{k+1}) + \ldots + \alpha_n T(x_n) \\ &= \alpha_{k+1} y_{k+1} + \ldots + \alpha_n y_n = \emptyset. \end{aligned}$$

It follows that $\alpha_{k+1} x_{k+1} + \ldots + \alpha_n x_n$ belongs to $\mathfrak{N}(T)$ and is thus a linear combination of $x_1, x_2, \ldots, x_k$. If at least one of the α's, say α_j, were not zero, then x_j would be a linear combination of $x_1, \ldots, x_{j-1}, x_{j+1}, \ldots, x_n$, contradicting the fact that a basis is linearly independent. Hence $\alpha_{k+1} = \ldots = \alpha_n = 0$ and $y_{k+1}, \ldots, y_n$ are independent.

Next, we must show that every member of $\mathfrak{R}(T)$ is a linear combination of $y_{k+1}, \ldots, y_n$. Let $y \in \mathfrak{R}(T)$. Then $y = T(x)$ for some $x \in X$. But x is a linear combination of $x_1, x_2, \ldots, x_n$, say

$$x = \alpha_1 x_1 + \ldots + \alpha_n x_n.$$

Then $y = T(x) = \alpha_1 y_1 + \ldots + \alpha_n y_n$. Recall that $x_i \in \mathfrak{N}(T)$ for $i = 1, 2, \ldots, k$; thus $y_i = T(x_i) = \emptyset$, $i = 1, 2, \ldots, k$. Hence $y = \alpha_{k+1} y_{k+1} + \ldots + \alpha_n y_n$ and $y_{k+1}, \ldots, y_n$ is a basis for $\mathfrak{R}(T)$.

The dimension of a subspace is the number of elements in a basis. Thus $\dim \mathcal{R}(T) = n - k$ and the result follows.

Now let us prove an existence theorem.

THEOREM 5.4. *Let X and Y be finite dimensional linear spaces and suppose T is a linear transformation from X into Y. If*

$$(5.4) \qquad \dim Y \leqslant \dim X - \dim \mathcal{N}(T),$$

then equation (5.1) *has at least one solution for each* $y \in Y$.

Proof. Note first that $\mathcal{R}(T) \subset Y$ and hence

$$(5.5) \qquad \dim \mathcal{R}(T) \leqslant \dim Y.$$

From (5.4), (5.5), and Theorem 5.3, we infer that

$$\dim \mathcal{R}(T) \leqslant \dim Y \leqslant \dim X - \dim \mathcal{N}(T) = \dim \mathcal{R}(T).$$

Thus $\dim \mathcal{R}(T) = \dim Y$. It follows (see Problem 4.10) that $\mathcal{R}(T) = Y$ and the proof is complete.

EXAMPLE 5.1. In order to illustrate the application of the theory developed in this section, let us outline briefly some techniques which are used for difference equations. An equation of the form

$$(5.6) \qquad a_2 x_{n+2} + a_1 x_{n+1} + a_0 x_n = y_n \qquad (a_2 \neq 0)$$

is referred to as a *linear second order difference equation with constant coefficients.* We want to find an infinite sequence $x = (x_1, x_2, \ldots, x_n, \ldots)$ of numbers which satisfies (5.6) for all positive integers n. Let X be the set of all complex-valued sequences $x = (x_1, x_2, \ldots, x_n, \ldots)$, and let addition and scalar multiplication be defined on X by the equations

$$x + y = (x_1 + y_1, x_2 + y_2, \ldots, x_n + y_n, \ldots),$$

$$\alpha x = (\alpha x_1, \alpha x_2, \ldots, \alpha x_n, \ldots),$$

whenever $x, y \in X$ and $\alpha \in K$. For each $x \in X$, let $T(x)$ be the sequence whose nth element is given by

$$T(x)_n = a_2 x_{n+2} + a_1 x_{n+1} + a_0 x_n.$$

Then T is a linear transformation from X into X, and thus (5.6) can be written in the form

$$T(x) = y.$$

With each such transformation T we associate a so-called characteristic polynomial p defined by

$$p(z) = a_2 z^2 + a_1 z + a_0.$$

Note that if $x_n = z^n$, $n = 1, 2, \ldots$, where $z \in K$,

$$T(x)_n = a_2 z^{n+2} + a_1 z^{n+1} + a_0 z^n \qquad (5.7)$$
$$= z^n (a_2 z^2 + a_1 z + a_0)$$
$$= p(z) z^n.$$

Let z_1 and z_2 be the zeros of p, and let $u_n = z_1^n$, $v_n = z_2^n$, $n = 1, 2, \ldots$. It follows from (5.7) that u and v are two solutions to the homogeneous equation. If $z_1 \neq z_2$, then u and v form a basis for $\mathfrak{N}(T)$. If $z_1 = z_2$, we let $u_n = z_1^n$, $v_n = nz_2^n$, and again obtain a basis for $\mathfrak{N}(T)$. The proofs of these statements are left as exercises (Problem 5.2).

There is a systematic way to obtain a particular solution to (5.6) (see Henrici [11]). However, for our purposes it will suffice to use a trial and error procedure known as the *method of undetermined coefficients.* This procedure consists of the following steps:

1. Guess at the form of a solution, i.e., $x_n = An + B$, $x_n = AB^n$, etc.
2. Attempt to determine A and B by substituting into the difference equation.
3. If successful in (2), check to see whether the solution is valid. If a solution has not been found, repeat (1) and (2).

For simple problems, a particular solution can usually be found within one or two tries.

To see more specifically how all of this works, consider the equation

$$x_{n+2} - 5x_{n+1} + 6x_n = 2n + 3. \qquad (5.8)$$

The characteristic polynomial p is given by

$$p(z) = z^2 - 5z + 6.$$

The zeros of p are 2 and 3; thus the complementary solution x^* is given by

$$x_n^* = c_1 2^n + c_2 3^n.$$

Let us guess that a particular solution w will be of the form $w_n = An + B$. Substituting w_n for x_n in (5.8), we have

$$A(n+2) + B - 5[A(n+1) + B] + 6(An + B) = 2n + 3.$$

Collecting like terms we obtain the equation

$$2An + 2B - 3A = 2n + 3.$$

This will be an identity in n if we set $2A = 2$ and $2B - 3A = 3$, i.e., $A = 1$ and $B = 3$. After checking that $w_n = n + 3$ is a valid solution, we can apply Theorem 5.2 and state that every solution x to (5.8) is given by an equation of the form

$$x_n = n + 3 + c_1 2^n + c_2 3^n,$$

where c_1 and c_2 are constants.

PROBLEMS

5.1. A transformation T is said to be one-to-one (1-1) iff $x \neq y$ implies that $T(x) \neq T(y)$. Let T be a linear transformation. Show that:
(a) T is 1-1 if and only if $\dim \mathfrak{N}(T) = \emptyset$.
(b) The inverse of T, T^{-1}, is linear if T is 1-1. (For each $y \in \mathfrak{R}(T)$, $T^{-1}(y)$ is the unique solution to the equation $T(x) = y$.)

5.2. Let z_1 and z_2 be the zeros of the characteristic polynomial p associated with the difference equation (5.6). If $z_1 = z_2$, let $u_n = z_1^n$, $v_n = nz_2^n$; otherwise let $u_n = z_1^n$ and $v_n = z_2^n$. Show that:
(a) u and v are solutions to the homogeneous equation.
(b) u and v are linearly independent.
(c) u and v span $\mathfrak{N}(T)$. Hint: Let $x \in \mathfrak{N}(T)$. Show that constants c_1 and c_2 exist which satisfy the equations $x_1 = c_1 u_1 + c_2 v_1$, $x_2 = c_1 u_2 + c_2 v_2$.

5.3. Prove that every difference equation of the form (5.6) has at least one solution. Hint: Construct a solution by induction.

5.4. Find the general form of the solutions to the following difference equations:
(a) $x_{n+2} - 5x_{n+1} + 4x_n = 2^n$
(b) $x_{n+2} + x_n = 5n$

5.5. Find the solution in Problem 5.4(a) which satisfies the initial conditions $x_1 = 1, x_2 = 6$.

6. NORMED LINEAR SPACES

A *norm* provides a measure of the distance between objects in a linear space. A norm will frequently be used to compare the quality of approximations. Consider the functions from $C[a,b]$ shown in the figure on the next page. Which function, g or h, provides the better approximation to f on $[a,b]$? This question has no absolute answer—it depends on the means employed to measure the quality of approximations. For approximating the integral of f over $[a,b]$, g would be better. For some other purpose, h might well be better.

DEFINITION 6.1. *A norm on a linear space X is a function which associates a real number, denoted by $\|x\|$, with each element $x \in X$*

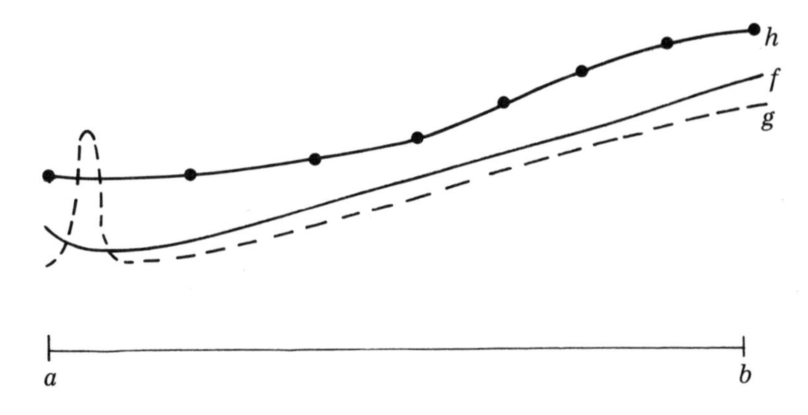

FIGURE 6.1. Comparison of Approximations.

in such a way that the following properties hold:

(i) $\|x\| \geqslant 0$ *for all* $x \in X$
(ii) $\|x\| = 0$ *iff* $x = \emptyset$
(iii) $\|\alpha x\| = |\alpha|\ \|x\|$ *for all* $\alpha \in F$ *and* $x \in X$
(iv) $\|x + y\| \leqslant \|x\| + \|y\|$ *whenever* x *and* $y \in X$

We call $\|x\|$ the norm or length of x and $\|x - y\|$ the distance between x and y. Property (iv) is referred to as the *triangle inequality*. Note that norms are a generalization of the concept of absolute value for real numbers.

EXAMPLE 6.1. In $C[a,b]$, let

$$\|f\| = \max_{t \in [a,b]} |f(t)|.$$

Let us show that this defines a function which possesses the four properties of a norm. The first property holds because $|f(t)| \geqslant 0$ for all t. The second property is then obvious. Now

$$\|f\| = \max_{t \in [a,b]} |f(t)|$$

means that

$$(1) \qquad |f(t)| \leqslant \|f\| \quad \text{for all} \quad t \in [a,b]$$

and

$$(2) \qquad |f(t_0)| = \|f\| \quad \text{for some} \quad t_0 \in [a,b].$$

Thus

$$|(\alpha f)(t)| = |\alpha f(t)| = |\alpha|\ |f(t)| \leqslant |\alpha|\ \|f\|$$

for all t, and

$$|(\alpha f)(t_0)| = |\alpha f(t_0)| = |\alpha|\,|f(t_0)| = |\alpha|\,\|f\|$$

for some t_0. It follows that

$$\max_{t\in[a,b]} |(\alpha f)(t)| = |\alpha|\,\|f\|$$

and property (iii) is satisfied. To prove that (iv) holds we note that

$$|(f+g)(t)| = |f(t)+g(t)| \leqslant |f(t)| + |g(t)| \leqslant \|f\| + \|g\|.$$

This follows from the triangle inequality for real numbers. Thus

$$\|f+g\| = \max_{t\in[a,b]} |(f+g)(t)| \leqslant \|f\| + \|g\|.$$

The norm defined in this example is frequently called the *uniform* or *Chebyshev norm.* If this norm is used for measuring the quality of an approximation, then the function h in Figure 6.1 provides a better approximation to f than does the function g.

Now that a notion of distance is available, it is possible to talk about limits in a normed linear space. The definitions are obtained by replacing absolute values by norms in the corresponding definitions concerning real numbers.

DEFINITION 6.2. *Let x_n be a sequence of elements from a normed linear space X. We say that the sequence converges to the element $a \in X$ iff for each positive number ϵ there exists a positive number N_ϵ such that*

$$\|x_n - a\| < \epsilon$$

whenever $n > N_\epsilon$. In this case we write

$$\lim_{n\to\infty} x_n = a.$$

EXAMPLE 6.2. For each $f \in C[0,1]$, let

$$\|f\|_2 = [\textstyle\int_0^1 f^2(t)dt]^{1/2}.$$

We will show in Section 7 that this is a valid norm. Let $f_n(t) = t^n$ and $\emptyset(t) = 0$ for each $t \in [0,1]$. Then

$$\|f_n - \emptyset\|_2 = \|f_n\|_2 = [\textstyle\int_0^1 t^{2n}dt]^{1/2} = \frac{1}{\sqrt{2n+1}}.$$

Let ϵ be an arbitrary positive number. Then with $N_\epsilon = \max\,[0, (\epsilon^{-2}-1)/2]$ and $n > N_\epsilon$, we have

$$\|f_n - \emptyset\|_2 = \frac{1}{\sqrt{2n+1}} < \frac{1}{\sqrt{2(\epsilon^{-2}-1)/2+1}} = \epsilon$$

if $\epsilon < 1$; and

$$\|f_n - \emptyset\|_2 = \frac{1}{\sqrt{2n+1}} < 1 < \epsilon$$

if $\epsilon > 1$. Thus

$$\lim_{n\to\infty} f_n = \emptyset$$

relative to the given norm.

Note that when $f_n(t) = t^n$, $\|f_n - \emptyset\|$ is always 1 when the uniform norm is used. Thus $\{f_n\}$ does not converge to $\emptyset$ in the uniform norm. In terms of approximation theory, this means that for a large value of n, f_n is a relatively good approximation to $\emptyset$ if $\|\ \|_2$ is used, but a relatively poor approximation if the uniform norm is used.

DEFINITION 6.3. *Let X and Y be normed linear spaces and let T be a transformation mapping X into Y. Then*

$$\lim_{x\to a} T(x) = b$$

means that for every positive number ϵ there exists a positive number δ_ϵ such that

$$\|T(x) - b\| < \epsilon$$

whenever $0 < \|x - a\| < \delta_\epsilon$.

Note that $T(x)$ and $b \in Y$ while x and $a \in X$. Technically different norm symbols should be used for X and Y, but we will do so only when confusion would otherwise be likely.

DEFINITION 6.4. *Let X and Y be normed linear spaces and let T be a transformation mapping X into Y. We say that T is continuous at the point $a \in X$ iff*

$$\lim_{x\to a} T(x) = T(a).$$

The phrase "T is continuous on X" means that T is continuous at each element of X.

EXAMPLE 6.3. For each $f \in C[a,b]$, let

$$T(f) = \int_a^b f(t)dt.$$

Then T is a transformation from $C[a,b]$ into R. Now it is easy to check that R is a linear space and that the equation $\|t\| = |t|$ defines a norm on R. We will show that T is continuous on $C[a,b]$ with respect to the uniform norm. Let f be an arbitrary member of $C[a,b]$ and ϵ an arbitrary positive number. Then

$$\begin{aligned}\|T(g) - T(f)\| &= \left|\int_a^b [g(t) - f(t)]\,dt\right| \\ &\leqslant \int_a^b |g(t) - f(t)|\,dt \\ &\leqslant \int_a^b \|g - f\|\,dt \\ &= (b - a)\,\|g - f\|.\end{aligned}$$

If we pick $\delta_\epsilon = \epsilon/(b - a)$, we will have

$$\|T(g) - T(f)\| < \epsilon$$

whenever $\|g - f\| < \delta_\epsilon$. Thus

$$\lim_{g \to f} T(g) = T(f)$$

and T is continuous on $C[a,b]$.

Let us consider some continuity properties of linear transformations. First, the word "bounded" is given a special meaning when used in connection with linear transformations.

DEFINITION 6.5. *Let X and Y be normed linear spaces and let T be a linear transformation from X into Y. Then T is said to be bounded iff there is a constant M such that*

$$\|T(x)\| \leqslant M\|x\|$$

for every $x \in X$.

Note that this definition does not conform to the usual meaning of the word "bounded," where one would expect $\|T(x)\| \leqslant M$ for all $x \in X$. However, it can be shown that $\|T(x)\| \leqslant M$ for all $x \in X$ if and only if $T(x) = \emptyset$ for all $x \in X$ (see Problem 6.7). The following theorem shows that boundedness and continuity are equivalent for linear transformations.

THEOREM 6.1. *Let X and Y be normed linear spaces and T a linear transformation from X into Y. Then T is bounded iff it is continuous on X.*

Proof. Suppose T is bounded. Then for some constant M, $\|T(x)\| \leqslant M\,\|x\|$ for all $x \in X$. Let ϵ be an arbitrary positive number and a an arbitrary member of X. Let $\delta_\epsilon = \epsilon/M$. Then

$$\|T(x) - T(a)\| = \|T(x - a)\| \leqslant M\,\|x - a\| < \epsilon$$

whenever $\|x - a\| < \delta_\epsilon$. Thus T is continuous on X.

Conversely, suppose T is continuous. In particular, T is continuous at $\emptyset$. Note that $T(\emptyset) = \emptyset$ (see Problem 2.5). Letting $\epsilon = 1$ in the continuity definition, we infer the existence of a number δ such that

$$\|T(x) - T(\emptyset)\| = \|T(x)\| < 1$$

whenever $\|x - \emptyset\| = \|x\| < \delta$. Now let z be an arbitrary member of X other than $\emptyset$, and let

$$x = \frac{\delta}{2\,\|z\|}\, z.$$

Then $\|x\| = \delta/2 < \delta$; so $\|T(x)\| < 1$. Thus

$$\|T(z)\| = \|T(2\,\|z\|\,\delta^{-1}x)\| = \frac{2\,\|z\|}{\delta}\,\|T(x)\| < \frac{2}{\delta}\,\|z\|.$$

Hence T is bounded and the proof is complete.

EXAMPLE 6.4. Consider $C[a,b]$ with the uniform norm, and suppose $K(t,u)$ is continuous for t and $u \in [a,b]$. For each $f \in C[a,b]$, let $T(f)$ be the function g defined by

$$g(t) = \int_a^b K(t,u)f(u)du.$$

Now it is easy to see that T is linear and bounded since

$$\begin{aligned} \|g\| &= \max_{t\in[a,b]} |g(t)| \\ &\leqslant \max_{t\in[a,b]} \int_a^b |K(t,u)f(u)|du \\ &\leqslant (b-a)M\,\|f\|, \end{aligned}$$

where

$$M = \max_{t,u\in[a,b]} |K(t,u)|.$$

From Theorem 6.1 we can infer that T is continuous. This type of linear transformation is important in the study of integral equations.

PROBLEMS

6.1. Let $f(t) = t^3$, $g(t) = 3t/4$, and $h(t) = 3t/2$ for each $t \in [-1,1]$. Using the uniform norm on $C[-1,1]$, decide which function, g or h, provides the better approximation to f.

6.2. Prove:
(a) $\|x - y\| \geqslant \|x\| - \|y\|$
(b) $\|x - y\| \geqslant |\,\|x\| - \|y\|\,|$
Hint: $x = x - y + y$.

6.3. Let $f(t) = e^t$ and $g(t) = 1 + t$ for each $t \in [0,1]$. Compute the distance between f and g relative to the norm defined in Example 6.2.

6.4. A metric for a set X is a function d which associates a real number with each pair of elements from X in such a way that:
(i) $d(x,y) \geqslant 0$
(ii) $d(x,y) = 0$ iff $x = y$
(iii) $d(x,y) = d(y,x)$
(iv) $d(x,z) \leqslant d(x,y) + d(y,z)$
Show that the equation $d(x,y) = \|x - y\|$ defines a metric in a normed linear space.

6.5. Suppose that the sequences $\{x_n\}$ and $\{y_n\}$ both converge in a normed linear space. Show that

$$\lim_{n\to\infty} (x_n + y_n) = \lim_{n\to\infty} x_n + \lim_{n\to\infty} y_n.$$

6.6. Let X be a normed linear space and define $T(x) = 20x$ for each $x \in X$. Show that T is a continuous transformation.

6.7. Let T be a linear transformation defined on X and let M be a constant. Show that if $\|T(x)\| \leqslant M$ for all $x \in X$, then $T(x) = \emptyset$ for all $x \in X$.

6.8. Let X and Y be normed linear spaces and let T be a linear transformation mapping X into Y. Show that if T is continuous at $\emptyset$, then T is continuous on all of X.

6.9. Let $T(f) = f'$ for each $f \in C'[0,1]$ (see Example 2.1). Show that T is not bounded with respect to the uniform norm. Hint: Consider the sequence defined by $f_n(t) = t^n$.

7. INNER PRODUCTS

An *inner product* is a multiplicative type of operation between elements of a linear space. Let F be the number system under consideration, X a linear space, and $(\cdot)$ an operation which associates a number in F with each pair of elements of X.

DEFINITION 7.1. *The operation $(\cdot)$ is said to be an inner product iff the following properties hold for all x, y, $z \in X$ and $\alpha \in F$:*

(i) $x \cdot x \geqslant 0$ *and* $x \cdot x = 0$ *iff* $x = \emptyset$
(ii) $x \cdot y = \overline{y \cdot x}$ ($\overline{y \cdot x}$ *is the complex conjugate of* $y \cdot x$)
(iii) $(\alpha x) \cdot y = \alpha(x \cdot y)$
(iv) $x \cdot (y+z) = x \cdot y + x \cdot z$

Note that (ii) can be written as $x \cdot y = y \cdot x$ when $F = R$, since the conjugate of a real number is equal to the number itself. The symbols (x,y) and $\langle x,y \rangle$ are quite frequently used in place of $x \cdot y$ to represent an inner product.

EXAMPLE 7.1. For f and $g \in C[a,b]$, let

$$f \cdot g = \int_a^b f(t)g(t)dt.$$

It follows easily from the properties of the integral that this defines an inner product on $C[a,b]$.

EXAMPLE 7.2. An inner product on E_n is usually defined by the equation

$$(7.1) \qquad x \cdot y = x_1 y_1 + x_2 y_2 + \ldots + x_n y_n,$$

where $x = (x_1, x_2, \ldots, x_n)$ and $y = (y_1, y_2, \ldots, y_n)$. In EK_n, the inner product is usually defined by

$$(7.2) \qquad x \cdot y = x_1 \overline{y_1} + x_2 \overline{y_2} + \ldots + x_n \overline{y_n}.$$

This agrees with (7.1) if the components of x and y are real.

It is always possible to define a norm in terms of an inner product.

DEFINITION 7.2. *Let X be an inner product space. The inner product norm is defined by the equation* $\|x\| = \sqrt{x \cdot x}$.

In order to have Definition 7.2 make sense, we must show that the equation $\|x\| = \sqrt{x \cdot x}$ defines a function which satisfies the conditions for a norm. Axioms (i) and (ii) for a norm (see Definition 6.1) are obviously satisfied in view of the first axiom for an inner product. (We use the convention that the symbol $\sqrt{a}$ or $a^{1/2}$ always refers to the non-negative square root of a.) Now

$$\begin{aligned} \|\alpha x\|^2 &= (\alpha x) \cdot (\alpha x) = \alpha [x \cdot (\alpha x)] \\ &= \alpha [\overline{(\alpha x) \cdot x}] = \alpha \overline{\alpha (x \cdot x)} \\ &= \alpha \overline{\alpha} (x \cdot x) = |\alpha|^2 \, \|x\|^2 \end{aligned}$$

since $x \cdot x$ is real. Hence $\|\alpha x\| = |\alpha| \, \|x\|$ and axiom (iii) is satisfied. In order to show that the triangle inequality holds we will use the inequality $|x \cdot y| \leqslant \|x\| \, \|y\|$ which is proved below. Using this inequality and the properties of an inner product, we obtain

$$\begin{aligned} \|x + y\|^2 &= (x+y) \cdot (x+y) = (x+y) \cdot x + (x+y) \cdot y \\ &= {}^*x \cdot x + y \cdot x + x \cdot y + y \cdot y \\ &= \|x\|^2 + 2\mathrm{Re}(x \cdot y) + \|y\|^2 \\ &\leqslant \|x\|^2 + 2|x \cdot y| + \|y\|^2 \\ &\leqslant \|x\|^2 + 2\,\|x\|\,\|y\| + \|y\|^2 \\ &= (\|x\| + \|y\|)^2. \end{aligned}$$

*It is easy to verify that $(x+y) \cdot z = x \cdot z + y \cdot z$.

The triangle inequality then follows by taking square roots on both sides of this inequality.

THEOREM 7.1. (The Schwarz Inequality.) *If x and y belong to an inner product space, then*

$$|x \cdot y| \leqslant \|x\| \, \|y\|,$$

where $\|\ \|$ represents the inner product norm.

Proof. If $\|y\| = 0$, then $y = \emptyset$ and the inequality holds. Otherwise let $\alpha = x \cdot y / \|y\|^2$. Then, using the result of Problem 7.6, we have

$$\begin{aligned} 0 \leqslant \|x - \alpha y\|^2 &= (x - \alpha y) \cdot (x - \alpha y) \\ &= \|x\|^2 - \alpha(\overline{x \cdot y}) - \overline{\alpha}(x \cdot y) + |\alpha|^2 \, \|y\|^2 \\ &= \|x\|^2 - |x \cdot y|^2 / \|y\|^2. \end{aligned}$$

The result now follows by straightforward manipulations on this inequality.

The concept of *orthogonality* is quite important throughout mathematics. For example, orthogonality plays a crucial role in the theory of Fourier series, in linear algebra, and in certain parts of the theory of differential and integral equations.

DEFINITION 7.3. *The elements x and y of an inner product space are said to be orthogonal iff $x \cdot y = 0$. In this case we write $x \perp y$.*

DEFINITION 7.4. *The subset A of an inner product space is said to be an orthogonal set iff $x \perp y$ whenever x and y are distinct elements of A. To say that A is an orthonormal set means that A is orthogonal and each member of A has unit length in the inner product norm.*

EXAMPLE 7.3. In E_3, let

$$u = \left(\frac{1}{\sqrt{2}}, \frac{1}{\sqrt{2}}, 0\right), \ v = \left(\frac{1}{\sqrt{2}}, -\frac{1}{\sqrt{2}}, 0\right), \text{ and } w = (0,0,1).$$

By computing all possible inner products involving u, v, w, one can readily see that these three vectors form an orthonormal set.

PROBLEMS

7.1. Let an inner product on $C[0,1]$ be defined by

$$f \cdot g = \int_0^1 f(t) g(t) dt.$$

Let $f(t) = 2t$ and $g(t) = t^2$ for every $t \in [0,1]$.
(a) Compute $f \cdot g$.
(b) Compute the distance between f and g relative to the inner product norm.

7.2. Let $x = (1,1,7,5)$ and $y = (2,0,1,6)$. Using the inner product (7.1) compute:
(a) $x \cdot y$
(b) $x \cdot (y - x)$
(c) $x \cdot y - x \cdot x$
(d) $\|x\|$
(e) $\|y - x\|$

7.3. Prove that equation (7.1) defines an inner product on E_n.

7.4. Let $\{x_1, x_2, \ldots, x_n\}$ be an orthonormal set of elements from an inner product space. Show that this set is linearly independent.

7.5. Show that if $x \perp y$, then $\|x + y\|^2 = \|x\|^2 + \|y\|^2$ in the inner product norm. This is called the *Pythagorean Theorem.*

7.6. Show that $x \cdot (\alpha y) = \overline{\alpha}(x \cdot y)$ if x and y are elements of an inner product space and $\alpha \in F$.

7.7. Let $x_1, \ldots, x_n$ be orthonormal and suppose

$$y = \alpha_1 x_1 + \ldots + \alpha_n x_n.$$

(a) Show that $\alpha_j = y \cdot x_j$, $j = 1, 2, \ldots, n$.
(b) Show that $\|y\|^2 = |\alpha_1|^2 + \ldots + |\alpha_n|^2$.

7.8. Prove that $x \cdot \emptyset = \emptyset \cdot x = 0$ for all x in an inner product space.

8. LEAST SQUARES APPROXIMATION

The *least squares method* is an important technique for obtaining approximations. Let $\{\varphi_1, \varphi_2, \ldots, \varphi_n\}$ be an orthonormal subset of an inner product space X. Given an element $f \in X$, we seek an approximation to f from the linear subspace spanned by $\{\varphi_1, \varphi_2, \ldots, \varphi_n\}$. The following theorem explains how to compute the best such approximation relative to the inner product norm.

THEOREM 8.1. *Let $\{\varphi_1, \varphi_2, \ldots, \varphi_n\}$ be an orthonormal subset of X. For each $f \in X$, the best approximation to f of the form*

$$\alpha_1 \varphi_1 + \alpha_2 \varphi_2 + \ldots + \alpha_n \varphi_n$$

relative to the inner product norm is obtained by setting

$$\alpha_j = f \cdot \varphi_j, \qquad j = 1, 2, \ldots, n.$$

Proof. Suppose $\alpha_j = f \cdot \varphi_j$ and $\beta_1 \varphi_1 + \ldots + \beta_n \varphi_n$ is a typical element of span $\{\varphi_1, \varphi_2, \ldots, \varphi_n\}$. Then

$$\left\|f-\sum_{j=1}^{n}\beta_j\varphi_j\right\|^2 = \left(f-\sum_{j=1}^{n}\beta_j\varphi_j\right)\cdot\left(f-\sum_{j=1}^{n}\beta_j\varphi_j\right)$$

$$(8.1) \qquad = \|f\|^2 - \sum_{j=1}^{n}\beta_j\varphi_j\cdot f - \sum_{j=1}^{n}\overline{\beta}_j f\cdot\varphi_j + \sum_{j=1}^{n}|\beta_j|^2$$

$$= \|f\|^2 - \sum_{j=1}^{n}(\beta_j\overline{\alpha}_j + \overline{\beta}_j\alpha_j) + \sum_{j=1}^{n}|\beta_j|^2.$$

Next note that

$$\sum_{j=1}^{n}|\alpha_j-\beta_j|^2 = \sum_{j=1}^{n}(\alpha_j-\beta_j)(\overline{\alpha}_j-\overline{\beta}_j)$$

$$= \sum_{j=1}^{n}|\alpha_j|^2 - \sum_{j=1}^{n}(\beta_j\overline{\alpha}_j + \overline{\beta}_j\alpha_j) + \sum_{j=1}^{n}|\beta_j|^2.$$

By subtracting and adding $\Sigma_{j=1}^{n}|\alpha_j|^2$ from the right side of equation (8.1), we obtain

$$(8.2) \qquad \left\|f-\sum_{j=1}^{n}\beta_j\varphi_j\right\|^2 = \|f\|^2 - \sum_{j=1}^{n}|\alpha_j|^2 + \sum_{j=1}^{n}|\alpha_j-\beta_j|^2.$$

From this equation we see that the minimum value of $\|f-\Sigma_{j=1}^{n}\beta_j\varphi_j\|$ is obtained if and only if $\beta_j=\alpha_j$ for each j.

As a corollary we obtain an inequality known as *Bessel's inequality*, in addition to an expression for $\|f-\Sigma_{j=1}^{n}\alpha_j\varphi_j\|$.

COROLLARY 8.2. *If* $\alpha_j = f\cdot\varphi_j$, $j=1,2,\ldots,n$, *then*

$$\text{(i)} \qquad \sum_{j=1}^{n}|\alpha_j|^2 \leqslant \|f\|^2,$$

$$\text{(ii)} \qquad \left\|f-\sum_{j=1}^{n}\alpha_j\varphi_j\right\|^2 = \|f\|^2 - \sum_{j=1}^{n}|\alpha_j|^2.$$

Proof. Set $\beta_j=\alpha_j$ in equation (8.2) and observe that

$$\left\|f-\sum_{j=1}^{n}\alpha_j\varphi_j\right\|^2 \geqslant 0.$$

EXAMPLE 8.1. Let $CK[-\pi,\pi]$ be the collection of all continuous complex-valued functions defined on the interval $[-\pi,\pi]$. For f and $g\in CK[-\pi,\pi]$ define

$$f\cdot g = \int_{-\pi}^{\pi} f(t)\overline{g(t)}dt.$$

Let

$$\varphi_j(t) = \frac{e^{ijt}}{\sqrt{2\pi}}, \qquad j=0,\pm1,\pm2,\ldots,\pm n,$$

where $i = \sqrt{-1}$. Then $\{\varphi_{-n}, \varphi_{-(n-1)}, \ldots, \varphi_0, \varphi_1, \ldots, \varphi_n\}$ is an orthonormal set. For a function $f \in CK[-\pi,\pi]$, the coefficients of the best approximation are given by the equations

$$\alpha_j = f \cdot \varphi_j = \frac{1}{\sqrt{2\pi}} \int_{-\pi}^{\pi} f(t) e^{-ijt} dt, \qquad j = 0, \pm 1, \ldots, \pm n.$$

The approximation to f is the nth partial sum of its Fourier series (see Problem 8.1).

EXAMPLE 8.2. For f and $g \in C[-1,1]$, let

$$f \cdot g = \int_{-1}^{1} \frac{f(t)g(t)dt}{\sqrt{1-t^2}}.$$

Let $\theta(x) = \text{arc cos } x$ and $T_j(x) = \cos[j\theta(x)]$. Then $\{T_0, T_1, \ldots, T_n\}$ forms an orthogonal set for any n. The coefficients of the best approximation to f from span $\{T_0, T_1, \ldots, T_n\}$ are given by the equations

$$\alpha_j = c_j \int_{-1}^{1} \frac{f(t)T_j(t)dt}{\sqrt{1-t^2}} = c_j \int_0^{\pi} f(\cos\theta) \cos j\theta \, d\theta,$$

where $c_0 = 1/\pi$ and $c_j = 2/\pi$ for $j > 0$. The function T_j is called the jth *Chebyshev polynomial of the first kind*, and the series $\Sigma_{j=0}^{\infty} \alpha_j T_j$ is known as the *Fourier-Chebyshev series for f*. Partial sums of this series frequently provide very good approximations to f relative to the uniform norm.

Examples 8.1 and 8.2 will be discussed further in Chapter 7.

PROBLEMS

In Problems 8.1–8.3 use the inner product and notation from Example 8.1.

8.1. Using the identity $e^{ijt} = \cos jt + i \sin jt$, write

$$\frac{1}{\sqrt{2\pi}} \sum_{j=-n}^{n} \alpha_j e^{ijt}$$

in the form

$$\sum_{j=0}^{n} (a_j \cos jt + b_j \sin jt).$$

Give formulas for a_j and b_j in terms of

$$\int_{-\pi}^{\pi} f(t) \cos jt \, dt \qquad \text{and} \qquad \int_{-\pi}^{\pi} f(t) \sin jt \, dt.$$

8.2. Let $f(t) = t$ for every $t \in [-\pi,\pi]$. Find the best approximation to f from span $\{\varphi_{-1}, \varphi_0, \varphi_1\}$ relative to the inner product norm. Also compute the norm of the error using Corollary 8.2.

8.3. Repeat Problem 8.2 with $f(t) = t^2$.

In Problems 8.4–8.9 use the inner product and notation given in Example 8.2.

8.4. Show that $f \cdot g = \int_0^\pi f(\cos\theta)g(\cos\theta)\,d\theta$.

8.5. Let $f(t) = t^3$ for every $t \in [-1,1]$. Use Bessel's inequality to find an upper bound for $|\alpha_0|^2 + |\alpha_1|^2 + |\alpha_2|^2$. Then compute α_0, α_1, and α_2 and see if Bessel's inequality really works in this case.

8.6. The integral $\int_{-1}^{1} f(t)g(t)(1-t^2)^{-1/2}\,dt$ is improper. Show that it converges whenever f and g are continuous on $[-1,1]$.

8.7. Compute $\|T_n\|$.

8.8. Let $f(t) = t^2 + 2t + 6$ for each $t \in [-1,1]$.
   (a) Compute the best approximation to f from span $\{T_0, T_1\}$ relative to the inner product norm.
   (b) Compute the norm of the error curve.
   (c) Approximate f by two terms of its Maclaurin series.
   (d) Compare the *uniform* norms of the error curves in (a) and (c).

8.9. Show that the Chebyshev polynomial T_n is really a polynomial.

*9. THE THEORY OF LINEAR PROGRAMMING

Linear programming provides a good illustration of the application of linear space concepts to a practical problem. In this section we will consider only the theoretical aspects of linear programming—a study of the computational aspects will be postponed until after the discussion of matrix operations in Chapter 3. A complete development of the theory would take us outside the scope of this course; so we will quote without proof several results from higher analysis which we hope the reader will find intuitively reasonable.

Problem 9.1 gives an example of an industrial resource allocation problem which can be solved by linear programming. A number of case histories on applications of linear programming can be found in various texts such as Gass [8], Garvin [7], and Dantzig [4].

The linear programming problem is an optimization problem with constraints. Let T be a linear functional defined on the Euclidean space E_n, and let K be the set of points $x \in E_n$ satisfying the constraints:

1. All coordinates of x are non-negative.
2. The coordinates of x satisfy the system of inequalities

$$\begin{aligned} a_{11}x_1 + a_{12}x_2 + \dots + a_{1n}x_n &\geqslant b_1, \\ a_{21}x_1 + a_{22}x_2 + \dots + a_{2n}x_n &\geqslant b_2, \\ &\vdots \\ a_{m1}x_1 + a_{m2}x_2 + \dots + a_{mn}x_n &\geqslant b_m, \end{aligned} \tag{9.1}$$

where the a_{ij}'s and b_i's are given real numbers.

The linear programming problem is to find a vector x^* which minimizes T over K, that is, to find $x^* \in K$ such that

$$T(\mathrm{x}^*) = \min_{x \in K} T(x).$$

The problem of maximizing T over K is also defined to be a linear programming problem.

The set K is called the *feasibility set* and T is referred to as the *objective function* or *cost function*. The cost function can always (see Problem 9.2) be written in the form

$$T(x) = c_1 x_1 + c_2 x_2 + \ldots + c_n x_n,$$

where $c_1, c_2, \ldots, c_n$ are real numbers which are referred to as the *cost coefficients*. It should be noted that the maximization and minimization problems are essentially the same since

$$\min_{x \in K} T(x) = -\max_{x \in K} [-T(x)].$$

EXAMPLE 9.1. For each $x \in E_2$, let $T(x) = 2x_1 + x_2$. Let K be the set of points $x \in E_2$ satisfying the constraints: $x_1 \geqslant 0$, $x_2 \geqslant 0$, and $x_1 + x_2 \leqslant 1$. [Note that the third inequality can be written as $-x_1 -x_2 \geqslant -1$ to conform with (9.1).] The problem is to find a vector $x^* \in K$ which maximizes T.

It will be instructive to look at the geometric interpretation of this problem. The feasibility set is the shaded triangular region in Figure 9.1.

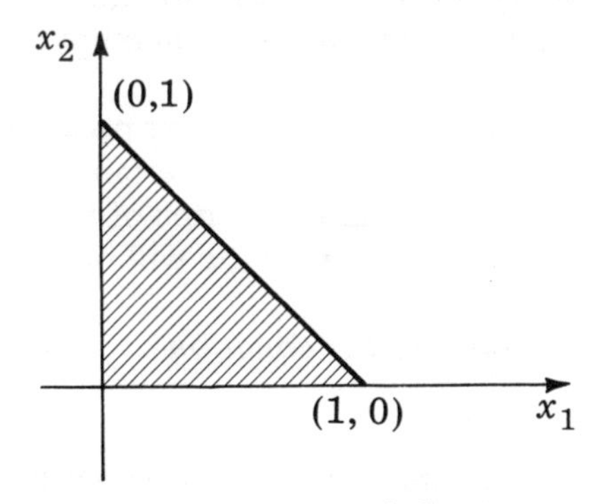

FIGURE 9.1. The Feasibility Set.

The equation $T(x) = 2x_1 + x_2$ describes a plane. The problem is to find the highest point on the plane above the feasibility set (see Figure 9.2 on the next page). From the figure it appears that the highest point on the cost plane occurs at (1,0). We can show that this is indeed the case, since the inequalities $x_1 + x_2 \leqslant 1$, $x_1 \geqslant 0$, $x_2 \geqslant 0$

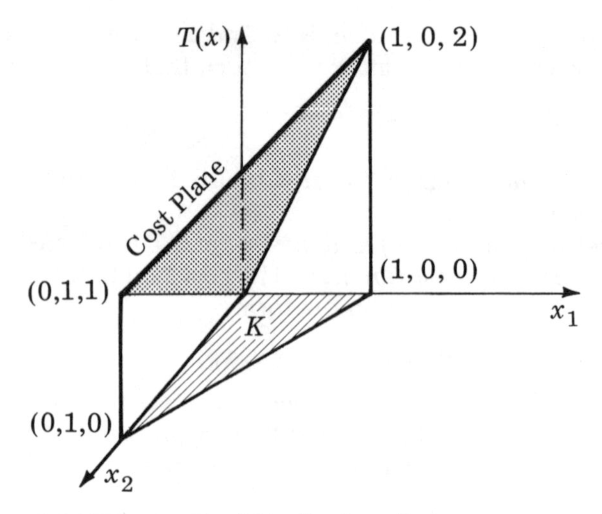

FIGURE 9.2. The Objective Function.

imply that $x_1 \leqslant 1$. Thus

$$T(x) = 2x_1 + x_2 = x_1 + (x_1 + x_2) \leqslant x_1 + 1 \leqslant 2 = T(1,0).$$

Thus $x^* = (1,0)$ is a solution to this particular linear programming problem, and the maximum value of T over K is $T(1,0) = 2$.

Note that the maximum occurs at a vertex of the triangular region K. It turns out that this is the typical behavior of linear programming problems—if there is a maximum value, then there is a vertex of the feasibility set K where the maximum value is assumed. This allows us to restrict our search for a maximum to the vertices of K.

In order to discuss the theory of linear programming, it will be convenient to introduce some geometric terminology. (Higher dimensional problems are best visualized in terms of two- or three-dimensional pictures.) In the sequel x, y, and z will be members of E_n.

DEFINITION 9.1. *Consider the equation*

$$(9.2) \qquad z = \alpha x + (1 - \alpha)y,$$

where α is a real number and x and y are distinct points of E_n.

(i) *The set of all points z satisfying (9.2) with $0 \leqslant \alpha \leqslant 1$ is called the closed line segment joining x and y. This is denoted by $[x,y]$.*
(ii) *The set of all points z satisfying (9.2) with $0 < \alpha < 1$ is called the open line segment joining x and y. This is denoted by (x,y).*
(iii) *The set of all points z satisfying (9.2) with $\alpha > 0$ is called the ray from y through x. This will be denoted by* Ray $[x,y]$.

(iv) *The set of all points z satisfying* (9.2) *(with no restriction on* α*) is called the line through x and y. This will be denoted by* Line $[x,y]$.

DEFINITION 9.2. *The set* $A \subset E_n$ *is said to be convex iff* $[x,y] \subset A$ *whenever x and* $y \in A$.

EXAMPLE 9.2. Figure 9.3 shows six regions in E_2, three of which are convex and three of which are not. Each of the non-convex regions shows a line segment which does not lie within the set.

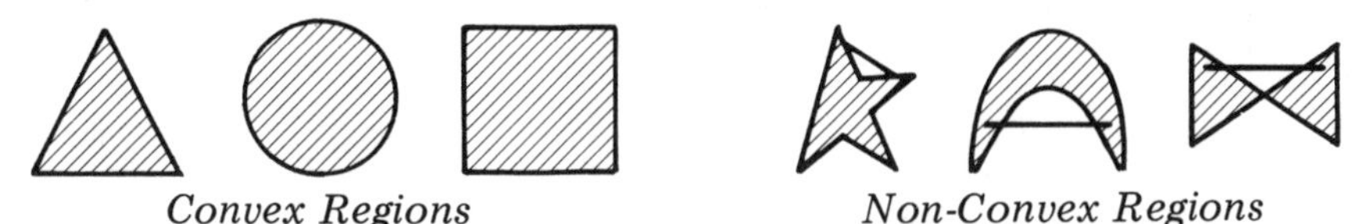

FIGURE 9.3. Convexity.

The convexity of the feasibility set plays a fundamental role in the theory of linear programming.

THEOREM 9.1. *The feasibility set K in the linear programming problem is convex.*

Proof. Suppose that x and y have non-negative coordinates and satisfy the inequalities (9.1). Let $z = \alpha x + (1-\alpha)y$ for some real number α, $0 \leqslant \alpha \leqslant 1$. Clearly z has non-negative coordinates, and for any i, $1 \leqslant i \leqslant m$,

$$\begin{aligned} a_{i1}z_1 &+ \ldots + a_{in}z_n \\ &= a_{i1}[\alpha x_1 + (1-\alpha)y_1] + \ldots + a_{in}[\alpha x_n + (1-\alpha)y_n] \\ &= \alpha(a_{i1}x_1 + \ldots + a_{in}x_n) + (1-\alpha)(a_{i1}y_1 + \ldots + a_{in}y_n) \\ &\geqslant \alpha b_i + (1-\alpha)b_i = b_i. \end{aligned}$$

Thus z satisfies (9.1) and $[x,y] \subset K$.

The notion of an extreme point is an extension of the idea of a vertex.

DEFINITION 9.3. *Let* $C \subset E_n$ *be convex. A point* $x \in C$ *is said to be an extreme point of C iff it does not lie on an open line segment joining any two distinct points of C.*

EXAMPLE 9.3. The extreme points of a triangular region are the vertices; the extreme points of a circular region are the points lying on the circle.

The fact we wish to develop is that the search for solutions to the linear programming problem can be restricted to the set of extreme points of K. The following theorem is the first step in this direction.

THEOREM 9.2. *If x^* is a solution to the linear programming problem which lies on an open line segment joining two distinct points x and $y \in K$, then every point on the closed line segment $[x,y]$ is a solution.*

Proof. By hypothesis, $x^* = \alpha x + (1 - \alpha)y$, where $x,y \in K$ and $0 < \alpha < 1$. Let

$$m = \min_{z \in K} T(z).$$

From the linearity of T we must have $m = T(x^*) = \alpha T(x) + (1 - \alpha) T(y)$. If either x or y is not a solution, then $T(x) > m$ or $T(y) > m$ and hence

$$m = \alpha T(x) + (1 - \alpha)T(y) > \alpha m + (1 - \alpha)m = m.$$

This contradiction shows that both x and y are solutions. The fact that each point on $[x,y]$ is a solution now follows readily from the linearity of T.

An easy corollary of this theorem is as follows:

COROLLARY 9.3. *If x^* is a unique solution to the linear programming problem, then x^* is an extreme point of the convex set K.*

Before proceeding further with our discussion of linear programming, we need to introduce several topological concepts. Let

$$\|x\| = \left[\sum_{i=1}^{n} x_i^2\right]^{1/2}$$

be the usual Euclidean norm.

DEFINITION 9.4. *Let r be a positive number. The open ball of radius r centered at the point $x \in E_n$ is the set*

$$N(x,r) = \{z: \|z - x\| < r\}.$$

DEFINITION 9.5. *The set $A \subset E_n$ is said to be open iff for each $x \in A$ there is a ball $N(x,r)$ such that $N(x,r) \subset A$.*

DEFINITION 9.6. *The set $A \subset E_n$ is said to be closed iff $E_n \sim A$ is open.*

A set is closed iff it contains all of its boundary points (see Problem 9.8). In most cases, sets defined only in terms of weak inequalities ($\leqslant$) are closed, while those defined only in terms of strict inequalities ($<$) are open if they are n-dimensional. For example, the closed line segment $[x,y]$ is closed.

In addition to being convex, the feasibility set in the linear programming problem is closed.

THEOREM 9.4. *The feasibility set K is closed.*

Proof. Since the intersection of closed sets is closed (Problem 9.9), it suffices to show that a set B of the form

$$B = \{x \in E_n \colon a_1 x_1 + \ldots + a_n x_n \geqslant b\}$$

is closed. To do this we will show that $E_n \sim B$ is open. If $x \in E_n \sim B$, then $a \cdot x < b$, where $a \cdot x = a_1 x_1 + \ldots + a_n x_n$ is the usual inner product for E_n. Using the Schwarz inequality (Theorem 7.1) we have for any $z \in E_n$,

$$|a \cdot z - a \cdot x| \leqslant \|a\| \, \|z - x\|.$$

Now let $\delta = b - a \cdot x$ and $r = \delta / \|a\|$. If $z \in N(x,r)$, we have

$$a \cdot z = a \cdot z - a \cdot x + a \cdot x \leqslant |a \cdot z - a \cdot x| + a \cdot x$$

$$\leqslant \|a\| \, \|z - x\| + a \cdot x < \|a\| r + a \cdot x = b.$$

Thus $z \in E_n \sim B$ and $N(x,r) \subset E_n \sim B$. It follows that $E_n \sim B$ is open and the proof is complete.

Next, let us quote a theorem from higher analysis.

THEOREM 9.5. *If K is a non-empty closed convex subset of E_n which contains no line, then K has at least one extreme point.*

An outline of the proof is given in Grunbaum [9], p. 18. This leads us to a fundamental fact about linear programming.

THEOREM 9.6. *If a linear programming problem has a solution, then it has an extreme point solution.*

Proof. Let x^* be a solution and let

$$m = T(x^*) = \min_{x \in K} T(x).$$

Let H be defined by the equation

$$H = \{x \in K \colon T(x) = m\}. \tag{9.3}$$

Now H is closed and convex (Problem 9.11). Note that $x^* \in H$; so H is non-empty. Furthermore, any line in E_n contains some points having negative coordinates (Problem 9.4). Thus H cannot contain any lines since one of the constraints is that all coordinates must be non-negative. Now H satisfies the hypotheses of Theorem 9.5 and must have an extreme point z. We claim that z is also an extreme point of K. If this were not the case, z would lie on an open line segment joining two distinct points of K, say $z \in (x,y)$. Then Theorem 9.2 implies that every point in $[x,y]$ is a solution, i.e., $[x,y] \subset H$. But this would contradict the fact that z is an extreme point of H. Thus z is an extreme point of K and the proof is complete.

Let us now state conditions which will guarantee the existence of a solution to a linear programming problem.

DEFINITION 9.7. *The set $A \subset E_n$ is said to be bounded iff there exists a constant R such that $\|x\| < R$ for all $x \in A$.*

THEOREM 9.7. *If the feasibility set K is non-empty and bounded, then the linear programming problem has a solution.*

An outline of the proof is as follows: The function T is continuous (Problem 9.12), and K is closed by Theorem 9.4. A famous theorem from higher analysis [1] states that if a real-valued function is continuous on a non-empty closed and bounded subset of E_n, then it is bounded on that set and takes on its maximum and minimum values. Thus there is a solution.

To show that K is non-empty one must exhibit at least one vector with non-negative coordinates which satisfies (9.1). For example, if $b_1, b_2, \ldots, b_n$ in (9.1) were all negative, then $\emptyset$ would satisfy (9.1). Two conditions which would imply the boundedness of K are:

1. An upper bound on each coordinate.
2. An inequality of the form

$$a_{i1}x_1 + \ldots + a_{in}x_n \leqslant b_i$$

with $a_{i1}, \ldots, a_{in}$ all positive.

In summary, we have shown that the search for solutions to a linear programming problem can be restricted to extreme points of the feasibility set. The simplex method, which we will study in Chapter 3, is a method for stepping from extreme point to extreme point in such a way as to always decrease the value of T in the case of a minimization problem, or to increase T for a maximization problem.

PROBLEMS

9.1. A furniture company manufactures colonial and Danish chairs. Each colonial chair requires 2 man-hours to build including 1 hour on a lathe, while the Danish chairs require 4 man-hours including half an hour on a lathe. The material for each chair costs $5.00. The company has 10 lathes and 40 carpenters available up to 40 hours per week, and a line of credit which allows the purchase of at most $2500 worth of materials per week. The profit per chair is $4.00 for the colonial and $3.00 for the Danish. We wish to determine the number of each type of chair that should be produced in order to maximize profits. Express these statements in the form of a linear programming problem.

9.2. Let T be a real-valued linear functional defined on E_n. Show that there is a vector $c \in E_n$ such that $T(x) = c \cdot x$ for all x.

9.3. Write parametric equations for the line segment $[a,b] \subset E_2$. Compare these equations with (9.2).

9.4. Show that if x and $y \in E_n$ and $x \neq y$, then Line $[x,y]$ contains some points which have negative coordinates.

9.5. Prove or disprove:
(a) The intersection of two convex sets is convex.
(b) The union of two convex sets is convex.
(c) Repeat (a), replacing "two" by "a family of."

9.6. Let $\overline{N}(x,r) = \{z: \|z - x\| \leqslant r\}$ be the closed ball about a point $x \in E_n$ (with the usual norm). Show that z is an extreme point of $\overline{N}(x,r)$ iff $\|z - x\| = r$.

9.7. Show that an open ball $N(x,r)$ in E_n is open and convex. Show that a closed ball is closed and convex.

9.8. Let X be a normed linear space. A point x is said to be on the boundary of a set $A \subset X$ iff every open ball about x intersects both A and $X \sim A$. Show that A is closed iff it contains all of its boundary points.

9.9. Prove or disprove:
(a) The union of a family of open sets is open.
(b) The intersection of a family of closed sets is closed.

9.10. Show that an extreme point of a convex set $K \subset E_n$ must be on the boundary of K.

9.11. Show that the set H defined by (9.3) is closed and convex.

9.12. Prove that every linear functional defined on E_n is continuous.

9.13. Give an example of a linear programming problem which does not have a solution.

9.14. Solve the linear program from Problem 9.1. Which resources are being used to capacity? How large a line of credit could be used without increasing the number of carpenters or machines? How much would that improve profits?

9.15. Suppose the furniture company in Problem 9.1 has increased its credit for materials to \$3500 and has 1200 square feet available for expansion. Each additional carpenter requires 200 square feet, while each lathe requires 100 square feet. We would like to determine what mixture of additional carpenters and machines would provide the most profit. Can this be formulated in terms of linear programming? Explain.

9.16. A technique for finding the maximum of a real-valued function T of n variables is to set all of the partial derivatives equal to zero and solve the resulting system of equations. Discuss the applicability of this technique to linear programming.

*9.17. Discuss the applicability of Lagrange multipliers to linear programming.

9.18. A theorem from higher analysis (see Grunbaum [9], p. 18) states that if $K \subset E_n$ is closed, convex, and bounded, then every point $x \in K$ can be expressed in the form

$$x = \sum_{i=1}^{m} \alpha_i z_i,$$

where $\alpha_i > 0$ for all i, $\sum_{i=1}^{m} \alpha_i = 1$, and $z_1, \ldots, z_m$ are extreme points of K. Use this result to prove the existence of an extreme point solution to the linear programming problem in case the feasibility set is non-empty and bounded.

9.19. A functional T on E_n is said to be convex iff

$$T(\alpha x + (1 - \alpha)y) \leqslant \alpha T(x) + (1 - \alpha)T(y)$$

whenever $x, y \in E_n$ and $0 \leqslant \alpha \leqslant 1$. Show that if a convex functional T has a maximum on a closed convex set K, then there is an extreme point z of K such that

$$T(z) = \max_{x \in K} T(x).$$

REFERENCES

1. T. M. Apostol, *Mathematical Analysis*, Addison-Wesley, 1957.
2. A. Charnes, W. W. Cooper, and A. Henderson, *An Introduction to Linear Programming*, Wiley, 1953.
3. E. W. Cheney, *Introduction to Approximation Theory*, McGraw-Hill, 1966.
4. G. B. Dantzig, *Linear Programming and Extensions*, Princeton University Press, 1963.
5. P. J. Davis, *Interpolation and Approximation*, Blaisdell, 1963.
6. D. K. Faddeev and V. N. Faddeeva, *Computational Methods of Linear Algebra*, Freeman, 1963.
7. W. W. Garvin, *Introduction to Linear Programming*, McGraw-Hill, 1960.
8. S. I. Gass, *Linear Programming*, McGraw-Hill, 1958.

9. B. Grunbaum, *Convex Polytopes*, Interscience, 1967.
10. P. R. Halmos, *Finite-Dimensional Vector Spaces*, Van Nostrand, 1958.
11. P. Henrici, *Elements of Numerical Analysis*, Wiley, 1964.
12. B. Kreko, *Lehrbuch der Linearen Optimierung*, VEB Deutscher Verlag der Wissenschaften, Berlin, 1965.
13. M. Marcus and H. Minc, *Introduction to Linear Algebra*, Macmillan, 1965.
14. W. A. Porter, *Modern Foundations of Systems Engineering*, Macmillan, 1966.
15. M. Simonnard, *Programmation Linearen*, Dunod, Paris, 1962.

CHAPTER 3

Matrix Operations

1. INTRODUCTION

Many mathematical problems require the solution of a linear system of equations or the computation of certain numbers associated with the coefficients of such a system. For example, a differential or integral equation is often replaced by an approximate linear system of equations. Such systems are conveniently discussed in terms of matrices. As we will see in Section 2, matrices are very closely related to linear transformations.

Roughly speaking, a matrix is a rectangular array of numbers. More precisely, let $L(m,n) = \{(i,j)\colon i \text{ and } j \text{ are integers, } 1 \leqslant i \leqslant m, \text{ and } 1 \leqslant j \leqslant n\}$.

DEFINITION 1.1. *An m by n matrix is a function which associates a real or complex number with each member of $L(m,n)$.*

We will use the symbol $M(m,n)$ for the collection of all m by n matrices. A matrix is customarily denoted by a capital letter, and the value of the matrix at (i,j) is denoted by the corresponding lowercase letter with double subscripts. For example, if A represents a matrix, then a_{ij} represents the value of the matrix at (i,j). (The notation $A(i,j)$ instead of a_{ij} would be more consistent with the usual functional notation; in fact, this notation is used in many programming languages.) It is customary to refer to a_{ij} as the (i,j)th element of A, and to display the matrix A in rows and columns as follows:

$$A = \begin{bmatrix} a_{11} & a_{12} & \cdots & a_{1n} \\ a_{21} & a_{22} & \cdots & a_{2n} \\ \vdots & & & \\ a_{m1} & a_{m2} & \cdots & a_{mn} \end{bmatrix}.$$

We call m the number of rows, n the number of columns, and we say that A has dimensions m by n. For brevity the equation $A = (a_{ij})$ is sometimes used to indicate that a_{ij} is the (i,j)th element of A. In discussions involving a given matrix A, we will always assume, in the absence of any contrary statement, that a_{ij} is the (ij)th element of A.

EXAMPLE 1.1. The equation

$$A = \begin{bmatrix} 2 & 1 & 3 \\ 7 & 4 & 6 \end{bmatrix}$$

defines a matrix A with two rows ($m = 2$) and three columns ($n = 3$). Here $a_{11} = 2$, $a_{12} = 1$, $a_{13} = 3$, $a_{21} = 7$, $a_{22} = 4$, and $a_{23} = 6$.

The matrices A and B are said to be equal iff A and B have the same number of rows (m) and columns (n) and $a_{ij} = b_{ij}$ for all $(i,j) \in L(m,n)$.

DEFINITION 1.2. *Let $A \in M(m,n)$. The transpose of A is the matrix $B \in M(n,m)$ defined by*

$$b_{ij} = a_{ji} \quad \text{for all} \quad (i,j) \in L(n,m).$$

The transpose of A is denoted by A' or A^T.

EXAMPLE 1.2. The transpose of the matrix A in Example 1.1 is given by

$$A' = \begin{bmatrix} 2 & 7 \\ 1 & 4 \\ 3 & 6 \end{bmatrix}.$$

Note that the rows of A form the columns of A'.

DEFINITION 1.3. *Let $A \in M(m,n)$. The conjugate of A is the matrix $B \in M(m,n)$ defined by*

$$b_{ij} = \overline{a}_{ij},$$

where $\overline{a}_{ij}$ is the complex conjugate of a_{ij}. The conjugate of A is denoted by $\overline{A}$.

DEFINITION 1.4. *The Hermitian transpose (adjoint) of the matrix A, denoted by A^*, is defined by the equation*

$$A^* = (\overline{A})'.$$

The Hermitian transpose is also denoted by A^H in the literature. We should note that the word "adjoint" has a different meaning in many algebra texts.

EXAMPLE 1.3. Define a 3 by 2 matrix by the equation

$$A = \begin{bmatrix} 3 & 1 \\ 1+i & i \\ 2+3i & 6 \end{bmatrix}.$$

Then

$$\overline{A} = \begin{bmatrix} 3 & 1 \\ 1-i & -i \\ 2-3i & 6 \end{bmatrix},$$

$$A^* = \begin{bmatrix} 3 & 1-i & 2-3i \\ 1 & -i & 6 \end{bmatrix}.$$

PROBLEMS

1.1. Let

$$A = \begin{bmatrix} 1 & 2+i & 3 \\ 4-i & 2i & 6 \\ 4 & 5i & 6+3i \end{bmatrix}.$$

Compute A', $\overline{A}$, and A^*.

1.2. Show that $(\overline{A})' = \overline{(A')}$.

1.3. Write a computer program to read in a 5 by 4 matrix, print out the matrix, and then print out its transpose. Assume the elements of the matrix are real numbers with two digits after the decimal point.

2. MATRIX ARITHMETIC

We will now consider the basic arithmetic operations which are defined for matrices. First, let us consider addition and multiplication by scalars.

DEFINITION 2.1. *Let A and $B \in M(m,n)$ and $\alpha \in K$ (the set of complex numbers). The sum $C = A + B$ and the product $D = \alpha A$ are defined by the equations*

$$c_{ij} = a_{ij} + b_{ij}$$

$$d_{ij} = \alpha a_{ij}$$

for all $(i,j) \in L(m,n)$.

These are element-by-element operations similar to those defined for E_n. Note that A and B must have the same dimensions in order to form the sum.

EXAMPLE 2.1. Let A and B be defined by the equations

$$A = \begin{bmatrix} 2 & 1 & 5 \\ 6 & 0 & 7 \end{bmatrix}, \qquad B = \begin{bmatrix} 3 & 4 & 2 \\ 1 & 8 & 5 \end{bmatrix}.$$

Then

$$A + B = \begin{bmatrix} 5 & 5 & 7 \\ 7 & 8 & 12 \end{bmatrix} \quad \text{and} \quad 5A = \begin{bmatrix} 10 & 5 & 25 \\ 30 & 0 & 35 \end{bmatrix}.$$

The set $M(m,n)$ forms a linear space under these two operations. That is, the eight algebraic laws listed in Section 1 of Chapter 2 hold for matrices. The proof of this fact is straightforward and is therefore omitted here.

If $A \in M(m,n)$, then the symbol $a_{i\cdot}$ will be used to represent the vector in EK_n formed from the elements of the ith row of A. Similarly, $a_{\cdot j}$ will represent the vector in EK_m formed by the jth column of A. For example, if A is the matrix in Example 2.1, then $a_{1\cdot} = (2,1,5)$, $a_{2\cdot} = (6,0,7)$, $a_{\cdot 1} = (2,6)$, $a_{\cdot 2} = (1,0)$, and $a_{\cdot 3} = (5,7)$. This notation will be helpful in discussing matrix multiplication.

DEFINITION 2.2. *Let $A \in M(m,n)$ and $B \in M(n,q)$. The product $C = AB$ is the member of $M(m,q)$ defined by the equation*

$$(2.1) \qquad c_{ij} = \sum_{k=1}^{n} a_{ik} b_{kj}$$

for all $(i,j) \in L(m,q)$.

An easy way to remember (2.1) is to replace it by the equivalent equation

$$(2.2) \qquad c_{ij} = a_{i\cdot} \cdot \overline{b}_{\cdot j} = (i\text{th row of } A) \cdot \overline{(j\text{th column of } B)},$$

where the inner product in (2.2) is the standard one for EK_n (see Example 7.2 in Chapter 2). If all the elements of A and B are real, then one can ignore the conjugates in (2.2).

EXAMPLE 2.2. Consider the three matrices

$$A = \begin{bmatrix} 1 & 2 & 3 \\ 4 & 5 & 6 \\ 7 & 8 & 9 \end{bmatrix}, \quad B = \begin{bmatrix} 1 & 0 & 1 \\ 2 & 1 & 3 \\ 4 & 2 & 1 \end{bmatrix}, \quad H = \begin{bmatrix} 2 & 1 \\ 1 & 1 \\ 3 & 2 \end{bmatrix}.$$

Let $C = AB$. Then:

$$c_{11} = a_{1\cdot} \cdot b_{\cdot 1} = (1,2,3)\cdot(1,2,4) = 1^2 + 2^2 + 3\cdot 4 = 17;$$

$$c_{23} = a_{2\cdot} \cdot b_{\cdot 3} = (4,5,6)\cdot(1,3,1) = 25.$$

The complete product is given by

$$C = AB = \begin{bmatrix} 17 & 8 & 10 \\ 38 & 17 & 25 \\ 59 & 26 & 40 \end{bmatrix}.$$

In a similar manner we obtain

$$BA = \begin{bmatrix} 8 & 10 & 12 \\ 27 & 33 & 39 \\ 19 & 26 & 33 \end{bmatrix},$$

$$AH = \begin{bmatrix} 13 & 9 \\ 31 & 21 \\ 49 & 33 \end{bmatrix},$$

$$BH = \begin{bmatrix} 5 & 3 \\ 14 & 9 \\ 13 & 8 \end{bmatrix}.$$

Note that $AB \neq BA$; thus matrix multiplication is not a commutative operation. In fact, the products HA and HB cannot even be formed because the number of columns in H is not the same as the number of rows in A or B.

Let us list some algebraic identities for matrix operations. For completeness, the list includes an identity and a remark concerning the matrix inverse, A^{-1}, which will be discussed later in this section. Those rules involving multiplication assume that the matrices are compatible for multiplication, while those involving inverses assume that the inverses exist, and so forth. Capital letters represent matrices and Greek letters denote numbers.

1. $A + B = B + A$
2. $(A + B) + C = A + (B + C)$, $(AB)C = A(BC)$, $(\alpha\beta)A = \alpha(\beta A)$
3. $A(B+C) = AB + AC$, $(B+C)A = BA + CA$
4. $\alpha(B+C) = \alpha B + \alpha C$, $(\beta + \gamma)A = \beta A + \gamma A$
5. $(A + B)' = A' + B'$
6. $(AB)' = B'A'$

7. $(A+B)^* = A^* + B^*$
8. $(AB)^* = B^*A^*$
9. $(AB)^{-1} = B^{-1}A^{-1}$

As is the case with numbers, the matrix $(A+B)^{-1}$ would seldom equal $A^{-1} + B^{-1}$, while in contrast to numbers, AB would seldom equal BA.

We now wish to show that there is a one-to-one correspondence between m by n matrices and linear transformations which map EK_n into EK_m. In discussions involving matrices, it is customary to write a vector $x \in EK_n$ as an n by 1 matrix (column vector):

$$x = \begin{bmatrix} x_1 \\ x_2 \\ \cdot \\ \cdot \\ \cdot \\ x_n \end{bmatrix}. \tag{2.3}$$

For convenience, single rather than double subscripts are used on the right side of (2.3) since the matrix has only one column. Let $A \in M(m,n)$ and let

$$y = \begin{bmatrix} y_1 \\ y_2 \\ \cdot \\ \cdot \\ \cdot \\ y_m \end{bmatrix} = Ax$$

represent the matrix product of A and x, where we again use single rather than double subscripts. Then $y \in EK_m$ and the equation

$$T(x) = Ax \quad \text{for each} \quad x \in EK_n \tag{2.4}$$

defines a transformation from EK_n into EK_m.

THEOREM 2.1. *Let $A \in M(m,n)$. Then the transformation defined by (2.4) is linear. Conversely, if T is a linear transformation from EK_n into EK_m, then there is a unique matrix $A \in M(m,n)$ which satisfies (2.4).*

Proof. Let T be defined by (2.4) and let $w = T(\alpha x + \beta z)$. From (2.1) we have for $i = 1,2,\ldots,m$,

$$w_i = \sum_{k=1}^{n} a_{ik}(\alpha x_k + \beta z_k) = \alpha \sum_{k=1}^{n} a_{ik}x_k + \beta \sum_{k=1}^{n} a_{ik}z_k.$$

Thus $w = \alpha T(x) + \beta T(z)$ and T is linear.

The proof of the converse is left for Problem 2.5.

A similar theorem holds for transformations from E_n into E_m. In this case the matrix A has real elements. It should be noted that under the above correspondence, composition of transformations corresponds to matrix multiplication (Problem 2.3).

In view of the one-to-one correspondence between matrices and linear transformations, it is possible to define properties of matrices in terms of the corresponding transformation. Let $A \in M(m,n)$ and let $T_A(x) = Ax$ for each $x \in EK_n$.

DEFINITION 2.3. *The rank and nullity of A are defined by the equations*

$$\textit{rank}\ (A) = \dim\ [\mathcal{R}(T_A)],$$

$$\textit{nullity}\ (A) = \dim\ [\mathcal{N}(T_A)],$$

where $\mathcal{R}(T_A)$ and $\mathcal{N}(T_A)$ represent the range and nullspace of T_A.

It follows from Theorem 5.3 in Chapter 2 that the sum of the rank and nullity of A must equal n. The rank is important in considerations involving the matrix inverse (defined below). Let us now restrict our attention to square n by n matrices.

DEFINITION 2.4. *The nth order identity matrix, denoted by I, is defined by the condition that its elements δ_{ij} must satisfy the equation*

$$\delta_{ij} = \begin{cases} 1 & \text{if } i = j. \\ 0 & \text{if } i \neq j. \end{cases}$$

The identity matrix is similar to the number 1; that is, $AI = IA = A$ for each $A \in M(n,n)$. The matrix inverse, which we will denote by A^{-1}, is similar to the reciprocal of a number in that $AA^{-1} = A^{-1}A = I$. However, not every matrix has an inverse. The theorems which follow lay the groundwork for a formal definition of the inverse.

DEFINITION 2.5. *The n by n matrix A is said to be singular iff rank $(A) < n$.*

THEOREM 2.2. *If $A \in M(n,n)$ is non-singular, then the equation*

$$(2.5) \qquad Ax = y$$

has a unique solution for each $y \in EK_n$.

Proof. Let $T(x) = Ax$. Now $\dim \mathfrak{N}(T) = 0$ and $\dim \mathfrak{R}(T) = n$. The result follows from Theorems 5.4 and 5.2 in Chapter 2.

THEOREM 2.3. *Let A and $X \in M(n,n)$. If $AX = I$, then both A and X have rank n.*

Proof. Suppose rank $(X) < n$. Then nullity $(X) \geqslant 1$ and there is a non-zero vector $z \in EK_n$ such that $Xz = \emptyset$. Using the associative law for matrix multiplication (Problem 2.3), we have

$$z = Iz = (AX)z = A(Xz) = A\emptyset = \emptyset.$$

This contradiction establishes the fact that rank $(X) = n$. If rank $(A) < n$, then there is a non-zero vector y such that $Ay = \emptyset$. By Theorem 2.2 there is a vector z such that $Xz = y$. Clearly $z \neq \emptyset$ because $y \neq \emptyset$. Thus

$$z = Iz = A(Xz) = Ay = \emptyset.$$

This contradiction shows that rank $(A) = n$ and the proof is complete.

From Theorem 2.3 we can see that there is no hope of finding a matrix A^{-1} such that $A^{-1}A$ or AA^{-1} is I if rank $(A) < n$. The following theorem shows that this will always be possible if rank $(A) = n$.

THEOREM 2.4. *Let A and $I \in M(n,n)$. The equations*

$$(2.6) \qquad AX = I$$

$$(2.7) \qquad XA = I$$

have unique solutions if rank $(A) = n$. Furthermore, the solution to both equations is the same.

Proof. For each $j = 1,2,\ldots,n$, let e_j be the jth column of the identity, and let $b_{.j}$ be the vector satisfying the condition $Ab_{.j} = e_j$. Let B be the matrix whose jth column is $b_{.j}$, $j = 1,2,\ldots,n$. It is easy to verify (see Problem 2.2) that $AB = I$ and thus (2.6) has a solution. Now rank $(B) = n$; so by the preceding argument there is a matrix C such that

$$(2.8) \qquad BC = I.$$

Multiplying both sides of (2.8) on the left by A, we see that $C = A$. Thus $AB = BA = I$. The uniqueness of the solution is left for Problem 2.7.

This leads us to the formal definition of an inverse.

DEFINITION 2.6. *Let $A \in M(n,n)$ be non-singular. The inverse of A, denoted by A^{-1}, is the unique matrix in $M(n,n)$ satisfying the conditions $AA^{-1} = A^{-1}A = I$.*

The solution to matrix equations can often be expressed in terms of the matrix inverse. For example, the equation

$$(A + B)X = C$$

has the solution

$$X = (A + B)^{-1}C,$$

providing that $A + B$ has an inverse.

PROBLEMS

2.1. Let A, B, C, and I be defined by the equations

$$A = \begin{bmatrix} 1 & 1 & 2 \\ 2 & 1 & 3 \\ 1 & 0 & 1 \end{bmatrix} \qquad B = \begin{bmatrix} 1 & 3 & 1 \\ 2 & 6 & 8 \\ 0 & 0 & 1 \end{bmatrix}$$

$$C = \begin{bmatrix} 1 & 0 & 0 & 1 \\ 2 & 0 & 1 & 2 \\ 1 & 1 & 0 & 5 \end{bmatrix} \qquad I = \begin{bmatrix} 1 & 0 & 0 \\ 0 & 1 & 0 \\ 0 & 0 & 1 \end{bmatrix}$$

Compute the following matrices:

(a) $A + B$
(b) $30A$
(c) $A - 3I$
(d) AB
(e) BA
(f) AC
(g) AI and IA
(h) $(A + B)'$ and $A'B'$

2.2. Let $C = AB$. Show that $c_{.j} = Ab_{.j}$, where $c_{.j}$ and $b_{.j}$ represent the jth columns of C and B. Also show that $c_{ij} = a_{i.} \cdot \bar{b}_{.j}$, where $a_{i.}$ is the ith row of A and the inner product is the usual one for EK_n [see equation (7.2) in Chapter 2].

2.3. Prove the *associative law* for matrix multiplication, that is, $(AB)C = A(BC)$. Use this result to show that composition of linear transformations corresponds to matrix multiplication.

2.4. Let $T(x) = Ax$, where

$$A = \begin{bmatrix} 1 & 0 & 2 \\ 0 & 1 & 0 \\ 2 & 0 & 1 \end{bmatrix}.$$

Compute the following:

(a) $T\begin{bmatrix} 1 \\ 5 \\ 10 \end{bmatrix}$ (b) $T\begin{bmatrix} 6 \\ 3 \\ 8 \end{bmatrix}$

(c) $T(x) + T(z)$
(d) $T(3x)$
(e) $T(x + z)$

2.5. Let T be a linear transformation mapping EK_n into EK_m. Show that there is a unique matrix $A \in M(m,n)$ such that $T(x) = Ax$ for all $x \in EK_n$.

2.6. Compute the inverse of the matrix A in Problem 2.4. Hint: One technique is suggested in the proof of Theorem 2.4.

2.7. Let $A \in M(n,n)$ be non-singular. Show that the solutions to equations (2.6) and (2.7) are unique, and that the solutions are the same for both equations.

2.8. Prove that $(AB)^{-1} = B^{-1}A^{-1}$, assuming that A and $B \in M(n,n)$ are non-singular.

2.9. Solve the following equations for X in terms of A, B, and C. Assume that A, B, C, and $X \in M(n,n)$, and that A and B have inverses.
(a) $AX + B = C$
(b) $BXA = C$
(c) $B^2X = AX + C$
(d) $ABX = C$

2.10. Let $A \in M(n,n)$. Show that

$$(Ax) \cdot y = x \cdot (A^* y)$$

for all $x,y \in EK_n$.

2.11. Let $\{u_1, u_2, \ldots, u_n\}$ be a basis for EK_n, and let $y_1, y_2, \ldots, y_n$ be the coordinates of $x \in EK_n$ with respect to this basis. Describe how to find a matrix A such that $y = Ax$.

2.12. Prove that $(AB)^* = B^*A^*$.

3. DETERMINANTS

The determinant of a square matrix is a certain number which is occasionally used in discussions involving linear systems of equations. It turns out that a square matrix A is non-singular if and only if the determinant A is not zero [12]. Thus by Theorem 2.2 the equation $Ax = y$ has a unique solution if the determinant of A is not zero. Also A has an inverse if and only if A has a non-zero determinant.

An inductive definition for the determinant of a square matrix of order n is given below. An inductive definition defines the determinant of a third order matrix in terms of determinants of second order matrices, fourth order in terms of third order, and so on.

Let A be a square matrix of order n. The $(n-1)$st order matrix formed by deleting the ith row and jth column of A is denoted by A_{ij}. The determinant of A is usually denoted by $|A|$ or $\det(A)$.

DEFINITION 3.1. *If A is of the second order, then*

$$|A| = a_{11}a_{22} - a_{12}a_{21}. \tag{3.1}$$

If A is of the nth order ($n > 2$), then

$$|A| = \sum_{i=1}^{n} (-1)^{1+i} a_{i1} |A_{i1}|. \tag{3.2}$$

EXAMPLE 3.1. Let

$$A = \begin{bmatrix} 3 & 1 \\ 4 & 6 \end{bmatrix}, \qquad B = \begin{bmatrix} 1 & 2 & 3 \\ 4 & 5 & 6 \\ 7 & 8 & 9 \end{bmatrix}.$$

By (3.1) we have

$$|A| = 3 \cdot 6 - 1 \cdot 4 = 14.$$

From (3.2) we obtain

$$\begin{aligned} |B| &= (-1)^2 b_{11}|B_{11}| + (-1)^3 b_{21}|B_{21}| + (-1)^4 b_{31}|B_{31}| \\ &= \begin{vmatrix} 5 & 6 \\ 8 & 9 \end{vmatrix} - 4\begin{vmatrix} 2 & 3 \\ 8 & 9 \end{vmatrix} + 7\begin{vmatrix} 2 & 3 \\ 5 & 6 \end{vmatrix} = 0. \end{aligned} \tag{3.3}$$

A rough estimate is sufficient to show that calculation of determinants directly from (3.2) is not feasible for matrices of large order. To get a lower bound on the time for such a calculation, let us count the number of multiplications required. For $n = 2$, we have 2 multiplications. Ignoring the multiplication by $(-1)^{i+1}$, the third order matrix requires evaluation of the determinants of 3 second order matrices, giving at least 6 multiplications. An easy inductive argument shows that at least $n!$ multiplications are required for the nth order matrix. Assuming 10 μ sec (i.e., $10 \cdot 10^{-6}$ sec) per multiplication, we see that $10 \cdot 10! \cdot 10^{-6} \doteq 36$ sec is a lower bound on the time required for a 10th order matrix. A similar calculation shows that nearly 5 months would be required for a 15th order matrix, while over half of a million years would be required for a matrix of the 20th order.

Actually, determinants are not normally used except for computations involving matrices of very low order. However, a more efficient scheme for calculation can be based on the following theorem.

THEOREM 3.1. *If a multiple of one row of a matrix is added to a different row, then the value of the determinant is unchanged. If two rows are interchanged, then the sign of the determinant is reversed.*

The proof of this theorem can be found in most books on linear algebra, matrix theory, or college algebra [12].

EXAMPLE 3.2. Let

$$A = \begin{bmatrix} -2 & 4 & 6 \\ 1 & 0 & 3 \\ 4 & 2 & -4 \end{bmatrix}.$$

We now add 0.5 times the first row to the second row, 2 times the first row to the third row, and −5 times the second row to the third row to obtain

$$(3.4) \qquad \begin{vmatrix} -2 & 4 & 6 \\ 1 & 0 & 3 \\ 4 & 2 & -4 \end{vmatrix} = \begin{vmatrix} -2 & 4 & 6 \\ 0 & 2 & 6 \\ 4 & 2 & -4 \end{vmatrix}$$

$$= \begin{vmatrix} -2 & 4 & 6 \\ 0 & 2 & 6 \\ 0 & 10 & 8 \end{vmatrix} = \begin{vmatrix} -2 & 4 & 6 \\ 0 & 2 & 6 \\ 0 & 0 & -22 \end{vmatrix}.$$

This process is called *triangularization of the matrix*, and the last matrix is said to be in *triangular form.* It follows from (3.2) that the determinant of a triangular matrix is the product of the diagonal elements. Thus $|A| = (-2)(2)(-22) = 88$. Note that this example requires 3 divisions, 7 multiplications, and 5 additions, compared with 0 divisions, 9 multiplications, and 5 additions required for the evaluation of $|B|$ by (3.2) in Example 3.1. In (3.4), the multiplication of (−2) by 0.5 and addition of the result to 1 was not counted since it is a foregone conclusion that the result will be zero, and thus a computer program would not have to include this operation. However, in (3.3), the implied multiplication of

$$\begin{vmatrix} 5 & 6 \\ 8 & 9 \end{vmatrix}$$

by 1 was counted since a general computer program based on (3.2) would have to include this operation. We will now show that the method described in this example is markedly faster than formula (3.2) for larger values of n.

DEFINITION 3.2. *The square matrix A is said to be in (upper) triangular form iff $a_{ij} = 0$ whenever $j < i$.*

DEFINITION 3.3. *The process of computing the determinant of a matrix by reducing it to triangular form is called the triangular method.*

THEOREM 3.2. *Let A be a square matrix of order n. The determinant of A can be evaluated by the triangular method using the following number of arithmetic operations:*

Additions $n(n-1)(2n-1)/6$
Multiplications $(n-1) + n(n-1)(2n-1)/6$
Divisions $n(n-1)/2$

Proof. The first step is to compute $m_1 = -a_{21}/a_{11}$. Then the quantities $m_1a_{12} + a_{22}$, $m_1a_{13} + a_{23}, \ldots, m_1a_{1n} + a_{2n}$ are computed. These calculations require 1 division and $n-1$ multiplications and additions. Thus the entire process of obtaining zeros in the first column requires $n-1$ divisions and $(n-1)^2$ multiplications and additions. Similarly, the process of obtaining zeros in the second column requires $n-2$ divisions and $(n-2)^2$ multiplications and additions. Formulas for the sum of the first $(n-1)$ integers and their squares can be found in the CRC Tables [18]. It follows that the total number of operations required to transform A into triangular form is given by:

Additions and multiplications
$$1 + 2^2 + 3^2 + \ldots + (n-1)^2 = n(n-1)(2n-1)/6.$$

Divisions
$$1 + 2 + 3 + \ldots + (n-1) = n(n-1)/2.$$

The theorem follows now from the fact that $n-1$ multiplications are required in computing the product of the diagonal elements.

PROBLEMS

3.1. Evaluate $|A|$ by means of (3.2) and also by means of the triangular method, where

$$A = \begin{bmatrix} 2 & 1 & 3 \\ 1 & 3 & 1 \\ 2 & 1 & 5 \end{bmatrix}.$$

3.2. Evaluate $|B|$ by the triangular method. Keep a tally on the number of operations which would be required by a computer program and compare with Theorem 3.2.

$$B = \begin{bmatrix} 1 & 0 & 1 & 1 \\ 2 & 1 & 3 & 1 \\ 1 & 2 & 0 & 1 \\ 3 & 1 & 1 & 1 \end{bmatrix}.$$

3.3. A hypothetical computer requires 5 μ sec per addition and 10 μ sec per multiplication or division. Compute the time required (considering only these operations) to calculate the determinant of an n by n matrix by the triangular method if
(a) $n = 5$
(b) $n = 10$
(c) $n = 15$
(d) $n = 20$
Compare these results with the lower bound of $n!$ multiplications required by (3.2).

4. SYSTEMS OF EQUATIONS

In this section we will describe a technique for solving linear systems of equations. Let us first see how such systems of equations can be expressed in matrix notation.

EXAMPLE 4.1. Consider the system of equations

$$\begin{aligned} x + y + z &= 9, \\ x - y + 2z &= 9, \\ 2x + 4y - z &= 14. \end{aligned} \tag{4.1}$$

Corresponding elements of equal matrices must be equal; thus the equations (4.1) can be written in the equivalent matrix form:

$$\begin{bmatrix} 1 & 1 & 1 \\ 1 & -1 & 2 \\ 2 & 4 & -1 \end{bmatrix} \begin{bmatrix} x \\ y \\ z \end{bmatrix} = \begin{bmatrix} 9 \\ 9 \\ 14 \end{bmatrix}.$$

Now consider the general system of n linear equations in n unknowns:

$$
\begin{aligned}
&a_{11}x_1 + a_{12}x_2 + \ldots + a_{1n}x_n = b_1, \\
&a_{21}x_1 + a_{22}x_2 + \ldots + a_{2n}x_n = b_2, \\
&\qquad \vdots \\
&a_{n1}x_1 + a_{n2}x_2 + \ldots + a_{nn}x_n = b_n.
\end{aligned}
\tag{4.2}
$$

This system of equations is equivalent to the matrix equation

$$Ax = b, \tag{4.3}$$

where

$$
A = \begin{bmatrix} a_{11} & a_{12} & \ldots & a_{1n} \\ a_{21} & a_{22} & \ldots & a_{2n} \\ \vdots & & & \\ a_{n1} & a_{n2} & \ldots & a_{nn} \end{bmatrix}, \quad x = \begin{bmatrix} x_1 \\ x_2 \\ \vdots \\ x_n \end{bmatrix}, \quad b = \begin{bmatrix} b_1 \\ b_2 \\ \vdots \\ b_n \end{bmatrix}.
$$

We will also consider matrix equations of the form

$$AX = B, \tag{4.4}$$

where $A \in M(n,n)$ and X and $B \in M(n,q)$. This type of equation is equivalent to the q independent matrix equations

$$Ax_{\cdot 1} = b_{\cdot 1}, Ax_{\cdot 2} = b_{\cdot 2}, \ldots, Ax_{\cdot q} = b_{\cdot q}, \tag{4.5}$$

where $x_{\cdot j}$ and $b_{\cdot j}$ represent the jth column vectors of X and B, respectively.

The following theorem concerns the existence and uniqueness of solutions to matrix equations. In this theorem it is assumed that $A \in M(n,n)$, and that X and $B \in M(n,q)$.

THEOREM 4.1. *If A is non-singular, then* (4.4) *has a unique solution, namely $X = A^{-1}B$. If A is singular, then either* (4.4) *has no solutions or it has infinitely many solutions.*

Proof. If A is non-singular, then A^{-1} exists by Theorem 2.4. Using the associative law, we have

$$A(A^{-1}B) = (AA^{-1})B = IB = B.$$

Thus $A^{-1}B$ is a solution. If X is another solution, then both sides of (4.4) can be multiplied on the left by A^{-1} to obtain $X = A^{-1}B$. Thus $A^{-1}B$ is the only solution.

If A is singular, then there is a non-zero vector $z \in EK_n$ such that $Az = \emptyset$. Let X be a solution to (4.4) and let Y be the matrix whose columns are given by

$$y_{.j} = x_{.j} + cz, \quad j = 1,2,\ldots,q,$$

where c is any number. Then

$$Ay_{.j} = Ax_{.j} + cAz = Ax_{.j} = b_{.j}, \quad j = 1,2,\ldots,q.$$

Thus Y is a solution to (4.4) for any value of c, and since there are infinitely many possible values for c, it follows that there are infinitely many solutions.

The Gauss Elimination Method. From Theorem 4.1 it might appear that the way to solve (4.4) is to compute A^{-1} and then multiply it by B. However, it turns out that this approach is not normally the most efficient. The method which we will now describe for solving such systems of equations is known as the *Gauss elimination method.* The basic idea of the method is probably already familiar to most readers. Let us first apply the method to the 3 by 3 system given by (4.1). The first step is to form the matrix

$$\begin{bmatrix} 1 & 1 & 1 & 9 \\ 1 & -1 & 2 & 9 \\ 2 & 4 & -1 & 14 \end{bmatrix},$$

where it is understood that the first row is just shorthand for the equation $x + y + z = 9$, the second row is shorthand for $x - y + 2z = 9$, etc. Now multiply the first equation (row) by -1 and add the result to the second equation to obtain an equation which is free of x. Then multiply the first equation by -2 and add the result to the third equation to obtain another equation which is free of x. At this stage, the shorthand notation gives

$$\begin{bmatrix} 1 & 1 & 1 & 9 \\ 0 & -2 & 1 & 0 \\ 0 & 2 & -3 & -4 \end{bmatrix},$$

representing a system equivalent to (4.1). Addition of the new second equation to the new third equation gives

$$\text{(4.6)} \qquad \begin{bmatrix} 1 & 1 & 1 & 9 \\ 0 & -2 & 1 & 0 \\ 0 & 0 & -2 & -4 \end{bmatrix}.$$

The third row of this matrix is shorthand for the equation $0x + 0y - 2z = -4$. Thus $z = 2$. Substitution of 2 for z in the second equation of (4.6) gives $0x - 2y + 1 \cdot 2 = 0$. Thus $y = 1$. Substituting 2 for z and 1 for y in the first equation, we get $1 \cdot x + 1 \cdot 1 + 1 \cdot 2 = 9$. Thus $x = 6$. Note that this procedure is essentially the same as that used to evaluate determinants by the triangular method.

The general description of the Gauss method for the system (4.2) now follows:

1. Form the n by $n + 1$ matrix
$$\begin{bmatrix} a_{11} & a_{12} & \cdots & a_{1n} & b_1 \\ a_{21} & a_{22} & \cdots & a_{2n} & b_2 \\ \vdots & & & & \\ a_{n1} & a_{n2} & \cdots & a_{nn} & b_n \end{bmatrix}.$$
(This matrix is called the *augmented matrix.*)
2. Let
$$m_2 = -a_{21}/a_{11},\ m_3 = -a_{31}/a_{11}, \ldots, m_n = -a_{n1}/a_{11}.$$
(If a_{11} is zero, switch rows so that a non-zero entry occupies the (1,1) position. If all elements of the first column are zero, proceed to step 4.)
3. Replace row i, $i = 2,3,\ldots,n$, by the row
$$0,\ m_i a_{12} + a_{i2},\ m_i a_{13} + a_{i3}, \ldots, m_i a_{1n} + a_{in},\ m_i b_1 + b_i.$$
4. Repeat steps 2 and 3 on the $(n - 1)$ by n submatrix formed by deleting the first row and first column of the augmented matrix. (Don't actually delete the first row and column when working on the submatrix.) Continue in this fashion until the matrix is in upper triangular form.
5. Solve the equation represented by the last row of the triangular matrix for x_n. If it should turn out that this equation is of the form $0 \cdot x_n = 0$, then choose $x_n = 1$ (or any other non-zero number). If this equation is of the form $0 \cdot x_n = c$ and c is not zero, then the system is inconsistent and has no solution.
6. Substitute the value of x_n in the next to last equation and solve for x_{n-1}. Continue in this fashion to obtain in succession $x_{n-2}, \ldots, x_1$.

The number of operations required to solve the system (4.2) by the Gauss method is:

$$
\begin{aligned}
&\text{Multiplications} \quad n(n-1)(2n+5)/6 \\
(4.7) \qquad &\text{Divisions} \quad n(n+1)/2 \\
&\text{Additions and subtractions} \quad n(n-1)(2n+5)/6
\end{aligned}
$$

The verification of (4.7) is left as an exercise (Problem 4.2).

When the matrix B in (4.4) has more than one column, the Gauss method can be used to solve each of the q systems of equations (4.5). In the section on matrix inversion we will discuss how to arrange the bookkeeping so as to avoid duplication in this process.

PROBLEMS

4.1. Solve the following systems of equations by the Gauss method. For part (b) keep a tally on the number of operations and compare with (4.7). How can you check your answers?

(a) $x + 2y = 14$
$2x - 3y = 14$

(b) $x + y + z + w = 10$
$2x - y - z + 3w = 9$
$4x + 3y + z + w = 17$
$x + y + z - w = 2$

(c) $x + y + z = 3$
$x - y + z = 1$
$2x + 2z = 7$

(d) Change the last equation in (c) to $2x + 2z = 4$.

4.2. Show that the number of operations given in (4.7) is correct. Calculate the computer time required for these operations for $n = 5, 10, 15, 20, 100$, if multiplications and divisions take 10 μ sec and additions and subtractions take 5 μ sec.

4.3. Write a flow diagram for the solution of three equations in three unknowns by the Gauss method.

4.4. Your computer installation probably has a program in its library for solving systems of linear equations. Use this program to solve the systems in Problem 4.1. Also try the program on a sixth order system with non-integer coefficients and solution. (How can you make up such a system with a known solution?)

4.5. Show that *Cramer's rule* does not compete with the Gauss method except for very small systems of equations. (Cramer's rule, which is also known as the *method of determinants*, is described in many algebra texts.)

5. ROUND-OFF AND STABILITY PROBLEMS

The Gauss method is somewhat unusual in that theoretically it produces an exact solution in a finite number of steps. More frequently the algorithms encountered in numerical analysis would in theory give an exact solution only after an infinite number of steps. Thus round-off is the only cause of error in the Gauss method (other than mistakes or input rounding). The next example shows that round-off can produce a serious error even in a system of small order.

EXAMPLE 5.1. Suppose that floating-point arithmetic with three significant digits is used on the system

$$\begin{aligned} x + 400y &= 801, \\ 200x + 200y &= 600. \end{aligned} \tag{5.1}$$

Multiplying the first equation by -200 and adding the result to the second equation, we get

$$-7.98 \cdot 10^4 y = -15.9 \cdot 10^4.$$

This gives the approximate value of 1.99 for y. Substitution of this value in the first equation yields the approximate value $x = 5$. The relative error in the value for x is 400% since the correct solution is (1,2). The problem here lies in the equation $x = 801 - 400y$. Any rounding error in y (in our case 0.01) gets magnified by a factor of 400, which can be quite significant because x is small.

There are a number of schemes which can be used to help control round-off error. These schemes, none of which is perfect, normally involve variations of some of the techniques listed below. With regard to the first two techniques, it can be shown [8] that when floating-point arithmetic is used, scaling by a power of the base has no influence on the result unless it is used in conjunction with partial or complete pivoting.

1. *Row Scaling.* Before starting the elimination process for the system of equations (4.2), divide each equation by an appropriate power of 10 so that the magnitude of the largest coefficient on the left side of the equation is between 0.1 and 1. More generally, if the base β is used for floating-point arithmetic in a particular computer, then divide each equation by a power of β so that the magnitude of the largest coefficient is between β^{-1} and 1. Another scheme is to divide each equation by the largest coefficient, so that the magnitude of this coefficient after scaling is 1.

EXAMPLE 5.2. Consider the system of equations

$$\begin{aligned} x + 230y + 3460z &= 20{,}000, \\ 30x + 5y + 0.1z &= 300, \\ 0.001x + 0.002y + 0.003z &= 7. \end{aligned} \tag{5.2}$$

In the decimal system, we would scale this system by dividing the first equation by 10^4, and the second equation by 10^2, and the third equation by 10^{-2}. This results in the system

$$0.0001x + 0.023y + 0.346z = 2,$$
$$0.3x + 0.05y + 0.001z = 3,$$
$$0.1x + 0.2y + 0.3z = 700.$$

Note that the right-hand side of each equation is ignored in the determination of the scale factor.

2. *Column Scaling.* Column scaling is similar to row scaling except that it alters the solution. For example, if the first column is divided by 10, then the new solution will be $10x_1$ instead of x_1.

EXAMPLE 5.3. In order to scale the system (5.2) by columns, we let $x = x^*/10^2$, $y = y^*/10^3$, and $z = z^*/10^4$. This results in the system

$$\begin{aligned} &0.01x^* + 0.230y^* + 0.3460z^* = 20{,}000,\\ (5.3)\qquad &0.3x^* + 0.005y^* + 0.00001z^* = 300,\\ &0.00001x^* + 0.000002y^* + 0.0000003z^* = 7. \end{aligned}$$

Of course, one must remember to replace x^*, y^*, and z^* by their values in terms of x, y, and z after solving the system (5.3).

3. *Partial Pivoting on the Largest Element.* Before obtaining zeros in the jth column of the augmented matrix, rows are interchanged so that the element in column j having the largest absolute value occupies the pivot [i.e., (j,j)th] position. Normally row scaling would be used along with this technique.

EXAMPLE 5.4. Scaling the system (5.1) by rows, we obtain

$$\begin{aligned} (5.4)\qquad &0.001x + 0.4y = 0.801,\\ &0.2x + 0.2y = 0.6. \end{aligned}$$

The technique of partial pivoting requires that we use the second equation as the pivot since $0.2 > 0.001$. Carrying three significant digits, we obtain the equation

$$0.399y = 0.798.$$

Thus $y = 2.00$ and $x = 1.00$. Note that a small round-off error in y would have produced only a small error in x, since the second equation is less sensitive than the first to an error in y.

4. *Complete Pivoting on the Largest Element.* Let the (i,j)th element of the partially triangularized augmented matrix be denoted by a'_{ij}. Interchange a row or column (or both) so that the element in the pivot position, a^*_{jj}, satisfies the condition

$$|a^*_{jj}| = \max_{\substack{j \leqslant k \leqslant n \\ j \leqslant m \leqslant n}} |a'_{k,m}|.$$

That is, a^*_{jj} is the largest element in the submatrix lying below and to the right of the $(j-1, j-1)$ position. Both row and column scaling are often used in conjunction with complete pivoting.

EXAMPLE 5.5. Let us use complete pivoting on the system (5.1). First, the rows are scaled to give (5.4). Now (5.4) is already properly scaled by columns since the largest coefficient in each column is in the interval [0.1,1]. Next, the columns are interchanged so that the largest coefficient (0.4) is in the pivot position:

$$0.4y + 0.001x = 0.801,$$

$$0.2y + 0.2x = 0.6.$$

Multiplying the first equation by -0.5 and adding the result to the second equation, we obtain (to three significant digits)

$$0.2x = 0.2,$$

$$x = 1.00,$$

$$y = 2.00.$$

It was only luck that the round-off errors compensated for each other in such a way that the solution is exact. Notice, however, that a small error in x would have produced only a small error in y.

5. *Increased Precision.* Some or all of the computations can be carried out in double precision. The use of complete double precision will double the storage requirements and may increase the computer time by a factor of four [6]. A significant gain in accuracy can be obtained by accumulating inner products (sums of products) in double precision and rounding back to single precision when the addition is complete. This requires only a few extra storage locations, and for some computers the increase in computing time is very small. Normally, a variation of Gaussian elimination known as the *Crout* or *Doolittle method* is used in conjunction with the double precision accumulation of inner products. This variation is described in Wilkinson [26] or Forsythe and Moler [8].

6. *Iteration.* An approximate solution, say x_0, is obtained by the Gauss method. Let $r_0 = b - Ax_0$ be the so-called *residual vector.*

Now solve the equation $Ay_0 = r_0$ and let $x_1 = x_0 + y_0$. Then let $r_1 = b - Ax_1$, solve the equation $Ay_1 = r_1$, set $x_2 = x_1 + y_1$, and so forth. This procedure is described more fully in Section 8 of Chapter 4.

These techniques, as well as certain variations of the Gauss method, are discussed in detail by Wilkinson [26, 27], Forsythe [6], and Forsythe and Moler [8]. Although the method of complete pivoting appears to be better from the round-off point of view, Wilkinson doubts that the benefits are worth the extra cost. A practical procedure which gives good results is to use row scaling and partial pivoting followed by iterative improvement. The residual vectors $r_0, r_1, \ldots$ in the iteration should be computed using double precision accumulation of inner products. A nearly singular matrix of coefficients is indicated if the residuals are nearly the same size as the solution vector. Algol, Fortran, and PL/I programs for this procedure can be found in Forsythe and Moler [8], while detailed flow charts can be found in Wilkinson [27]. Some numerical examples are given by Carnahan, Luther, and Wilkes [2], Wilkinson [25], and Forsythe and Moler [8].

The concept of stability is very important in the study of round-off errors. Some matrices are unstable in the sense that a small change in the elements of A or b leads to a relatively large change in the solution of the equation $Ax = b$. Such a matrix is said to be *ill-conditioned.*

EXAMPLE 5.6. Consider the system of equations $Ax = b$, where

$$A = \begin{bmatrix} 1.001 & 1 \\ 1 & 1 \end{bmatrix}, \qquad b = \begin{bmatrix} 5.002 \\ 5 \end{bmatrix}.$$

It is easy to see that $x_1 = 2$, $x_2 = 3$ is the solution.

Now let

$$b^* = \begin{bmatrix} 5.003 \\ 4.999 \end{bmatrix}.$$

The equation $Ax = b^*$ has the solution $x_1 = 4$, $x_2 = 0.999$. Thus a small change in b produces a relatively large change in x. The geometric interpretation is obvious—we are trying to find the intersection of two nearly parallel lines.

The *condition number*, which will be defined presently, is a quantitative measure of the stability of the system $Ax = b$. Let us first define a *matrix norm.*

DEFINITION 5.1. *Suppose A is an n by n matrix and $x \in EK_n$. The spectral norm of A is defined by the equation*

$$\|A\| = \max_{\|x\|=1} \|Ax\|,$$

where $\|x\| = [\Sigma_{i=1}^n |x_i|^2]^{1/2}$ is the usual norm for EK_n.

Let $T(x) = Ax$ define the linear transformation associated with A. The spectral norm is interpreted geometrically as the maximum distortion in length produced when T is applied to the unit sphere in EK_n. Of course, we should show that such a maximum exists, and that the norm satisfies the axioms for a norm. The fact that T has a maximum (as well as a minimum) on the unit sphere follows from the fact that T is continuous and the unit sphere is closed and bounded (see Section 9 in Chapter 2 and [1]). The verification of the norm conditions is left as an exercise (Problem 5.2).

DEFINITION 5.2. *Let A be a non-singular matrix. The condition number of A is defined by the equation*

$$\text{cond}\,(A) = \|A\|\ \|A^{-1}\|.$$

Let $T(x) = Ax$ as before. We will show that the condition number is the ratio of the largest to the smallest distortion in length caused by T on the unit sphere. The actual computation of cond (A) is based on the theory of quadratic and Hermitian forms (see Problems 7.8–7.10). Problem 5.7 shows one way to estimate cond (A), while Forsythe and Moler [8] have shown how to estimate this quantity from the results of the iterative correction process. The following theorem gives some of the fundamental inequalities involving the spectral norm and the condition number.

THEOREM 5.1. *Let $A \in M(n,n)$ and*

$$m = \min_{\|x\|=1} \|Ax\|.$$

Then:

$$\|Ax\| \leqslant \|A\|\ \|x\| \tag{5.5}$$

for all $x \in EK_n$.

$$\|A\| \leqslant \left[\sum_{i=1}^{n} \sum_{j=1}^{n} |a_{ij}|^2\right]^{1/2}. \tag{5.6}$$

If A is non-singular, then

$$\text{cond}\,(A) = \frac{\|A\|}{m}, \tag{5.7}$$

$$\text{cond}\,(A) \geqslant 1. \tag{5.8}$$

Proof. If $x = \emptyset$, then (5.5) obviously holds. Otherwise, let $y = x/\|x\|$. Then $\|y\| = 1$ and

$$\|Ay\| \leqslant \max_{\|z\|=1} \|Az\| = \|A\|.$$

But $\|Ay\| = \|Ax\|/\|x\|$ and (5.5) follows.

In order to prove (5.6), let $a_{i\cdot}$ denote the ith row of A and let x have unit length. Then:

$$Ax = \begin{bmatrix} a_{1\cdot} \cdot \overline{x} \\ \vdots \\ a_{n\cdot} \cdot \overline{x} \end{bmatrix}.$$

According to the Schwarz inequality,

$$|a_{i\cdot} \cdot \overline{x}| \leqslant \|a_{i\cdot}\| \, \|\overline{x}\| = \|a_{i\cdot}\| \, \|x\|.$$

Using the fact that $\|x\| = 1$, we have

$$\|Ax\|^2 = \sum_{i=1}^{n} |a_{i\cdot} \cdot \overline{x}|^2 \leqslant \sum_{i=1}^{n} \|a_{i\cdot}\|^2 = \sum_{i=1}^{n} \sum_{j=1}^{n} |a_{ij}|^2.$$

Thus

$$\|A\| = \max_{\|x\|=1} \|Ax\| \leqslant \left[\sum_{i=1}^{n} \sum_{j=1}^{n} |a_{ij}|^2\right]^{1/2}.$$

Next, suppose that A is non-singular. To say that the minimum of $\|Ax\|$ over the unit sphere exists means that there is a vector x^* of unit length such that $\|Ax^*\| = m$. It follows that $m > 0$; for otherwise $x^* = A^{-1}(Ax^*)$ would be zero. Let $w = (1/m)Ax^*$. Then $\|w\| = 1$ and $\|A^{-1}w\| = (1/m)\|x^*\| = 1/m$. Hence

$$\frac{1}{m} \leqslant \|A^{-1}\|. \tag{5.9}$$

On the other hand, if $\|y\| = 1$ and $x = A^{-1}y$, then

$$\frac{1}{\|A^{-1}y\|} = \frac{\|y\|}{\|x\|} = \frac{\|Ax\|}{\|x\|} = \left\|A\left(\frac{x}{\|x\|}\right)\right\| \geqslant m.$$

Hence $\|A^{-1}y\| \leqslant 1/m$ whenever $\|y\| = 1$, and $\|A^{-1}\| \leqslant 1/m$. Combining this with (5.9) we see that $\|A^{-1}\| = 1/m$ and (5.7) follows.

Finally, (5.8) follows from the fact that the minimum of $\|Ax\|$ over the unit sphere cannot exceed the maximum.

The next theorem indicates the sense in which the condition number is a measure of stability for the system of equations $Ax = b$.

THEOREM 5.2. *Suppose A is non-singular, x, Δx, b, $\Delta b \in EK_n$, and $b \neq \emptyset$.*

(i) *If $Ax = b$ and $A(x + \Delta x) = b + \Delta b$, then*

$$\frac{\|\Delta x\|}{\|x\|} \leqslant \operatorname{cond}(A) \frac{\|\Delta b\|}{\|b\|}. \tag{5.10}$$

(ii) *If $Ax = b$, $(A + \Delta A)(x + \Delta x) = b$, $x + \Delta x \neq \emptyset$, and $\rho = \operatorname{cond}(A)\,\|\Delta A\|/\|A\|$, then*

$$\frac{\|\Delta x\|}{\|x + \Delta x\|} \leqslant \operatorname{cond}(A) \frac{\|\Delta A\|}{\|A\|} = \rho. \tag{5.11}$$

If, in addition $\rho < 1$, then

$$\frac{\|\Delta x\|}{\|x\|} \leqslant \frac{\rho}{1-\rho}. \tag{5.12}$$

The proof of this theorem is left as an exercise (Problem 5.4).

Let us look carefully at the interpretation of this theorem. The inequality (5.10) states that if b is in error by an amount Δb, then cond (A) is a bound on the magnification of relative error that is induced in the solution x. Inequality (5.11) says essentially the same thing about errors in A. While cond (A) is only an upper bound for this magnification, one can show (Problems 5.5 and 5.6) that in fact there do exist choices of b, Δb, and ΔA for which equality holds in (5.10) or (5.11). In most practical problems there is some uncertainty in the values of A or b due to measurement errors, round-off errors upon entering values in a computer, and so forth. The point to be emphasized is this: If the condition number of A is quite large, then we would tend to have very little faith in the solution *even if the computation were carried out exactly*. Of course, the round-off errors which do occur in the solution process tend to aggravate an already bad situation. If floating-point arithmetic with base β and m significant digits is used, then a value of cond $(A) > \beta^m/2$ would indicate that the computed solution is likely to have little significance.

Wilkinson has made a thorough analysis of the rounding errors in Gaussian elimination. He has shown [25, 8] that the computed solution x satisfies exactly an equation of the form

$$(A + \Delta A)x = b.$$

Wilkinson gives bounds on the quantity

$$\|\Delta A\|_\infty = \max_{1 \leqslant i \leqslant n} \sum_{j=1}^{n} |\Delta a_{ij}|$$

which are roughly proportional to n^3. This is an example of the so-called *inverse type of error analysis*, in which it is shown that the computed solution is the exact solution to a slightly altered problem. If the bounds on $\|\Delta A\|$ are small compared to the inherent inaccuracy of A, then one can feel comfortable in knowing that the computed solution is as good as is warranted by the data. If cond (A) is known, then inequality (5.11) can be used to compute a bound on the difference between the computed and exact solutions.

PROBLEMS

5.1. Solve each of the following systems (1) without pivoting on largest elements or scaling, (2) with partial pivoting and scaling of the rows. Use floating-point arithmetic with m significant digits.

(a)
$$\begin{aligned} x + 10{,}000y &= 10{,}000 \\ x + y &= 2 \\ m &= 3 \end{aligned}$$

(b)
$$\begin{aligned} -2x + 1000y - z &= 1000 \\ x + y + z &= 0 \\ x + y - z &= 4 \\ m &= 2 \end{aligned}$$

5.2. Show that the norm given by Definition 5.1 does indeed have the properties required of a norm.

5.3. Compute the condition number of the matrix

$$A = \begin{bmatrix} 1 & 0 & 0 \\ 0 & 2 & 0 \\ 0 & 0 & 3 \end{bmatrix}.$$

5.4. Prove Theorem 5.2.

5.5. Show that Δb and b can be chosen so that equality holds in (5.10). Hint: Choose these vectors in the directions of maximum and minimum distortion for A^{-1}.

5.6. Show that ΔA and b can be chosen so that equality holds in (5.11). Hint: Choose ΔA to be a small multiple of the identity.

5.7. Let

$$\lambda = \max_{1 \leqslant j \leqslant n} \|a_{\cdot j}\| \qquad \text{and} \qquad \sigma = \min_{1 \leqslant j \leqslant n} \|a_{\cdot j}\|,$$

where $a_{\cdot j}$ represents the jth column of A. Show that cond $(A) \geqslant \lambda/\sigma$.

6. COMPUTATION OF THE INVERSE

We will consider four ways to compute the inverse of an n by n matrix: Cramer's rule, the Gauss method, the Jordan method, and the iterative method. Cramer's rule, which is based on the theory of determinants, is inefficient for moderate or large values of n. The iteration method is used primarily to improve on the accuracy of an approximate inverse computed by some other means. (Discussion of iteration techniques will be postponed until Chapter 4.)

Cramer's Rule. Let A be the n by n matrix whose inverse is desired. Let C be the n by n matrix defined by the equation

$$c_{ij} = (-1)^{i+j}|A_{ji}|/|A|,$$

where A_{ij} is the $(n-1)$ by $(n-1)$ matrix formed by deleting the ith row and jth column of A. It can be shown [12] that if $|A| \neq 0$, then $C = A^{-1}$. This formula is convenient for certain theoretical purposes and for small values of n. It is left as an exercise (Problem 6.1) to show that computation of the inverse by means of this formula requires too many operations to compete with the Gauss or Jordan method except when n is small.

The Gauss Method. Note that the matrix equation

$$AX = I$$

is equivalent to the n linear systems of equations

$$Ax_{.1} = e_1, \quad Ax_{.2} = e_2, \quad \ldots, \quad Ax_{.n} = e_n, \tag{6.1}$$

where $x_{.j}$ and e_j are the jth column vectors of X and I, respectively. Thus the inverse of A can be computed by applying Gaussian elimination to each of the n systems of equations in (6.1). However, before jumping headlong into this task, let us observe that in each of these systems the triangularized augmented matrix will differ only in the last column. Thus we can save a lot of work by solving all of the systems at the same time. To do this we form the augmented matrix

$$\begin{bmatrix} a_{11} & a_{12} & \cdots & a_{1n} & 1 & 0 & 0 & \cdots & 0 \\ a_{21} & a_{22} & \cdots & a_{2n} & 0 & 1 & 0 & \cdots & 0 \\ \vdots & & & & & & & & \\ a_{n1} & a_{n2} & \cdots & a_{nn} & 0 & 0 & 0 & \cdots & 1 \end{bmatrix}.$$

The matrix is then reduced by the Gauss method to triangular form:

$$\begin{bmatrix} a'_{11} & a'_{12} & \cdots & a'_{1n} & b_{11} & \cdots & b_{1n} \\ 0 & a'_{22} & \cdots & a'_{2n} & b_{21} & \cdots & b_{2n} \\ \vdots & & & & & & \\ 0 & \cdots & \cdots & a'_{nn} & b_{n1} & \cdots & b_{nn} \end{bmatrix}.$$

The last row is shorthand for the equations

$$a'_{nn}x_{n1} = b_{n1}, \quad a'_{nn}x_{n2} = b_{n2}, \quad \ldots, \quad a'_{nn}x_{nn} = b_{nn}.$$

Solving these equations and substituting in the next to last row we obtain the equations

$$a'_{n-1,n-1}x_{n-1,j} + a'_{n-1,n}x_{nj} = b_{n-1,j}, \qquad j = 1, 2, \ldots, n.$$

These equations are solved for $x_{n-1,j}$, the values are substituted in the $(n-2)$nd row, and so on. When this is completed we set $A^{-1} = X$.

EXAMPLE 6.1. Consider the matrix

$$A = \begin{bmatrix} 1 & 0 & 1 \\ 2 & 2 & 3 \\ 1 & 2 & 4 \end{bmatrix}.$$

In actual numerical work one should use a pivoting strategy in order to help control round-off error. To simplify the manipulations, we will reduce the augmented matrix to triangular form without pivoting on the largest element. This yields the matrix

$$\begin{bmatrix} 1 & 0 & 1 & 1 & 0 & 0 \\ 0 & 2 & 1 & -2 & 1 & 0 \\ 0 & 0 & 2 & 1 & -1 & 1 \end{bmatrix}.$$

Thus $x_{31} = 1/2$, $x_{32} = -1/2$, $x_{33} = 1/2$. These values lead to the equations

$$2x_{21} + 1/2 = -2, \quad 2x_{22} - 1/2 = 1, \quad 2x_{23} + 1/2 = 0,$$
$$x_{21} = -5/4, \quad x_{22} = 3/4, \quad x_{23} = -1/4.$$

Substituting in the first row, we get

$$x_{11} + 1/2 = 1, \quad x_{12} - 1/2 = 0, \quad x_{13} + 1/2 = 0,$$
$$x_{11} = 1/2, \quad x_{12} = 1/2, \quad x_{13} = -1/2.$$

Finally,

$$A^{-1} = \begin{bmatrix} 1/2 & 1/2 & -1/2 \\ -5/4 & 3/4 & -1/4 \\ 1/2 & -1/2 & 1/2 \end{bmatrix}.$$

The following theorem gives the number of arithmetic operations required by the Gauss method.

THEOREM 6.1. *The Gauss method for computing the inverse of an n by n matrix requires the following number of operations (not counting operations required for scaling and partial or complete pivoting):*

Multiplications $n^2(n-1)$

Additions and subtractions $n^2(n-1)$

Divisions $n(3n-1)/2$

Proof. Let $B = (A,I)$ be the augmented matrix associated with the n by n matrix A. The first step in the diagonalization process is to set $m_2 = -b_{21}/b_{11}$ and then replace b_{2j} by $m_2 b_{1j} + b_{2j}$, $j = 1, 2, \ldots, 2n$. However, $m_2 b_{11} + b_{21} = 0$; and for $j > n + 1$, $m_2 b_{1j} + b_{2j} = b_{2j}$ since $b_{1j} = 0$. Thus we need only carry out the arithmetic operations in $m_2 b_{1j} + b_{2j}$ for $j = 2, 3, \ldots, n + 1$. The first step of the process requires 1 division, n multiplications, and n additions. Using similar reasoning we obtain the following breakdown on the number of operations required for the computation of the inverse:

	Divisions	*Multiplications*	*Additions and Subtractions*
Obtaining zeros in 1st col.	$n-1$	$(n-1)n$	$(n-1)n$
Obtaining zeros in 2nd col.	$n-2$	$(n-2)n$	$(n-2)n$
⋮			
Obtaining zeros in $(n-1)$st col.	1	n	n
Subtotal (Triangularization)	$n(n-1)/2$	$n^2(n-1)/2$	$n^2(n-1)/2$
Solving for x_{nj}, $j=1,n$	n	0	0
Solving for $x_{n-1,j}$, $j=1,n$	n	n	n
Solving for $x_{n-2,j}$, $j=1,n$	n	$2n$	$2n$
⋮			
Solving for $x_{1,j}$, $j=1,n$	n	$(n-1)n$	$(n-1)n$
Subtotal (Back Substitution)	n^2	$n^2(n-1)/2$	$n^2(n-1)/2$*

The theorem follows by adding the subtotals given in the table and simplifying the resulting expressions.

*The submatrix occupying the original position of I still has zeros above the diagonal. If the program takes this into account there would be $n(n-1)/2$ fewer additions.

Operation counts such as this depend on the actual program used for the computations. For example, Fox [9] stores the reciprocals of the pivot elements and replaces some of our divisions by multiplications. This would be advantageous on a machine that had a slow division speed compared with multiplication.

The Jordan Method. The Jordan method differs from the Gauss method in that zeros are obtained above the diagonal as well as below. Also, each row is divided by the diagonal element so that when the process is complete, the identity is at the left of the augmented matrix and the inverse is at the right.

EXAMPLE 6.2. Let A be the matrix in Example 6.1. The first step is to divide the first row of the augmented matrix by a_{11} so that a 1 appears in the (1,1) position. (In this example $a_{11} = 1$ so this step does not alter the first row.) Next, obtain zeros in the (2,1) and (3,1) positions in the usual way. Then divide the second row by a'_{22} (2 in this case) to obtain a 1 in the (2,2) position. At this stage we have the matrix

$$\begin{bmatrix} 1 & 0 & 1 & 1 & 0 & 0 \\ 0 & 1 & 1/2 & -1 & 1/2 & 0 \\ 0 & 2 & 3 & -1 & 0 & 1 \end{bmatrix}.$$

Now add $-a'_{12}$ (0 in this case) times the second row to the first row to obtain a zero in the (1,2) position. Then add -2 times the second row to the third row, and divide the third row by a'_{33} to obtain

$$\begin{bmatrix} 1 & 0 & 1 & 1 & 0 & 0 \\ 0 & 1 & 1/2 & -1 & 1/2 & 0 \\ 0 & 0 & 1 & 1/2 & -1/2 & 1/2 \end{bmatrix}.$$

Finally, add $-1/2$ times the third row to the second row, and -1 times the third row to the first row to obtain

$$\begin{bmatrix} 1 & 0 & 0 & 1/2 & 1/2 & -1/2 \\ 0 & 1 & 0 & -5/4 & 3/4 & -1/4 \\ 0 & 0 & 1 & 1/2 & -1/2 & 1/2 \end{bmatrix}.$$

The submatrix consisting of the last three columns is A^{-1}.

THEOREM 6.2. *The Jordan method for computing the inverse of an n by n matrix requires the following number of arithmetic* *operations (not counting the operations required for scaling and partial or complete pivoting):*

Multiplications $n^2(n-1)$

Additions and subtractions $n^2(n-1)$

Divisions n^2

The proof of this theorem is left as an exercise (Problem 6.4).

The Gauss and Jordan methods would appear to require an n by $2n$ matrix for storage during the computations. However, the results of the computations can be stored in an n by n matrix. By the left and right submatrix of the transformed augmented matrix, we refer to the positions originally occupied by A and I, respectively. In the Gauss method the elements below the diagonal in the right submatrix can be stored below the diagonal in the left submatrix. This is possible because the elements below the diagonal in the left submatrix end up being zero and need not be stored. The elements on or above the diagonal in the right submatrix need not be stored because they are never altered in the reduction process. Of course, this technique requires some extra bookkeeping, especially when partial or complete pivoting is used.

In the Jordan method, the first column of the right submatrix is stored in the first column of the left submatrix, then the second column on the right is stored in the second column on the left, and so on.

PROBLEMS

6.1. Count the numbers of arithmetic operations required to compute the inverse of an n by n matrix using Cramer's rule. Compare your results with the Jordan method.

6.2. Use the Gauss method (without pivoting) to compute the inverse of the matrix

$$A = \begin{bmatrix} 1 & 1 & 1 \\ 2 & 3 & 2 \\ 1 & 0 & 2 \end{bmatrix}.$$

Check your answer by Cramer's rule. (How else could you check your answer?) Count the number of operations and compare with Theorem 6.1.

6.3. Compute the inverse of the matrix in Problem 6.2 by the Jordan method. Count the number of operations and compare with Theorem 6.2.

6.4. Prove Theorem 6.2.

6.5. The number of divisions given for the Jordan method is roughly 2/3 of the number for the Gauss method. By comparing the division counts, find a modification of the Gauss method which would require only n^2 divisions.

6.6. Let

$$A = \begin{bmatrix} 1 & 0 & 2 \\ 1 & 1 & 0 \\ 2 & 2 & 1 \end{bmatrix}.$$

(a) Compute A^{-1} by the Gauss method using only a 3 by 3 matrix for storage.
(b) Repeat (a) using the Jordan method.

7. COMPUTATION OF MATRIX EIGENVALUES

We will consider four methods for computing eigenvalues of matrices. The first method, which will be discussed in this section, is that of computing directly the zeros of the characteristic polynomial. A more complete discussion concerning the computation of zeros of polynomials will be given in Chapter 4. The power method, which is used primarily to approximate the largest eigenvalue, will also be covered in Chapter 4. The third method, introduced by Jacobi in 1846, is based on the idea of reducing a real symmetric matrix to diagonal form by means of orthogonal transformations. Jacobi's method will be treated in Section 8. The Givens-Householder method, covered in Section 9, is similar to Jacobi's method, but it is more efficient. Other techniques are discussed by Forsythe [6], Wilkinson [26], and Householder [13].

In this section we will review some terminology and theory needed in the discussion of methods for computing eigenvalues. Recall that EK_n is the collection of ordered n-tuples of complex numbers, and that the dot product for EK_n is defined by $x \cdot y = x_1\overline{y}_1 + x_2\overline{y}_2 + \ldots + x_n\overline{y}_n$.

DEFINITION 7.1. *Let A be an n by n matrix. The number λ is said to be an eigenvalue of A iff*

$$(7.1) \qquad Ax = \lambda x$$

for some non-zero (column) vector $x \in EK_n$. A non-zero vector x satisfying (7.1) is said to be an eigenvector of A corresponding to λ.

Let $T(x) = Ax$ be the linear transformation associated with the matrix A. Note that the eigenvalues and eigenvectors of A defined here are the same as those for T given by Definition 3.1 in Chapter 2.

A first step toward developing a method for computing eigenvalues would be to work directly with equation (7.1). Subtracting x from both sides of (7.1) we obtain the homogeneous equation

$$(A - \lambda I)x = \emptyset. \tag{7.2}$$

According to Theorem 4.1, the equation (7.2) has a non-trivial solution iff $|A - \lambda I| = 0$. This proves the next theorem.

THEOREM 7.1. *The number λ is an eigenvalue of A iff $|A - \lambda I| = 0$.*

DEFINITION 7.2. *The characteristic polynomial of A is the function f defined by the equation*

$$f(\lambda) = |A - \lambda I|.$$

It is an easy inductive argument to show that the function f so defined is a polynomial of degree n. Thus the problem of computing eigenvalues of a matrix can be reduced to the problem of computing the zeros of the characteristic polynomial. This gives us a practical method for computing eigenvalues when the size of the matrix is small. The computation of the coefficients of the characteristic polynomial is quite difficult for large matrices.

EXAMPLE 7.1. Let

$$A = \begin{bmatrix} 3 & 1 \\ 4 & 3 \end{bmatrix}.$$

Then:

$$f(\lambda) = \begin{vmatrix} 3-\lambda & 1 \\ 4 & 3-\lambda \end{vmatrix} = (3 - \lambda)^2 - 4.$$

The zeros of f (and hence the eigenvalues of A) are given by

$$\lambda_1 = 5, \qquad \lambda_2 = 1.$$

An eigenvector corresponding to λ_1 can be obtained by solving equation (7.2) with $\lambda = \lambda_1$:

$$\begin{bmatrix} -2 & 1 \\ 4 & -2 \end{bmatrix} \begin{bmatrix} x_1 \\ x_2 \end{bmatrix} = \emptyset,$$

$$-2x_1 + x_2 = 0,$$

$$4x_1 - 2x_2 = 0.$$

This system is dependent, so x_2 may be chosen arbitrarily. Choosing $x_2 = 2$, we find that (1,2) is an eigenvector corresponding to λ_1.

The following theorem of Gershgorin will be used later to determine error bounds for computed eigenvalues.

THEOREM 7.2 (Gershgorin's Theorem). *Let A be an n by n matrix,*

$$R_i = |a_{i1}| + |a_{i2}| + \ldots + |a_{in}| - |a_{ii}|$$

and

$$D_i = \{z \in K: |z - a_{ii}| \leqslant R_i\}.$$

Then:

(i) *Every eigenvalue of A lies in the set* $D = \cup_{i=1}^{n} D_i$.
(ii) *If the circular discs* D_i *are disjoint, then each disc contains exactly one eigenvalue.*

Proof. Let λ be an eigenvalue and x a corresponding eigenvector. Let k be the index of the coordinate of x which has the largest magnitude. Now $Ax = \lambda x$; so $(Ax)_k = \lambda x_k$. Computing $(Ax)_k$ we obtain the equation

$$a_{k1}x_1 + a_{k2}x_2 + \ldots + a_{kn}x_n = \lambda x_k.$$

Solving for $a_{kk} - \lambda$ and noting that $|x_j|/|x_k| \leqslant 1$, we have

$$|a_{kk} - \lambda| \leqslant \sum_{j \neq k} \frac{|a_{kj}|\,|x_j|}{|x_k|} \leqslant \sum_{j \neq k} |a_{kj}| = R_k.$$

For a proof of part (ii) see Faddeev and Faddeeva [5] and Ostrowski [17].

Matrices which are symmetric about the diagonal are very important in applications involving eigenvalues.

DEFINITION 7.3. *The matrix A is said to be symmetric iff* $A' = A$, *and Hermitian iff* $A^* = A$.

THEOREM 7.3. *If the matrix A is Hermitian, then its eigenvalues are real. Furthermore, eigenvectors corresponding to distinct eigenvalues are orthogonal.*

Proof. Let x and y be column vectors in EK_n. Note that $x \cdot y = x_1\bar{y}_1 + x_2\bar{y}_2 + \ldots + x_n\bar{y}_n = y^*x$, where $y^* = \bar{y}' = (\bar{y}_1, \bar{y}_2, \ldots, \bar{y}_n)$. (No distinction is made between the 1 by 1 matrix y^*x and its only element.) Let x be an eigenvector corresponding to λ. Then

$$\begin{aligned}\lambda(x \cdot x) &= (\lambda x) \cdot x = (Ax) \cdot x = x^*Ax \\ &= x^*A^*x = (Ax)^*x = x \cdot (Ax) \\ &= x \cdot (\lambda x) = \bar{\lambda}(x \cdot x).\end{aligned}$$

Thus $\lambda = \bar{\lambda}$, and λ is real.

Now let x and y be eigenvectors corresponding to λ_1 and λ_2, respectively, with $\lambda_1 \neq \lambda_2$. Then

$$\begin{aligned}(\lambda_1 - \lambda_2)(x \cdot y) &= \lambda_1(x \cdot y) - \lambda_2(x \cdot y) \\ &= (Ax) \cdot y - x \cdot (Ay) \\ &= y^*Ax - y^*A^*x = y^*Ax - y^*Ax = 0.\end{aligned}$$

It follows that $x \cdot y = 0$ since $\lambda_1 - \lambda_2 \neq 0$.

COROLLARY 7.4. *If A is real and symmetric, then its eigenvalues are real. Furthermore, eigenvectors corresponding to distinct eigenvalues are orthogonal.*

Proof. Note that a real symmetric matrix is Hermitian.

DEFINITION 7.4. *The matrix P is said to be orthogonal iff $P'P = I$.*

A linear transformation T from E_n into E_n is said to be orthogonal if and only if the corresponding matrix is orthogonal. Geometrically, orthogonal transformations are rotations or reflections; that is, they preserve angles and lengths. The proof of this fact is reserved for Problem 7.7.

The following theorem is the theoretical basis for Jacobi's method of computing eigenvalues which will be discussed in the next section. A proof may be found in [12] (see also Problem 7.6).

THEOREM 7.5. *If A is a real symmetric matrix, then there exists a real orthogonal matrix P such that*

$$P'AP = \begin{bmatrix} \lambda_1 & 0 & 0 & \cdots & 0 \\ 0 & \lambda_2 & 0 & \cdots & 0 \\ \vdots & & & & \\ 0 & \cdots & \cdots & \cdots & \lambda_n \end{bmatrix},$$

where $\lambda_1, \lambda_2, \ldots, \lambda_n$ are the eigenvalues of A.

EXAMPLE 7.2. Let

$$A = \begin{bmatrix} 1 & 2 \\ 2 & 1 \end{bmatrix} \quad \text{and} \quad P = \begin{bmatrix} \frac{1}{\sqrt{2}} & \frac{1}{\sqrt{2}} \\ \frac{1}{\sqrt{2}} & \frac{-1}{\sqrt{2}} \end{bmatrix}.$$

Then:

$$P'P = \begin{bmatrix} \frac{1}{\sqrt{2}} & \frac{1}{\sqrt{2}} \\ \frac{1}{\sqrt{2}} & \frac{-1}{\sqrt{2}} \end{bmatrix} \begin{bmatrix} \frac{1}{\sqrt{2}} & \frac{1}{\sqrt{2}} \\ \frac{1}{\sqrt{2}} & \frac{-1}{\sqrt{2}} \end{bmatrix} = \begin{bmatrix} 1 & 0 \\ 0 & 1 \end{bmatrix}.$$

Thus P is orthogonal. Furthermore

$$P'AP = \begin{bmatrix} \frac{3}{\sqrt{2}} & \frac{3}{\sqrt{2}} \\ \frac{-1}{\sqrt{2}} & \frac{1}{\sqrt{2}} \end{bmatrix} \begin{bmatrix} \frac{1}{\sqrt{2}} & \frac{1}{\sqrt{2}} \\ \frac{1}{\sqrt{2}} & \frac{-1}{\sqrt{2}} \end{bmatrix} = \begin{bmatrix} 3 & 0 \\ 0 & -1 \end{bmatrix}.$$

The reader may check that 3 and -1 are the eigenvalues of A. One way of constructing such an orthogonal matrix P is to obtain eigenvectors of A corresponding to 3 and -1, normalize the eigenvectors to length 1, and form the matrix P by using the normalized eigenvectors as columns. Jacobi's method provides a way to compute P and the eigenvalues of A at the same time.

PROBLEMS

7.1. Let P be a non-singular matrix (i.e., $|P| \neq 0$). Show that $P^{-1}AP$ and A have the same eigenvalues. Hint: Use Theorem 7.1 and the identity $|AB| = |A|\,|B|$.

7.2. Suppose $|A| \neq 0$. Show that AB and BA have the same eigenvalues.

7.3. Let

$$A = \begin{bmatrix} 3 & 0 & 1 \\ 1 & 1 & 1 \\ 2 & 0 & 5 \end{bmatrix}.$$

Find the characteristic polynomial of A.

7.4. Let

$$A = \begin{bmatrix} 1 & 3 \\ 3 & 1 \end{bmatrix}.$$

(a) Compute the eigenvalues of A.

(b) Find an eigenvector corresponding to each eigenvalue. Are they orthogonal?

(c) Find an orthogonal matrix P such that

$$P'AP = \begin{bmatrix} \lambda_1 & 0 \\ 0 & \lambda_2 \end{bmatrix}.$$

7.5. Show that the product of two orthogonal matrices is orthogonal.

7.6. Prove Theorem 7.5 for the case where A has distinct eigenvalues. Hint: Form P from normalized eigenvectors.

7.7. Let T be a transformation defined on E_n by the equation

$$T(x) = Px,$$

where P is a real orthogonal n by n matrix. Show that T preserves distances and inner products, i.e.,

$$\|T(x) - T(y)\| = \|x - y\|$$

and

$$T(x) \cdot T(y) = x \cdot y$$

for all $x,y \in E_n$. Conversely, show that if T is a linear transformation which preserves distances and inner products, then there exists an orthogonal matrix P such that $T(x) = Px$ for all $x \in E_n$.

7.8. A *quadratic form* is a function Q from E_n into R defined by an equation of the form

$$Q(x) = x'Ax, \tag{7.3}$$

where A is a real n by n matrix. Prove the following:

(a) We might as well assume that the matrix A in (7.3) is symmetric.

(b) Let λ_1 be the eigenvalue of A which has the largest magnitude. Then

$$|\lambda_1| = \max_{\|x\|=1} |Q(x)|.$$

Hint: Let P be the orthogonal matrix from Theorem 7.5 and let $x = Py$.

7.9. A matrix P is said to be *unitary* iff $P^*P = I$. Show that:

(a) The product of two unitary matrices is unitary.

(b) Unitary transformations preserve distances.

7.10. A theorem analogous to Theorem 7.5 states that a Hermitian matrix can be reduced to diagonal form by a unitary matrix. Use this theorem to show that

$$\|A\| = \sqrt{\lambda_1},$$

where λ_1 is the largest eigenvalue of A^*A. Hint: Note that $\|Ax\|^2 = (Ax)^*Ax$.

8. THE JACOBI METHOD

The Jacobi method can be applied to real symmetric matrices, or a modified version can be applied to Hermitian matrices [10]. We will first describe how the method works on a 2 by 2 real symmetric matrix A. An orthogonal matrix P will be found so that

$$P'AP = \begin{bmatrix} \lambda_1 & 0 \\ 0 & \lambda_2 \end{bmatrix},$$

where λ_1 and λ_2 are the eigenvalues of A. Let

$$P = \begin{bmatrix} \cos\varphi & -\sin\varphi \\ \sin\varphi & \cos\varphi \end{bmatrix}$$

and check that P is orthogonal for any value of φ. (Geometrically, the equation $y = Px$, $x \in E_2$, describes a rotation of axes through an angle φ.) Now let $B = P'A$ and $C = P'AP = BP$. Then the following equations must hold:

$$\begin{aligned} b_{11} &= a_{11}\cos\varphi + a_{21}\sin\varphi, \\ b_{12} &= a_{12}\cos\varphi + a_{22}\sin\varphi, \\ b_{21} &= a_{21}\cos\varphi - a_{11}\sin\varphi, \\ b_{22} &= a_{22}\cos\varphi - a_{12}\sin\varphi. \end{aligned} \tag{8.1}$$

$$\begin{aligned} c_{11} &= b_{11}\cos\varphi + b_{12}\sin\varphi, \\ c_{12} &= b_{12}\cos\varphi - b_{11}\sin\varphi, \\ c_{21} &= b_{21}\cos\varphi + b_{22}\sin\varphi, \\ c_{22} &= b_{22}\cos\varphi - b_{21}\sin\varphi. \end{aligned} \tag{8.2}$$

Now substitute (8.1) into (8.2), replace a_{21} by a_{12}, and simplify to obtain the equations

$$\begin{aligned} c_{11} &= a_{11}\cos^2\varphi + 2a_{12}\sin\varphi\cos\varphi + a_{22}\sin^2\varphi, \\ c_{12} = c_{21} &= a_{12}(\cos^2\varphi - \sin^2\varphi) + (a_{22} - a_{11})\sin\varphi\cos\varphi \\ &= a_{12}\cos 2\varphi + (a_{22} - a_{11})\frac{\sin(2\varphi)}{2}, \\ c_{22} &= a_{22}\cos^2\varphi - 2a_{12}\sin\varphi\cos\varphi + a_{11}\sin^2\varphi. \end{aligned}$$

We want to have $c_{12} = c_{21} = 0$. Thus we pick φ so that $\tan 2\varphi = 2a_{12}/(a_{11} - a_{22})$. If $a_{11} = a_{22}$, we choose $\varphi = 45°$.

EXAMPLE 8.1. Let

$$A = \begin{bmatrix} 7 & \sqrt{3} \\ \sqrt{3} & 5 \end{bmatrix}.$$

Then $\tan 2\varphi = 2\sqrt{3}/(7-5) = \sqrt{3}$. Thus $2\varphi = 60°$ and $\varphi = 30°$. Hence

$$P = \begin{bmatrix} \cos 30° & -\sin 30° \\ \sin 30° & \cos 30° \end{bmatrix} = \frac{1}{2}\begin{bmatrix} \sqrt{3} & -1 \\ 1 & \sqrt{3} \end{bmatrix},$$

$$P'A = \frac{1}{2}\begin{bmatrix} \sqrt{3} & 1 \\ -1 & \sqrt{3} \end{bmatrix}\begin{bmatrix} 7 & \sqrt{3} \\ \sqrt{3} & 5 \end{bmatrix} = \frac{1}{2}\begin{bmatrix} 8\sqrt{3} & 8 \\ -4 & 4\sqrt{3} \end{bmatrix},$$

and

$$P'AP = \begin{bmatrix} 2\sqrt{3} & 2 \\ -1 & \sqrt{3} \end{bmatrix}\begin{bmatrix} \sqrt{3} & -1 \\ 1 & \sqrt{3} \end{bmatrix} = \begin{bmatrix} 8 & 0 \\ 0 & 4 \end{bmatrix}.$$

Thus the eigenvalues of A are 8 and 4. The columns of P are eigenvectors of A. That is,

$$Ap_{\cdot 1} = \begin{bmatrix} 7 & \sqrt{3} \\ \sqrt{3} & 5 \end{bmatrix}\begin{bmatrix} \sqrt{3}/2 \\ 1/2 \end{bmatrix} = \begin{bmatrix} 4\sqrt{3} \\ 4 \end{bmatrix} = 8p_{\cdot 1},$$

and (by a similar computation)

$$Ap_{\cdot 2} = 4p_{\cdot 2}.$$

Before describing Jacobi's method in general, let us establish several facts. Let $e_1, e_2, \ldots, e_n$ be the columns, and δ_{ij} the (i,j)th element of the n by n identity matrix I.

THEOREM 8.1. *Let P be an n by n matrix whose columns $p_{\cdot j}$ are defined by the equations*

$$p_{\cdot j} = e_j \quad \text{if} \quad j \neq k \text{ or } m,$$

$$(8.3) \qquad p_{\cdot k} = (\cos\varphi)e_k + (\sin\varphi)e_m,$$

$$p_{\cdot m} = (-\sin\varphi)e_k + (\cos\varphi)e_m,$$

where φ is a real number. Then P is orthogonal.

Proof. Let $C = P'P$. Notice that C is symmetric and that the ith row of P' is the same as the ith column of P. In the equations below, which follow from (2.2), we use the convention that i and j never take on the values k or m.*

$$c_{ij} = p_{\cdot i} \cdot p_{\cdot j} = e_i \cdot e_j = \delta_{ij},$$

$$c_{jk} = c_{kj} = [(\cos\varphi)e_k + (\sin\varphi)e_m] \cdot e_j = 0 = \delta_{jk} = \delta_{kj},$$

$$c_{jm} = c_{mj} = [(-\sin\varphi)e_k + (\cos\varphi)e_m] \cdot e_j = 0 = \delta_{jm} = \delta_{mj},$$

*The reader who is familiar with block matrix diagrams might prefer a proof based on that concept.

$$c_{km} = c_{mk} = [(\cos\varphi)e_k + (\sin\varphi)e_m]\cdot[(-\sin\varphi)e_k + (\cos\varphi)e_m]$$
$$= -\sin\varphi\cos\varphi + \sin\varphi\cos\varphi = 0 = \delta_{km} = \delta_{mk},$$
$$c_{kk} = p_{\cdot k}\cdot p_{\cdot k} = \cos^2\varphi + \sin^2\varphi = 1 = \delta_{kk},$$
$$c_{mm} = p_{\cdot m}\cdot p_{\cdot m} = \cos^2\varphi + \sin^2\varphi = 1 = \delta_{mm}.$$

We have shown that $c_{ij} = \delta_{ij}$ for all i and j (including now k and m) and hence $P'P = I$.

THEOREM 8.2. *Let A be an n by n real symmetric matrix and let $\tan 2\varphi = 2a_{km}/(a_{kk} - a_{mm})$. With this value of φ let P be defined by (8.3) and suppose $C = P'AP$. Then $c_{km} = c_{mk} = 0$.*

Proof. Let $B = P'A$. Then

$$c_{mk} = b_{m\cdot}\cdot p_{\cdot k}$$
$$(8.4) \qquad = (b_{m1}, b_{m2}, \ldots, b_{mn})\cdot[(\cos\varphi)e_k + (\sin\varphi)e_m]$$
$$= b_{mk}\cos\varphi + b_{mm}\sin\varphi.$$

But

$$b_{mk} = p_{\cdot m}\cdot a_{\cdot k} = [(-\sin\varphi)e_k + (\cos\varphi)e_m]\cdot a_{\cdot k}$$
$$= -a_{kk}\sin\varphi + a_{mk}\cos\varphi,$$

and

$$b_{mm} = p_{\cdot m}\cdot a_{\cdot m} = -a_{km}\sin\varphi + a_{mm}\cos\varphi.$$

Combining these two equations with (8.4) we obtain

$$c_{mk} = -a_{kk}\sin\varphi\cos\varphi + a_{mk}\cos^2\varphi - a_{km}\sin^2\varphi + a_{mm}\sin\varphi\cos\varphi$$
$$= a_{km}(\cos^2\varphi - \sin^2\varphi) + (a_{mm} - a_{kk})\sin\varphi\cos\varphi$$
$$= a_{km}\cos 2\varphi + (a_{mm} - a_{kk})\sin(2\varphi)/2$$
$$= 0$$

since $\tan 2\varphi = 2a_{km}/(a_{kk} - a_{mm})$. It follows from the symmetry of C that $c_{km} = 0$.

Jacobi's method proceeds by the following steps:

1. Choose k and m so that $k \neq m$ and

$$|a_{km}| = \max_{i \neq j} |a_{ij}|.$$

2. Let φ be chosen so that $\tan 2\varphi = 2a_{km}/(a_{kk} - a_{mm})$. (If $a_{kk} = a_{mm}$ choose $\varphi = 45°$.)
3. Let P be the matrix defined by (8.3) with the value of φ obtained in step 2.
4. Compute the matrix $A_1 = P'AP$.
5. Repeat steps 1–4 on A_1 to obtain the matrix $A_2 = P_2'A_1P_2$. Continue cycling through steps 1–4 until the transformed matrix A_q has off-diagonal elements which are sufficiently small.

The approximate eigenvalues of A are the diagonal elements of A_q. The product $P_1P_2 \ldots P_q$ of all the orthogonal matrices found in step 3 has approximate eigenvectors of A for columns.

We will prove later that Jacobi's method converges. Since each matrix P_i formed in this process is orthogonal, one can see that

$$A_q = P_q' \, P_{q-1}' \ldots P_1' \, A \, P_1 \, P_2 \ldots P_q$$

has the same eigenvalues as A (see Problems 7.1 and 7.5). At each step, the method produces zeros in the (k,m) and (m,k) positions. It destroys some of the zeros obtained in previous steps, but the off-diagonal elements tend to zero as the process continues. The theorem of Gershgorin (Theorem 7.2) can be used to obtain an error estimate for the approximate eigenvalues. In using this theorem on the matrix A_q we are assuming no round-off error in the process. Thus we will use the term "error estimate" rather than "error bound" for our results. An analysis which takes round-off error into account is given in Wilkinson [23, 24].

EXAMPLE 8.2. Let

$$A = \begin{bmatrix} 3 & 0.01 & 0.1 \\ 0.01 & 2 & 0.01 \\ 0.1 & 0.01 & 1 \end{bmatrix}.$$

With $k = 1$, $m = 3$, we have

$$\tan 2\varphi = 2(0.1)/(3-1) = 0.1.$$

Thus $\varphi \doteq 2°51'$ and

$$P = \begin{bmatrix} 0.99876 & 0 & -0.04972 \\ 0 & 1 & 0 \\ 0.04972 & 0 & 0.99876 \end{bmatrix}.$$

This yields (to 5 decimal places) the matrix

$$A_1 = P'AP = \begin{bmatrix} 3.00497 & 0.01048 & 0.00019 \\ 0.01048 & 2 & 0.00949 \\ 0.00019 & 0.00949 & 0.99501 \end{bmatrix}.$$

We should have 0 instead of 0.00019 in the (1,3) and (3,1) positions. This discrepancy is due partly to round-off, but primarily to the approximation of φ. The error in φ does not affect the error estimate given below.

Now, in the notation of Theorem 7.2, $R_1 = 0.01067$, $R_2 = 0.01997$, and $R_3 = 0.00968$. The discs D_1, D_2, and D_3 are disjoint; thus (ignoring round-off error) we have

$$|\lambda_1 - 3.00497| \leqslant 0.01067,$$

$$|\lambda_2 - 2| \leqslant 0.01997,$$

$$|\lambda_3 - 0.99501| \leqslant 0.00968,$$

where λ_1, λ_2, and λ_3 are the eigenvalues of A.

Sample computer runs with Jacobi's method can be found in Carnahan, Luther, and Wilkes [2].

On a digital computer the search for the largest off-diagonal element consumes a relatively large amount of time. This search can be avoided by using one of the so-called *cyclic Jacobi methods*. Here the values of k and m in step 1 are chosen to run systematically through the above-diagonal elements. For example, in the row-cyclic method, the indices are chosen in the order

$$(1,2),(1,3),\ldots,(1,n), \quad (2,3),\ldots,(2,n),\ldots,(n-1,n), \quad (1,2),\ldots .$$

Forsythe and Henrici [7] have shown that the row-cyclic method converges if the angle φ in step 2 satisfies the additional condition

$$\frac{-\pi}{4} \leqslant \varphi \leqslant \frac{\pi}{4}.$$

The cyclic method has the disadvantage that the computation for the index (k,m) is carried out even if the element in that position is already sufficiently small. The *threshold method* is designed to avoid this difficulty. Here the computation is performed for the index (k,m) only if the magnitude of the corresponding element exceeds a threshold value α_1. When all off-diagonal elements have a magnitude less than α_1, the process is repeated with a smaller threshold α_2, and so on until the matrix is sufficiently close to being diagonal.

To conclude this section let us outline a proof of the fact that the off-diagonal elements tend to zero in Jacobi's method.

THEOREM 8.3. *Suppose $A_1, A_2, \ldots, A_q, \ldots$ are computed by steps 1–5 of Jacobi's method. Let a_{qij} represent the (i,j)th element of A_q. If $i \neq j$, then*

$$\lim_{q\to\infty} a_{qij} = 0. \tag{8.5}$$

Proof. For any n by n matrix A, let

$$D(A) = \sum_{i=1}^{n} a_{ii}^2,$$

$$T(A) = \sum_{i=1}^{n} \sum_{j=1}^{n} a_{ij}^2,$$

$$E(A) = T(A) - D(A).$$

We will show that $E(A_q) \to 0$ as $q \to \infty$ which will imply (8.5). Leaving most of the work to the reader, we have from Problem 8.5 that

$$T(A_{q+1}) = T(A_q).$$

Thus

$$E(A_{q+1}) = E(A_q) + D(A_q) - D(A_{q+1}).$$

Using part (d) of Problem 8.6, we have

$$E(A_{q+1}) = E(A_q) - 2a_{qkm}^2, \tag{8.6}$$

where

$$|a_{qkm}| = \max_{i\neq j} |a_{qij}|.$$

But

$$E(A_q) = \sum_{i\neq j} a_{qij}^2 \leqslant n(n-1)a_{qkm}^2,$$

and hence

$$-2a_{qkm}^2 \leqslant \frac{-2E(A_q)}{n(n-1)}. \tag{8.7}$$

Combining (8.6) and (8.7), we have

$$E(A_{q+1}) \leqslant E(A_q)\left(1 - \frac{2}{n(n-1)}\right). \tag{8.8}$$

The repeated application (8.8) implies that

$$E(A_q) \leqslant \left(1 - \frac{2}{n(n-1)}\right)^q E(A).$$

The proof is completed by noting that

$$0 \leqslant 1 - \frac{2}{n(n-1)} < 1,$$

and thus $E(A_q) \to 0$ as $q \to \infty$.

PROBLEMS

8.1. Apply Jacobi's method to the matrix

$$A = \begin{bmatrix} 1 & 2 \\ 2 & 1 \end{bmatrix}.$$

How can you check your result?

8.2. Let

$$A = \begin{bmatrix} 5 & 1 & .1 \\ 1 & 3 & 0 \\ .1 & 0 & 1 \end{bmatrix}.$$

(a) Compute $A_1 = P_1' A P_1$ using Jacobi's method.
(b) What are the approximate eigenvalues?
(c) Compute error estimates for the eigenvalues.
(d) Find the matrix P_2 which will be used for the next cycle of Jacobi's method.
(e) If your energy is holding up you might even compute $P_2' A_1 P_2$ and repeat parts (b) and (c).

8.3. Write a flow diagram for Jacobi's method.

8.4. Write a Fortran program for Jacobi's method.

8.5. Let A be a real matrix and

$$T(A) = \sum_{i=1}^{n} \sum_{j=1}^{n} a_{ij}^2 .$$

Show that if P is orthogonal, then

$$T(P'AP) = T(A).$$

Hint: Use Problem 7.7.

8.6. Using the notation of Theorem 8.2, prove the following:
(a) For all i except k and m, $c_{ii} = a_{ii}$.
(b)

$$\begin{bmatrix} c_{kk} & c_{km} \\ c_{mk} & c_{mm} \end{bmatrix} = Q' \begin{bmatrix} a_{kk} & a_{km} \\ a_{mk} & a_{mm} \end{bmatrix} Q,$$

where $Q = \begin{bmatrix} \cos\varphi & -\sin\varphi \\ \sin\varphi & \cos\varphi \end{bmatrix}$.

(c) $c_{kk}^2 + c_{mm}^2 = a_{kk}^2 + a_{mm}^2 + 2a_{km}^2$. Hint: Use Problem 8.5.

(d) $D(C) = D(A) + 2a_{km}^2$, where D is defined as in the proof of Theorem 8.3.

8.7. Prove that the off-diagonal elements converge to zero for the threshold Jacobi method if the threshold values $\alpha_1, \alpha_2, \ldots$ tend to zero. Hint: Show that after a finite number of steps we must have

$$|a_{qij}| \leqslant \alpha_1$$

for all $i \neq j$.

9. THE GIVENS-HOUSEHOLDER METHOD

The Givens-Householder method is another technique for computing the eigenvalues and eigenvectors of real symmetric matrices. This technique is somewhat more complicated but considerably more efficient than the Jacobi method. The basic idea is to use orthogonal transformations to reduce the given matrix A to tridiagonal form; that is, to a matrix which has non-zero elements only on and directly above and below the main diagonal. The characteristic polynomials of symmetric tridiagonal matrices have special properties which can be exploited to produce an efficient scheme for computation of the eigenvalues.

Originally, Givens devised both a method for the reduction to tridiagonal form as well as a method for computing the eigenvalues of such a matrix. A few years later Householder discovered a more efficient scheme for the reduction to tridiagonal form. The technique which we will describe consists of the Householder reduction coupled with the Givens method for computing the eigenvalues.

Let us now give a formal definition of a tridiagonal matrix.

DEFINITION 9.1. *The matrix A is said to be tridiagonal iff $a_{ij} = 0$ whenever $|i - j| > 1$. A position (i,j) in the matrix is called off-tridiagonal iff $|i - j| > 1$.*

EXAMPLE 9.1. The following matrix is tridiagonal:

$$\begin{bmatrix} 1 & 2 & 0 & 0 & 0 \\ 3 & 3 & 6 & 0 & 0 \\ 0 & 5 & 1 & 4 & 0 \\ 0 & 0 & 2 & 1 & 2 \\ 0 & 0 & 0 & 7 & 9 \end{bmatrix}.$$

The off-tridiagonal positions are the positions containing zero elements.

The orthogonal matrices used in the reduction process will be of the form

$$P = I - 2vv',$$

where v is a unit vector chosen in such a way that the matrix $P'AP$ will have zeros in certain off-tridiagonal positions. Let us show that such a matrix is orthogonal and symmetric.

THEOREM 9.1. *Suppose $v \in E_n$ is a column vector with unit length and $P = I - 2vv'$. Then P is orthogonal and symmetric.*

Proof. Note that

$$P' = I' - 2v''v' = I - 2vv' = P.$$

Thus P is symmetric. Also $\|v\|^2 = v'v = 1$. Hence

$$P'P = PP = I - 2vv' - 2vv' + 4v(v'v)v' = I,$$

and P is orthogonal.

Let A be the real symmetric matrix under consideration. For the first step of the reduction process we let

$$v' = \frac{1}{\alpha}(0, a_{12} + s, a_{13}, \ldots, a_{1n}), \tag{9.1}$$

where $s^2 = \sum_{j=2}^{n} a_{1j}^2$ and $\alpha^2 = 2s(a_{12} + s)$. Note that if $s = 0$, the first row of A is already in tridiagonal form and the first step in the reduction can be skipped.

THEOREM 9.2. *Suppose A is a real symmetric n by n matrix, v is defined by (9.1), and $P = I - 2vv'$. If $C = P'AP$, then*

$$c_{1k} = c_{k1} = 0$$

for $k = 3, 4, \ldots, n$.

Proof. Using the fact that $v_1 = 0$, we obtain

$$P = I - 2vv' = \begin{bmatrix} 1 & 0 & \ldots & 0 \\ 0 & 1-2v_2^2 & \ldots & -2v_2v_n \\ \vdots & & & \\ 0 & -2v_nv_2 & \ldots & 1-2v_n^2 \end{bmatrix}.$$

Let $B = AP$. A direct calculation shows that:

$$b_{11} = a_{11},$$

$$b_{12} = a_{12}(1 - 2v_2^2) - 2 \sum_{j=3}^{n} a_{1j}v_jv_2$$

$$= a_{12} - \frac{2(a_{12} + s)^2 a_{12}}{\alpha^2} - \frac{2}{\alpha^2}(a_{12} + s) \sum_{j=3}^{n} a_{1j}^2$$

$$= a_{12} - \frac{a_{12}}{s}(a_{12} + s) - \frac{1}{s}(s^2 - a_{12}^2) = -s.$$

For $k = 3, 4, \ldots, n$, we have

$$b_{1k} = -2a_{12}v_2v_k - 2a_{13}v_3v_k - \ldots + (1-2v_k^2)a_{1k} - \ldots - 2a_{1n}v_nv_k$$

$$= a_{1k} - 2v_k \sum_{j=2}^{n} a_{1j}v_j$$

$$= a_{1k} - \frac{2a_{1k}}{\alpha^2} a_{12}(a_{12} + s) - \frac{2a_{1k}}{\alpha^2} \sum_{j=3}^{n} a_{1j}^2$$

$$= a_{1k} - \frac{a_{1k}a_{12}}{s} - \frac{a_{1k}(s^2 - a_{12}^2)}{s(a_{12} + s)} = 0.$$

Now note that $C = P'B = PB$; hence

$$C = \begin{bmatrix} 1 & 0 & \ldots & 0 \\ 0 & 1-2v_2^2 & \ldots & -2v_2v_n \\ \vdots & & & \\ 0 & -2v_nv_2 & \ldots & 1-2v_n^2 \end{bmatrix} \begin{bmatrix} a_{11} & -s & 0 & \ldots & 0 \\ \vdots & & & & \\ b_{n1} & \cdot & \cdot & \cdot & b_{nn} \end{bmatrix}.$$

Another direct calculation shows that $c_{1k} = 0$ for $k \geqslant 3$. The proof is completed by noting that C is symmetric.

The matrix C in Theorem 9.2 has zeros in the off-tridiagonal positions of the first row and column. The next theorem shows that we can continue this process until a tridiagonal matrix is obtained.

THEOREM 9.3. *Suppose C is a real symmetric matrix which has zeros in the off-tridiagonal positions of the first $k-1$ rows and columns ($2 \leqslant k \leqslant n-2$). Let*

$$s^2 = \sum_{j=k+1}^{n} c_{kj}^2,$$

$$\alpha^2 = 2s(c_{k,k+1} + s),$$

$$v' = \frac{1}{\alpha}(0,\ldots,0,c_{k,k+1}+s,c_{k,k+2},\ldots,c_{k,n}),$$

$$P = I - 2vv',$$

$$D = P'CP.$$

Then D has zeros in the off-tridiagonal positions of the first k rows and columns.

The proof of this theorem is similar to that of Theorem 9.2. The details are left for an exercise (Problem 9.3).

From Theorem 9.3 it is clear that a sequence $P_1, P_2, \ldots, P_{n-2}$ of orthogonal matrices can be chosen in such a way that if $A_0 = A$, $A_k = P_k' A_{k-1} P_k$ for $k = 1, 2, \ldots, n-2$, then A_{n-2} will be tridiagonal. By Problem 7.1, A_{n-2} has the same eigenvalues as A.

It should be noted that there are two choices for the sign of the quantity s in Theorem 9.3. Theoretically it makes no difference how this sign is chosen, but from the computational point of view it is better to choose s so that it has the same sign as $c_{k,k+1}$. This avoids the possible loss of significance in computing $c_{k,k+1} + s$. The choice of the sign of α is unimportant.

EXAMPLE 9.2. Consider the matrix

$$A = \begin{bmatrix} 3 & 1 & 0 & 0 & 0 \\ 1 & 2 & 3 & 0 & 0 \\ 0 & 3 & 7 & 3 & 4 \\ 0 & 0 & 3 & 2 & 1 \\ 0 & 0 & 4 & 1 & 6 \end{bmatrix}.$$

The first two rows and columns of A are in tridiagonal form. Letting $k = 3$ in Theorem 9.3 we obtain

$$s^2 = \sum_{j=4}^{5} c_{3j}^2 = 9 + 16 = 25.$$

Choosing the sign of s to agree with c_{34} we have $s = 5$. Then $\alpha^2 = 10(3 + 5) = 80$ and $\alpha = 4\sqrt{5}$. This gives

$$v' = \frac{1}{4\sqrt{5}}(0,0,0,8,4),$$

and

$$P = \begin{bmatrix} 1 & 0 & 0 & 0 & 0 \\ 0 & 1 & 0 & 0 & 0 \\ 0 & 0 & 1 & 0 & 0 \\ 0 & 0 & 0 & -0.6 & -0.8 \\ 0 & 0 & 0 & -0.8 & 0.6 \end{bmatrix}.$$

It is left for the reader to check that $P'AP$ is in tridiagonal form.

The next step is to describe a procedure for computing the eigenvalues of a symmetric tridiagonal matrix. Suppose

$$C = \begin{bmatrix} c_1 & b_1 & 0 & \cdots & 0 \\ b_1 & c_2 & b_2 & \cdots & 0 \\ 0 & & & & \vdots \\ \vdots & & & & b_{n-1} \\ 0 & \cdots & & b_{n-1} & c_n \end{bmatrix}.$$

in such a matrix. Let us define a sequence of polynomials by the equations

$$\begin{aligned} &p_0(\lambda) = 1, \\ (9.2) \qquad &p_1(\lambda) = c_1 - \lambda, \\ &p_k(\lambda) = (c_k - \lambda)p_{k-1}(\lambda) - b_{k-1}^2 p_{k-2}(\lambda), \quad k = 2,3,\ldots n. \end{aligned}$$

We will now establish some properties of these polynomials which will be useful in the eigenvalue computation.

THEOREM 9.4. *The polynomial p_k defined by (9.2) is the characteristic polynomial of the k by k tridiagonal matrix*

$$C_k = \begin{bmatrix} c_1 & b_1 & 0 & \cdots & 0 \\ b_1 & c_2 & b_2 & \cdots & 0 \\ \vdots & & & & \vdots \\ \vdots & & & & b_{k-1} \\ 0 & 0 & \cdots & b_{k-1} & c_k \end{bmatrix}.$$

Consequently the polynomial p_n is the characteristic polynomial of $C_n = C$.

The proof of this theorem is a good exercise in the use of mathematical induction. The details are left for Problem 9.5.

In view of the fact that the eigenvalues of a real symmetric matrix are real, we infer from Theorem 9.4 that the zeros of each polynomial p_k are real. In the discussions which follow we will assume that none of the elements b_i, $i = 1,2,\ldots,n-1$ is zero. If one of the b_i's were zero, then the eigenvalue problem could be reduced to two smaller problems (see Problem 9.9).

THEOREM 9.5. *If $k \geqslant 2$ and z is a zero of p_{k-1}, then*

$$p_k(z)p_{k-2}(z) < 0.$$

Proof. If $k = 2$ and $p_1(z) = 0$, then (9.2) implies that

$$p_2(z)p_0(z) = p_2(z) = -b_1^2 p_0(z).$$

Thus the theorem holds for $k = 2$ in view of our assumption that none of the b_i's is zero. Proceeding by induction, we suppose that for $j = 2,3,\ldots,k-1$

$$p_j(z)p_{j-2}(z) < 0 \tag{9.3}$$

if $p_{j-1}(z) = 0$. Now suppose that $p_{k-1}(z) = 0$. Then $p_{k-2}(z) \neq 0$; otherwise the inequality (9.3) would fail with $j = k-1$. From (9.2) we have

$$p_k(z)p_{k-2}(z) = -b_{k-1}^2 p_{k-2}^2(z) < 0,$$

which completes the proof.

THEOREM 9.6. *Suppose $k \geqslant 2$ and let w_i and z_i represent the ith zeros of p_{k-1} and p_k, respectively, indexed in increasing order. Then*

$$z_1 < w_1 < z_2 < w_2 < \ldots < w_{k-1} < z_k. \tag{9.4}$$

Proof. Let $\psi_k(\lambda) = \operatorname{sgn} p_k(\lambda)$. It follows readily from (9.2) that

$$\begin{aligned} \psi_k(-\infty) &= 1, \\ \psi_k(\infty) &= (-1)^k. \end{aligned} \tag{9.5}$$

Now let $k = 2$ in order to start an inductive proof of (9.4). In this case $w_1 = c_1$. Also, from Theorem 9.5, $\psi_2(w_1) = -\psi_0(w_1) = -1$. Combining this with (9.5) we have

$$\psi_2(-\infty) = 1,$$

$$\psi_2(w_1) = -1,$$

$$\psi_2(\infty) = 1.$$

It follows that p_2 has exactly one zero in each of the intervals $(-\infty, w_1)$ and (w_1, ∞); so (9.4) holds when $k = 2$.

Turning to the inductive step, let $v_1, v_2, \ldots, v_{k-2}$ represent the zeros of p_{k-2} and suppose that

$$w_1 < v_1 < w_2 < v_2 < \ldots < v_{k-2} < w_{k-1}.$$

Now p_{k-2} must change sign at each of the points $v_1, \ldots, v_{k-2}$ because these zeros are distinct. It follows that $\psi_{k-2}(w_i) = (-1)^{i-1}$, $i = 1,2,\ldots,k-1$. Theorem 9.5 then imples that $\psi_k(w_i) = (-1)^i$.

Combining this with (9.5) we have

$$\psi_k(-\infty) = 1, \ \psi_k(w_1) = -1,$$
$$\psi_k(w_2) = 1, \ldots, \psi_k(w_{k-1}) = (-1)^{k-1},$$
$$\psi_k(\infty) = (-1)^k.$$

Thus p_k has a zero in each of the intervals $(-\infty, w_1)$, (w_1, w_2), $\ldots$, (w_{k-1}, ∞); and the proof is complete.

The number of signs variations in the vector $(p_0(\lambda), p_1(\lambda), \ldots, p_k(\lambda))$ contains information about the zeros of p_k. Let us formally describe the number of sign variations, and then show how this can be used to compute the eigenvalues of C.

Let $(a_0, a_1, \ldots, a_k)$ be a vector with real components. For $i = 1,2,\ldots,k$, let $j(i)$ be the largest index less than i such that $a_{j(i)} \neq 0$. Define b_i by the equation

$$b_i = \begin{cases} 0 & \text{if} \quad a_i a_{j(i)} \geqslant 0 \\ 1 & \text{if} \quad a_i a_{j(i)} < 0 \end{cases}.$$

The number of variations in sign of the vector $(a_0, a_1, \ldots, a_k)$ is the quantity $\Sigma_{i=1}^{k} b_i$. Less formally, it is the number of sign changes when zero components are ignored.

EXAMPLE 9.3. On the left are several vectors and on the right are their numbers of sign variation.

$(1, -1, 2, 3, -5)$	3
$(2, 1, -1, -2)$	1
$(2, 0, 2, -1)$	1
$(2, 0, -2, 1)$	2

For each real number λ, let $\varphi_k(\lambda)$ represent the number of sign variations in the sequence $(p_0(\lambda), p_1(\lambda), \ldots, p_k(\lambda))$. The next theorem shows the relationship between φ_k and the zeros of p_k.

THEOREM 9.7. *For each real number λ, the number of zeros of p_k which are less than λ is $\varphi_k(\lambda)$. Consequently, $\varphi_n(\lambda)$ is the number of eigenvalues of C which are less than λ.*

Proof. As usual we will use mathematical induction. It is left as an exercise (Problem 9.7) to show that the theorem holds for $k = 1$. For the inductive step let $\nu_j(\lambda)$ be the number of zeros of p_j which are less than λ, and suppose that

$$\varphi_{k-1}(\lambda) = \nu_{k-1}(\lambda)$$

for each real number λ. We wish to show that $\varphi_k(\lambda) = \nu_k(\lambda)$ for all λ. Let w_i and z_i represent the ith zeros of p_{k-1} and p_k, respectively, indexed in increasing order. Also, for notational convenience, let $z_0 = w_0 = -\infty$ and $z_{k+1} = w_k = \infty$. With this notation we see from (9.4) that for an arbitrary real number λ there exists a unique index i for which one of the following holds:

(9.6) $\qquad w_i < \lambda \leqslant z_{i+1},$

(9.7) $\qquad z_i < \lambda < w_i,$

(9.8) $\qquad \lambda = w_i.$

As before, let $\psi_j(\lambda) = \operatorname{sgn} p_j(\lambda)$. From (9.4) and (9.5) one can deduce that if $w_i < \lambda < w_{i+1}$, then

(9.9) $\qquad \psi_{k-1}(\lambda) = (-1)^i, \quad \nu_{k-1}(\lambda) = i;$

while if $z_i < \lambda < z_{i+1}$, then

(9.10) $\qquad \psi_k(\lambda) = (-1)^i, \quad \nu_k(\lambda) = i.$

If (9.6) holds, then (9.9) and (9.10) imply that there is no sign change between $p_{k-1}(\lambda)$ and $p_k(\lambda)$. Thus

$$\varphi_k(\lambda) = \varphi_{k-1}(\lambda) = \nu_{k-1}(\lambda) = i = \nu_k(\lambda).$$

If (9.7) holds, then $\psi_{k-1}(\lambda) = (-1)^{i-1}$ and $\psi_k(\lambda) = (-1)^i$. Now there is a sign change between $p_{k-1}(\lambda)$ and $p_k(\lambda)$. Thus $\varphi_k(\lambda) = 1 + \varphi_{k-1}(\lambda)$ and

$$\varphi_k(\lambda) = 1 + (i-1) = i = \nu_k(\lambda).$$

If (9.8) holds, then we infer from Theorem 9.5 that there is a sign change between $p_{k-2}(\lambda)$ and $p_k(\lambda)$. Thus

$$\varphi_k(\lambda) = \varphi_{k-1}(\lambda) + 1 = (i-1) + 1 = i = \nu_k(\lambda).$$

In any case, $\varphi_k(\lambda) = \nu_k(\lambda)$ and the proof is complete.

We are now in a position to compute the eigenvalues of C. The first step will be to isolate the eigenvalues by means of Theorem 9.7. To start the process, note that Gershgorin's theorem (Theorem 7.2) implies that all the eigenvalues of C lie in the interval $[a,b]$, where

$$a = \min_{1 \leqslant i \leqslant n} (c_i - R_i), \quad b = \max_{1 \leqslant i \leqslant n} (c_i + R_i),$$

and R_i is the sum of the magnitudes of the off-diagonal elements in the ith row. Suppose, for example, that we wish to isolate the smallest eigenvalue λ_1. Let m_1 be the midpoint of $[a,b]$. If $\varphi_n(m_1) = 1$, then Theorem 9.7 implies that λ_1 lies in the interval $[a,m_1]$. In addition, λ_1 is the only eigenvalue in this interval; so the eigenvalue λ_1 has been isolated and the search is stopped. Otherwise, Theorem 9.7 implies that $\lambda_1 \in [m_1,b]$ if $\varphi_n(m_1) = 0$, or $\lambda_1 \in [a,m_1]$ if $\varphi_n(m_1) > 1$. Now let m_2 be the midpoint of the interval containing λ_1 and continue the search. In this way we obtain an interval which contains the eigenvalue λ_1 and none of the others. The remaining eigenvalues of C can be isolated in a similar fashion. It is left to the reader (Problem 9.8) to work out a scheme whereby all of the eigenvalues would be isolated in one search, so that information obtained in the search for λ_1 could be applied to λ_2, and so forth.

The process just described produces for each i an interval $[a_i, b_i]$ which contains λ_i and none of the other eigenvalues. The value a_i or b_i can be used as an initial approximation for one of the iteration procedures, such as Newton's method, which will be described in Chapter 4. The technique which was used to isolate the roots, known as the *method of bisection*, could also be used to approximate λ_i to within the desired accuracy.

If x is an eigenvector of C, then Px is an eigenvector of A, where P is the product of the matrices $I - 2vv'$ used to reduce A to tridiagonal form (Problem 9.10). Thus eigenvectors of A can be computed from eigenvectors of C. Let λ^* be close to the computed value of λ_i determined by the method described above. To compute an eigenvector corresponding to λ_i we let x_0 be an arbitrary vector in E_n and solve the system of equations

$$(9.11) \qquad (C - \lambda^* I)x_k = x_{k-1}$$

for $k = 1, 2, \ldots$. After several iterations x_k is normally a good approximation to an eigenvector of C corresponding to λ_i. This technique is based on the *power method*, which will be described in Chapter 4, applied to the matrix $(C - \lambda^* I)^{-1}$. Problem 9.11 shows the relationship between the eigenvectors of $(C - \lambda^* I)^{-1}$ and C.

Flow diagrams, Fortran programs, and a sample problem for the Givens-Householder method can be found in Ortega [15]. Algol programs and further examples have been published by Wilkinson [21, 22, 26].

PROBLEMS

9.1. For the matrix A given below, find an orthogonal matrix P such that $P'AP$ will be tridiagonal. Check your result by computing $P'AP$.

$$A = \begin{bmatrix} 3 & 5 & 0 & 0 \\ 5 & 6 & 1 & 2 \\ 0 & 1 & 4 & 3 \\ 0 & 2 & 3 & 5 \end{bmatrix}.$$

9.2. Write a computer program for reducing a real symmetric matrix to tridiagonal form.

9.3. Prove Theorem 9.3.

9.4. Find the coefficients of the characteristic polynomial for the matrix C.

$$C = \begin{bmatrix} 1 & 2 & 0 & 0 \\ 2 & 3 & 4 & 0 \\ 0 & 4 & 5 & 6 \\ 0 & 0 & 6 & 7 \end{bmatrix}.$$

9.5. Prove Theorem 9.4.

9.6. Isolate the eigenvalues of the matrix C in Problem 9.4. Then determine the largest eigenvalue to the nearest tenth.

9.7. Verify that Theorem 9.7 holds for $k = 1$.

9.8. Prepare a flow diagram for a search technique to isolate all the eigenvalues of a real symmetric tridiagonal matrix.

9.9. Suppose the off-diagonal element b_k is zero in the tridiagonal matrix C. Show that $p_n(\lambda) = p_k(\lambda)q(\lambda)$, where q is the characteristic polynomial of the tridiagonal matrix

$$\begin{bmatrix} c_{k+1} & b_{k+1} & 0 & \cdots & 0 \\ b_{k+1} & c_{k+2} & b_{k+2} & \cdots & 0 \\ \vdots & & & & \\ 0 & \cdot & \cdot & b_{n-1} & c_n \end{bmatrix}.$$

9.10. Let $C = P'AP$, where P is orthogonal. Show that if x is an eigenvector of C, then Px is an eigenvector of A.

9.11. Let λ_i be an eigenvalue of the symmetric matrix C. Suppose the number λ^* is not equal to λ_i, but that λ^* is closer to λ_i than any other eigenvalue of C. Prove the following:

(a) The number $\lambda_i - \lambda^*$ is an eigenvalue of $C - \lambda^* I$.

(b) The eigenvalue of $(C - \lambda^* I)^{-1}$ which has the greatest magnitude is $1/(\lambda_i - \lambda^*)$.

(c) If x is an eigenvector of $(C - \lambda^* I)^{-1}$ corresponding to $1/(\lambda_i - \lambda^*)$, then x is also an eigenvector of C corresponding to λ_i.

*10. COMPUTATIONAL ASPECTS OF LINEAR PROGRAMMING

The discussion in Chapter 2 showed that the search for a solution to the linear programming problem could be restricted to the extreme points of the feasibility set. In this section we will describe a scheme, known as the *simplex method*, for stepping from extreme point to extreme point in such a way as to always decrease the value of the linear functional T until a minimum is found. The first step will be to transform the problem slightly so that the feasibility set is described in terms of equalities rather than inequalities. Then we will develop the following tools which form the basis of the simplex method:

1. A test to determine when a given point x of the feasibility set is an extreme point.
2. A test to determine when a given extreme point x minimizes the functional T. This test will also detect situations where T has no minimum on the feasibility set.
3. An algorithm for finding an extreme point which gives a lower value of T than the extreme point x in case x does not already minimize T.

In the simplex method an initial extreme point is found with the aid of the first test. Then the second test and the algorithm are applied repeatedly until either a minimum value of T is found or it is shown that the problem has no solution.

For computational purposes it is convenient to replace the inequalities in (9.1) of Chapter 2 by equalities. Consider the inequality

$$a_{11}x_1 + a_{12}x_2 + \ldots + a_{1n}x_n \geqslant b_1.$$

This is equivalent to the statement

$$a_{11}x_1 + a_{12}x_2 + \ldots + a_{1n}x_n - x_{n+1} = b_1$$

for some number $x_{n+1} \geqslant 0$. The quantity x_{n+1} is referred to as a *slack variable*. By introducing slack variables into each of the other inequalities in (9.1) and replacing $m + n$ by n in the result, we can restrict our attention to problems of the following form:

Given vectors $c \in E_n$, $b \in E_m$ and m by n matrix A, find the minimum of T over the set K, where

$$T(x) = c_1x_1 + c_2x_2 + \ldots + c_nx_n,$$

and K is the set of points $x \in E_n$ which satisfy the constraints

$$x_j \geqslant 0, \quad j = 1, 2, \ldots, n$$

$$Ax = b.$$

The only requirements which were used in developing the theory of Chapter 2 were that T be linear and K be closed and convex; so that theory still applies to the new formulation of the problem.

The columns of the matrix A play an important role in the theory of extreme points. For that reason it will be convenient to write the system of equations $Ax = b$ in the form

$$(10.1) \qquad x_1A_1 + x_2A_2 + \ldots + x_nA_n = b,$$

where $A_1, A_2, \ldots, A_n$ represent the columns of A. The following theorem gives a necessary and sufficient condition for a non-negative solution of (10.1) to be an extreme point of K.

THEOREM 10.1. *Let $x \in E_n$ have all non-negative coordinates and satisfy* (10.1). *Then x is an extreme point of K if and only if the positive coordinates of x correspond to linearly independent columns of A.*

Proof. For notational convenience let us assume that the first k coordinates of x are positive and the remaining $n - k$ coordinates are zero. Suppose x is an extreme point of K and $A_1, A_2, \ldots, A_k$ are dependent. Then there exist constants $\alpha_1, \alpha_2, \ldots, \alpha_k$, not all zero, such that

$$\alpha_1 A_1 + \alpha_2 A_2 + \ldots + \alpha_k A_k = \emptyset.$$

Now let

$$y = (x_1 - \lambda\alpha_1, x_2 - \lambda\alpha_2, \ldots, x_k - \lambda\alpha_k, 0, \ldots, 0),$$
$$z = (x_1 + \lambda\alpha_1, x_2 + \lambda\alpha_2, \ldots, x_k + \lambda\alpha_k, 0, \ldots, 0),$$

where λ is a positive number which is chosen to be sufficiently small so that the first k coordinates of y and z are positive. Now it is easy to see that y and z satisfy (10.1); so y and $z \in K$. But $x = (1/2)y + (1/2)z$. This contradicts the fact that x is an extreme point of K. It follows that x is an extreme point of K only if $A_1, A_2, \ldots, A_k$ are linearly independent.

Conversely, suppose that $A_1, A_2, \ldots, A_k$ are independent and x is not an extreme point of K. Then there exist distinct points $y,z \in K$ and a number $\alpha \in (0,1)$ such that

$$x = \alpha y + (1 - \alpha)z.$$

Since α and $1 - \alpha$ are positive, the last $n - k$ coordinates of y and z must be zero; otherwise x_i would be positive for some $i > k$. From the fact that y and z satisfy (10.1) we have

$$\sum_{j=1}^{k} y_j A_j = \sum_{j=1}^{k} z_j A_j = b.$$

This implies that

$$\sum_{j=1}^{k} (y_j - z_j) A_j = \emptyset,$$

which contradicts the independence of $A_1, A_2, \ldots, A_k$ because $y \neq z$. Thus x is an extreme point and the proof is complete.

The next step will be to describe a test for determining when a given extreme point is a solution to the linear programming problem. Suppose that we have found an extreme point $x \in K$ which has exactly m non-zero coordinates. For notational convenience we will again assume that these are the first m coordinates, so that $x = (x_1, x_2, \ldots, x_m, 0, \ldots, 0)$ and $x_j > 0, j = 1, 2, \ldots, m$. By Theorem 10.1, $A_1, A_2, \ldots, A_m$ are linearly independent; so they span all of E_m. Thus each vector $A_j, j = 1, 2, \ldots, n$, has a unique representation as a linear combination of the first m columns. Let α_{ij} be the coefficient of A_i in the representation of A_j, so that

$$(10.2) \qquad A_j = \sum_{i=1}^{m} \alpha_{ij} A_i.$$

For each j we let

$$(10.3) \qquad \gamma_j = \sum_{i=1}^{m} \alpha_{ij} c_i.$$

The following theorem gives a test for determining whether or not $T(x) = \min_{y \in K} T(y)$.

THEOREM 10.2. *The extreme point x is a solution to the linear programming problem if and only if $\gamma_j \leqslant c_j$ for each $j, j = 1, 2, \ldots, n$.*

Proof. Suppose $\gamma_j \leqslant c_j$ for each j. Let $y \in K$; then y has non-negative coordinates and satisfies (10.1). Substituting in (10.1) the value of A_j given by (10.2), we have

$$\sum_{j=1}^{n} y_j \sum_{i=1}^{m} \alpha_{ij} A_i = b.$$

Changing the order of summation and using the fact that x satisfies (10.1), we obtain

$$\sum_{i=1}^{m} \left(\sum_{j=1}^{n} y_j \alpha_{ij} \right) A_i = b = \sum_{i=1}^{m} x_i A_i.$$

We infer from the independence of $A_1, A_2, \ldots, A_m$ that

$$x_i = \sum_{j=1}^{n} y_j \alpha_{ij}, \quad i = 1, 2, \ldots, m.$$

Thus

$$T(x) = \sum_{i=1}^{m} c_i x_i = \sum_{i=1}^{m} \sum_{j=1}^{n} c_i \alpha_{ij} y_j = \sum_{j=1}^{n} \left(\sum_{i=1}^{m} c_i \alpha_{ij} \right) y_j.$$

This together with (10.3) implies that

$$T(x) = \sum_{j=1}^{n} \gamma_j y_j \leqslant \sum_{j=1}^{n} c_j y_j = T(y).$$

It follows that

$$T(x) = \min_{y \in K} T(y).$$

To prove the converse suppose that $\gamma_k > c_k$ for some index k. Now $\gamma_j = c_j$ for $j = 1, 2, \ldots, m$; so $k > m$. Let

$$(10.4) \qquad z = (-\alpha_{1k}, -\alpha_{2k}, \ldots, -\alpha_{mk}, 0, \ldots, 1, \ldots, 0).$$

where the 1 occurs in the kth coordinate. Note that

$$(10.5) \qquad \sum_{j=1}^{n} z_i A_i = A_k - \sum_{i=1}^{m} \alpha_{ik} A_i = \emptyset.$$

From (10.5) it follows that $x + \lambda z$ satisfies equation (10.1) for any real number λ. But

$$(10.6) \qquad x + \lambda z = (x_1 - \lambda\alpha_{1k}, \ldots, x_m - \lambda\alpha_{mk}, 0, \ldots, \lambda, 0, \ldots, 0).$$

Thus $x + \lambda z \in K$ if λ is chosen to be a positive number which is sufficiently small so that $x_i - \lambda\alpha_{ik} \geqslant 0$ for $i = 1, 2, \ldots, m$. From (10.3) and the hypothesis we have

$$T(z) = c_k - \sum_{i=1}^{m} \alpha_{ik} c_i = c_k - \gamma_k < 0.$$

Thus when $\lambda > 0$ is sufficiently small, $x + \lambda z \in K$ and

$$T(x + \lambda z) = T(x) + \lambda T(z) < T(x).$$

It follows that x is not a solution to the minimization problem.

The following corollary gives a condition under which the problem has no solution.

COROLLARY 10.3. *If $\gamma_k > c_k$ for some k, and if $\alpha_{ik} \leqslant 0$ for $i = 1, 2, \ldots, m$, then T has no lower bound on K.*

Proof. From (10.6) one can see that for any $\lambda > 0$, $x + \lambda z$ now has non-negative coordinates and hence belongs to K. But $T(z) < 0$; so

$$\lim_{\lambda \to \infty} T(x + \lambda z) = \lim_{\lambda \to \infty} [T(x) + \lambda T(z)] = -\infty.$$

Thus T has no lower bound on K.

Now suppose $\gamma_k > c_k$ for some k, and at least one α_{ik} is positive. In the computational scheme, which will be described presently, we would like to go along the ray $x + \lambda z$ as far as possible without leaving K. From (10.6) we see that in order for $x + \lambda z$ to be in K we must have for $i = 1, 2, \ldots, m$

$$(10.7) \qquad x_i - \lambda\alpha_{ik} \geqslant 0.$$

If $\alpha_{ik} \leqslant 0$, then (10.7) is automatically satisfied. Let $I = \{i: 1 \leqslant i \leqslant m$ and $\alpha_{ik} > 0\}$. For each $i \in I$ we must have $\lambda \leqslant x_i/\alpha_{ik}$. Thus we choose

$$\lambda = \min_{i \in I} \frac{x_i}{\alpha_{ik}}. \tag{10.8}$$

Let us now show that this choice of λ yields an extreme point of K.

THEOREM 10.4. *If $\gamma_k > c_k$ for some k, $\alpha_{ik} > 0$ for some i, z is given by (10.4), and λ is given by (10.8), then $x + \lambda z$ is an extreme point of K and $T(x + \lambda z) < T(x)$.*

Proof. It has already been shown that $x + \lambda z$ satisfies (10.1) and $T(x + \lambda z) < T(x)$. It remains to show that $x + \lambda z$ is an extreme point of K. From (10.8) it is clear that $x + \lambda z$ has non-negative coordinates. Let p be an index between 1 and m such that

$$\lambda = \frac{x_p}{\alpha_{pk}}. \tag{10.9}$$

Then

$$(x + \lambda z)_p = x_p - \frac{x_p}{\alpha_{pk}} \alpha_{pk} = 0.$$

Thus the positive coordinates of $x + \lambda z$ include at most those indexed with $1, 2, \ldots, p-1, p+1, \ldots, m, k$. Let us now show that the corresponding columns of A are linearly independent. Suppose

$$\beta_1 A_1 + \ldots + \beta_{p-1} A_{p-1} + \beta_{p+1} A_{p+1} + \ldots + \beta_m A_m + \beta_k A_k = \emptyset.$$

If $\beta_k \neq 0$, then α_{pk} would be zero, which is contrary to the choice of p. Given that $\beta_k = 0$, the remaining β_i's are zero because $A_1, A_2, \ldots, A_m$ are independent. Thus the columns of A corresponding to positive coordinates of $x + \lambda z$ are independent and $x + \lambda z$ is an extreme point of K.

We are now in a position to describe the simplex algorithm for solving linear programs. The steps in the algorithm are as follows:

1. Obtain an extreme point $x \in K$ which has exactly m positive coefficients. (On p. 128 we will describe a general method for finding such an extreme point.)
2. Compute the quantities γ_j given by (10.3). If $\gamma_j \leqslant c_j$ for all j, then $T(x) = \min_{y \in K} T(y)$ and the process is stopped. Otherwise, proceed to step 3.
3. Let k be an index with $\gamma_k > c_k$. If $\alpha_{ik} \leqslant 0$ for all i, then the problem has no solution and the process is stopped. Repeat this test for all indices k such that $\gamma_k > c_k$. If for each such value of k at least one α_{ik} is positive, proceed to step 4.

4. Choose k so that $\gamma_k - c_k = \max_{1 \leqslant j \leqslant n} (\gamma_j - c_j)$. Replace x by $x + \lambda z$, where λ and z are defined by (10.8) and (10.4), respectively. Cycle back to step 2.

In Theorems 10.2 through 10.4 we assumed that the first m coordinates of x were the positive ones. If this is not the case, then a formula such as (10.2) would be replaced by

$$A_j = \sum_{i \in \mathcal{I}} \alpha_{ij} A_i,$$

where $\mathcal{I} = \{i: x_i > 0\}$.

The solution is said to be *degenerate* when the new x in step 4 has fewer than m non-zero coordinates. In this case there is a possibility that the simplex algorithm will get caught in an infinite loop. The occurrence of such a loop has been extremely rare in practice, but techniques for handling degeneracy are described in Gass [11] and Dantzig [4]. For our purposes we will simply print out a warning that cycling might occur and return to step 2. The columns to be used in computing the new α's are the same as in the previous step, except that A_k replaces A_p, where p is given by (10.9).

If no degeneracy occurs and a solution exists, then the simplex method necessarily produces a solution because K has at most a finite number of extreme points and $T(x)$ decreases at each stage. Similarly, the algorithm will eventually stop during step 3 if no degeneracy occurs and T has no lower bound on K.

EXAMPLE 10.1. Compute $\max_{x \in K} T(x)$, where $T(x) = 5x_1 + 6x_2$ and K is the set of points $x \in E_2$ which satisfy the constraints:

$$x_1 \geqslant 0, \quad x_2 \geqslant 0,$$

$$4x_1 + x_2 \leqslant 20,$$

$$x_1 + x_2 \leqslant 8,$$

$$x_1 + 2x_2 \leqslant 14.$$

The addition of slack variables gives the system of equations

$$\begin{aligned} 4x_1 + x_2 + x_3 \qquad\qquad &= 20, \\ x_1 + x_2 \qquad + x_4 \qquad &= 8, \\ x_1 + 2x_2 \qquad\qquad + x_5 &= 14. \end{aligned} \tag{10.10}$$

Because this is a maximization problem we minimize $-T(x) = -5x_1 - 6x_2 - 0x_3 - 0x_4 - 0x_5$. An obvious extreme point is given by

$$x = (0, 0, 20, 8, 14).$$

Since A_3, A_4, and A_5 are unit vectors it is easy to see that

$$A_1 = \begin{bmatrix} 4 \\ 1 \\ 1 \end{bmatrix} = 4A_3 + A_4 + A_5.$$

In fact the α_{ij}'s are the matrix of coefficients from (10.10), provided that the rows are indexed with 3, 4, and 5. This gives $\gamma_1 = c_3\alpha_{31} + c_4\alpha_{41} + c_5\alpha_{51} = 0$. Similarly, $\gamma_2, \ldots, \gamma_5$ are all zero. Thus

$$c_1 - \gamma_1 = -5, \quad c_2 - \gamma_2 = -6,$$

and

$$c_i - \gamma_i = 0, \quad i = 3, 4, 5.$$

The algorithm is not stopped during step 3. The index k in step 4 is chosen to be 2 because $\gamma_2 - c_2 = 6$ is the maximum. Then

$$\lambda = \min\left(\frac{x_3}{\alpha_{32}}, \frac{x_4}{\alpha_{42}}, \frac{x_5}{\alpha_{52}}\right) = \min\left(\frac{20}{1}, \frac{8}{1}, \frac{14}{2}\right) = 7,$$

and $p = 5$. Thus $z = (0, 1, -\alpha_{32}, -\alpha_{42}, -\alpha_{52}) = (0, 1, -1, -1, -2)$. The new x is given by

$$x = (0, 0, 20, 8, 14) + 7z = (0, 7, 13, 1, 0),$$

and the process cycles back to step 2.

The new values of α_{ij} are computed by replacing A_5 by A_2 in the set of basis vectors. From the equation $A_2 = A_3 + A_4 + 2A_5$ it follows that

$$(10.11) \quad A_5 = \tfrac{1}{2}(A_2 - A_3 - A_4).$$

Using (10.11) we see that

$$A_1 = 4A_3 + A_4 + A_5 = \tfrac{1}{2}A_2 + \tfrac{7}{2}A_3 + \tfrac{1}{2}A_4.$$

This, together with (10.11) gives

$$\alpha_{21} = \tfrac{1}{2}, \quad \alpha_{25} = \tfrac{1}{2}$$

$$\alpha_{31} = \tfrac{7}{2}, \quad \alpha_{35} = -\tfrac{1}{2}$$

$$\alpha_{41} = \tfrac{1}{2}, \quad \alpha_{45} = -\tfrac{1}{2}$$

$$\alpha_{ij} = \begin{cases} 1 & \text{if } i = j \\ 0 & \text{if } i \neq j \end{cases} \quad \text{for } i,j = 2, 3, 4.$$

Proceeding as before, we now obtain the following:

$$c_1 - \gamma_1 = -2, \quad c_5 - \gamma_5 = 3,$$
$$c_i - \gamma_i = 0, \quad i = 2, 3, 4,$$
$$k = 1, \quad \lambda = 2, \quad p = 4,$$
$$z = (1, -1/2, -7/2, -1/2, 0),$$
$$x = (2, 6, 6, 0, 0).$$

This time step 2 gives $c_j - \gamma_j \geqslant 0$ for all j and hence $x = (2, 6, 6, 0, 0)$ is a solution. In terms of the original problem this shows that

$$\max_{x \in K} T(x) = T(2,6) = 46.$$

The simplex procedure can be represented in terms of tables like those given below for the first two stages in this example. It should be noted that the values of α_{ij} and x_i in the second table can be obtained from those in the first table by row operations which are essentially the same as those used in the Jordan method for computing the matrix inverse.

	$i \backslash j$	c_i	α_{i1}	α_{i2}	α_{i3}	α_{i4}	α_{i5}	x_i	x_i/α_{ik}
			1	2	3	4	5		
	3	0	4	1	1	0	0	20	20
	4	0	1	1	0	1	0	8	8
	5	0	1	(2)	0	0	1	14	7
c_j			−5	−6	0	0	0		
γ_j			0	0	0	0	0		
$c_j - \gamma_j$			−5	−6	0	0	0		

TABLE 10.1. Initial Stage of the Simplex Method.

	$i \backslash j$	c_i	α_{i1}	α_{i2}	α_{i3}	α_{i4}	α_{i5}	x_i	x_i/α_{ik}
			1	2	3	4	5		
	3	0	7/2	0	1	0	−1/2	13	26/7
	4	0	(1/2)	0	0	1	−1/2	1	2
	2	−6	1/2	1	0	0	1/2	7	14
c_j			−5	−6	0	0	0		
γ_j			−3	−6	0	0	−3		
$c_j - \gamma_j$			−2	0	0	0	3		

TABLE 10.2. Second Stage of the Simplex Method.

In this example the circled element, α_{52}, is the pivot element; so the third row of the first table (corresponding to $i = 5$) is divided by 2 to obtain a 1 in the pivot position. Then −1 times the new third row (now corresponding to $i = 2$) is added to the first two rows to obtain zeros in the rest of the pivot column. The columns labeled c_i and x_i/α_{ik}, as well as the rows labeled c_j, γ_j, and $c_j - \gamma_j$, are not transformed by row operations. These values are calculated after the row operations have been completed.

Let us now outline a general technique for obtaining an initial extreme point to start the simplex algorithm. The first step is to multiply equations in (10.1) by -1 if necessary so that none of the components of b is negative. Then a new system of equations is obtained by introducing an extra variable in each of the equations; namely,

$$(10.12) \quad x_1A_1 + \ldots + x_nA_n + x_{n+1}e_1 + \ldots + x_{n+m}e_m = b,$$

where e_j is the jth column of the m by m identity matrix. Let

$$T_1(x) = x_{n+1} + \ldots + x_{n+m},$$

and define K_1 to be the set of points in E_{n+m} which have non-negative coordinates and satisfy (10.12). The next theorem shows that an extreme point of the original feasibility set can be found by minimizing T_1 over K_1.

THEOREM 10.5. *Let $x^* = (x_1, \ldots, x_n, x_{n+1}, \ldots, x_{n+m})$ be an extreme point of K_1 and suppose*

$$T_1(x^*) = \min_{z \in K_1} T_1(z).$$

If $x_{n+1} = x_{n+2} = \ldots = x_{n+m} = 0$, then $x = (x_1, \ldots, x_n)$ is an extreme point of the original feasibility set K associated with (10.1). If $x_j > 0$ for some $j > n$, then K is empty.

The proof of this theorem is left for Problem 10.4.

The reason for introducing the extra variables is that K_1 has an obvious extreme point, namely

$$(0, 0, \ldots, 0, b_1, b_2, \ldots, b_m).$$

This can be used as the initial extreme point for computing

$$\min_{z \in K_1} T_1(z)$$

by the simplex method. Assuming that the process does not cycle, an initial extreme point of K is obtained, or else K is shown to be empty.

In actual practice it is usually not necessary to introduce extra variables into every equation of (10.1) since some of the columns of A may already be columns of the m by m identity. In fact, if each equation in (10.1) contains a slack variable, then it can be shown that only one extra variable is needed (see Problem 10.5).

EXAMPLE 10.2. Let K be the set of points in E_5 having non-negative coordinates and satisfying the system of equations

$$\begin{aligned} x_1 + 4x_2 + x_3 &= 9, \\ 2x_1 + x_2 + x_4 &= 7, \\ 3x_1 - x_2 - x_5 &= 1. \end{aligned}$$

The third and fourth columns are already equal to e_1 and e_2; so it is only necessary to introduce one extra variable. The third equation is replaced by

$$3x_1 - x_2 - x_5 + x_6 = 1.$$

With $T_1(x) = x_6$, we obtain the following simplex table:

	$i \backslash j$	c_i	α_{i1}	α_{i2}	α_{i3}	α_{i4}	α_{i5}	α_{i6}	x_i	x_i/α_{ik}
			1	2	3	4	5	6		
	3	0	1	4	1	0	0	0	9	9
	4	0	2	1	0	1	0	0	7	7/2
	6	1	(3)	−1	0	0	−1	1	1	1/3
c_j			0	0	0	0	0	1		
γ_j			3	−1	0	0	−1	1		
$c_j - \gamma_j$			−3	1	0	0	1	0		

TABLE 10.3. First Stage in the Computation of an Initial Extreme Point.

The next step gives the table:

	$i \backslash j$	c_i	α_{i1}	α_{i2}	α_{i3}	α_{i4}	α_{i5}	α_{i6}	x_i	
			1	2	3	4	5	6		
	3	0	0	13/3	1	0	1/3	−1/3	26/3	
	4	0	0	5/3	0	1	2/3	−2/3	19/3	
	1	0	1	−1/3	0	0	−1/3	1/3	1/3	
c_j			0	0	0	0	0	1		
γ_j			0	0	0	0	0	0		
$c_j - \gamma_j$			0	0	0	0	0	1		

TABLE 10.4. Final Stage.

Thus (1/3, 0, 26/3, 19/3, 0) is an extreme point of K.

PROBLEMS

10.1. Show that the feasibility set K associated with (10.1) has at most a finite number of extreme points.

10.2. Use the simplex method to solve:
(a) Problem 9.1 in Chapter 2.
(b) Problem 9.15 in Chapter 2.

10.3. Find the minimum of T if
(a) $T(x) = x_1 + x_2 - 4x_3$

$$x_1 \geq 0, \quad x_2 \geq 0, \quad x_3 \geq 0$$
$$x_1 + x_2 - x_3 \leq 5$$
$$5x_1 - 5x_2 - 3x_3 \leq 10$$
$$6x_1 + x_2 \leq 20$$

(b) $T(x) = x_1 - x_2$

$$x_1 \geq 0, \quad x_2 \geq 0$$
$$3x_1 + x_2 \leq 5$$
$$-x_1 + x_2 \leq -1$$
$$x_1 + 5x_2 \leq 20$$

10.4. Prove Theorem 10.5.

10.5. Suppose that each equation in (10.1) contains a slack variable. Show that it is necessary to introduce at most one extra variable in order to obtain an initial extreme point. Hint: Subtract one of the equations from some of the others.

REFERENCES

1. T. M. Apostol, *Mathematical Analysis*, Addison-Wesley, 1957.
2. B. Carnahan, H. A. Luther, and J. O. Wilkes, *Applied Numerical Methods*, Wiley, 1969.
3. L. Cooper and D. I. Steinberg, *Methods and Applications of Linear Programming*, Saunders, 1974.
4. G. B. Dantzig, *Linear Programming and Extensions*, Princeton University Press, 1963.
5. D. K. Faddeev and V. N. Faddeeva, *Computational Methods of Linear Algebra*, Freeman, 1963.
6. G. E. Forsythe, "Today's computational methods of linear algebra," *SIAM Review*, 9 (1967), 489–515.
7. G. E. Forsythe and P. Henrici, "The cyclic Jacobi method for computing the principal values of a complex matrix," *Trans. Amer. Math. Soc.*, 94 (1960), 1–23.
8. G. Forsythe and C. B. Moler, *Computer Solution of Linear Algebraic Systems*, Prentice-Hall, 1967.
9. L. Fox, *An Introduction to Numerical Linear Algebra*, Oxford, Eng., 1965

10. C.-E. Fröberg, *Introduction to Numerical Analysis*, Addison-Wesley, 1969.
11. S. I. Gass, *Linear Programming*, 3rd Ed., McGraw-Hill, 1969.
12. F. E. Hohn, *Elementary Matrix Algebra*, Macmillan, 1964.
13. A. S. Householder, *The Theory of Matrices in Numerical Analysis*, Blaisdell, 1964.
14. B. Noble, *Applied Linear Algebra*, Prentice-Hall, 1969.
15. J. M. Ortega, "The Givens-Householder method for symmetric matrices," in *Mathematical Methods for Digital Computers*, Vol. 2, edited by A. Ralston and H. S. Wilf, Wiley, 1967.
16. J. M. Ortega, *Numerical Analysis: A Second Course*, Academic Press, 1972.
17. A. Ostrowski, "Über die Stetigkeit von charakteristischen Wurzeln in Abhängigkeit von Matrizenelementen," *Deutsche Mathematiker Vereinigung Jahresbericht*, 60 (1957), 40–42.
18. S. M. Selby (Ed.), *C.R.C. Standard Mathematical Tables*, 19th Ed., Chemical Rubber Publishing Co., Cleveland, 1971.
19. D. I. Steinberg, *Computational Matrix Algebra*, McGraw-Hill, 1974.
20. J. E. Strum, *Introduction to Linear Programming*, Holden-Day, 1972.
21. J. H. Wilkinson, "Calculation of the eigenvalues of a symmetric tridiagonal matrix by the method of bisection," *Numer. Math.*, 4 (1962), 362–367.
22. J. H. Wilkinson, "Calculation of the eigenvectors of a symmetric tridiagonal matrix by inverse iteration," *Numer. Math.*, 4 (1962), 368–376.
23. J. H. Wilkinson, "Error analysis of eigenvalue techniques based on orthogonal transformations," *SIAM J.*, 10 (1962), 162–195.
24. J. H. Wilkinson, "Plane rotations in floating-point arithmetic," *Proceedings of Symposia in Applied Mathematics*, 15 (1963), 185–198 (Published by the Amer. Math. Soc.).
25. J. H. Wilkinson, *Rounding Errors in Algebraic Processes*, Prentice-Hall, 1963.
26. J. H. Wilkinson, *The Algebraic Eigenvalue Problem*, Clarendon Press, Oxford, Eng., 1965.
27. J. H. Wilkinson, "The solution of ill-conditioned linear equations," in *Mathematical Methods for Digital Computers*, Vol. 2, edited by A. Ralston and H. S. Wilf, Wiley, 1967.

CHAPTER 4

Iteration

1. INTRODUCTION

An iteration procedure for approximating a solution to an equation usually proceeds along the following lines: First, an initial approximation is obtained by one means or another. Then the initial approximation is refined to obtain a better approximation. This second approximation is refined to obtain an even better approximation, and so on. The process is stopped when the approximate solution is sufficiently close to the true solution.

In this chapter we will consider three types of equations:

$$(1.1) \qquad T(x) = x,$$

$$(1.2) \qquad T(x) = \emptyset,$$

$$(1.3) \qquad T(x) = \lambda x.$$

Here T is a transformation, x and $\emptyset$ are elements of a linear space, and λ is a number. A solution to equation (1.1) is called a *fixed point of the transformation T*. If T is a real or complex-valued function of a real or complex variable, then a solution to equation (1.2) is called a *zero of T*. The third equation represents the eigenvalue problem which has been considered previously (Chapter 2, Section 3, and Chapter 3, Sections 7–9).

In this section we will consider only equations of the type (1.1). A common technique used to approximate solutions to this type of equation is to define an iteration process by the equation

$$(1.4) \qquad x_{n+1} = T(x_n) \qquad n = 0, 1, 2, \ldots.$$

That is, an initial approximation (or guess) x_0 is obtained by independent means, then $x_1 = T(x_0)$, $x_2 = T(x_1)$, etc., form successive approximations to a solution of (1.1). We will refer to this as the *Picard iteration method*, although in the literature this term frequently refers to a specific method of the type (1.4) which is used for differential and integral equations.

EXAMPLE 1.1. Consider the equation

$$(1.5) \qquad x = (x^3 + 1)/4,$$

where x is a real number. This equation is of the form (1.1) with $T(x) = (x^3 + 1)/4$. By trial and error, we find that $0 < T(0)$, while $1/2 > T(1/2)$. Thus (1.5) must have a solution between 0 and 1/2 (why?). Taking $x_0 = 0$ as our initial guess and using fixed-point arithmetic truncated to eight decimal places, we obtain the following approximations:

$$x_0 = 0,$$

$$x_1 = T(x_0) = 0.2500\ 0000,$$

$$x_2 = T(x_1) = 0.2539\ 0625,$$

$$x_3 = T(x_2) = 0.2540\ 9223,$$

$$x_4 = T(x_3) = 0.2541\ 0123,$$

$$x_5 = T(x_4) = 0.2541\ 0166,$$

$$x_6 = T(x_5) = 0.2541\ 0168,$$

$$x_7 = T(x_6) = 0.2541\ 0168.$$

EXAMPLE 1.2. Consider the equation

$$(1.6) \qquad x = 3/x.$$

Equation (1.6) is of the form (1.1) with $T(x) = 3/x$. It is easy to see that $\sqrt{3}$ is a solution; so we choose $x_0 = 1.7$ as a first approximation. This gives

$$x_0 = 1.7000\ 0000,$$

$$x_1 = 1.7647\ 0588,$$

$$x_2 = 1.7000\ 0000,$$

$$x_3 = 1.7647\ 0588,$$

$$x_4 = 1.7000\ 0000.$$

The immediate objective in presenting Examples 1.1 and 1.2 is to illustrate the mechanics of the iteration process defined by (1.4). However, Example 1.2 shows that the method does not always work. One might wonder how to tell whether or not the procedure will provide a good approximation. In order to provide insight into this question we will now consider the theoretical basis for Picard's method.

THEOREM 1.1. *Let T be a transformation which maps a space X into itself. Suppose $x_{n+1} = T(x_n)$ for all non-negative integers n and $\lim_{n\to\infty} x_n = v$. If T is continuous at v, then $T(v) = v$.*

Proof.

$$v = \lim_{n\to\infty} x_n = \lim_{n\to\infty} x_{n+1} = \lim_{n\to\infty} T(x_n) = T(\lim_{n\to\infty} x_n) = T(v).$$

The reader should decide which step in the proof depends on the continuity of T at v.

Roughly speaking, Theorem 1.1 implies that there is no harm in trying the Picard method even if convergence cannot be guaranteed in advance. If the sequence settles down to a certain point, then that point is a fixed point of T. Of course, from a strictly logical point of view, one cannot tell from a finite number of terms whether or not a sequence is convergent. However, practical mathematicians must be willing to gamble a little and most would agree that the sequence in Example 1.1 appears to be convergent while the one in Example 1.2 clearly is not convergent. Error estimates for certain Picard iteration problems will be discussed in Theorems 1.3, 2.1, 2.3, and 2.4.

A general sufficient condition for convergence of the Picard method will be given below. First, some definitions are needed.

DEFINITION 1.1. *Let $\{x_n\}$ be a sequence of members of a normed linear space X. The sequence is said to be a Cauchy sequence iff for each positive number ϵ, there exists a number N so that*

$$\|x_m - x_n\| < \epsilon$$

whenever n and m are greater than N.

DEFINITION 1.2. *The normed linear space X is said to be complete iff every Cauchy sequence in X has a limit in X. A complete normed linear space is also called a Banach space.*

Examples of Banach spaces are R, K, E_n, and EK_n under their usual norms, and $C[a,b]$ or $CK[a,b]$ under the uniform norm. The proofs of these facts can be found in books on modern analysis or topology [1, 2].

DEFINITION 1.3. *Let T be a transformation between the normed linear spaces X and Y (usually $X = Y$). Then T is said to be a contraction mapping on the set $A \subset X$ iff there is a constant $c \in [0,1)$ such that*

$$\|T(x) - T(y)\| \leqslant c\,\|x - y\|$$

for all x, $y \in A$.

The connection between contraction mappings and convergence of the Picard iterates is illustrated by the following theorem.

THEOREM 1.2. *Let T be a transformation from the Banach space X into itself. If T is a contraction mapping on X, $x_0 \in X$, and $x_{n+1} = T(x_n)$ for $n = 0, 1, 2, \ldots$, then $\lim_{n\to\infty} x_n$ exists. This limit is the only fixed point of T.*

Proof. Since T is a contraction mapping, there exists a constant $c \in [0,1)$ such that

$$\|T(x_1) - T(x_0)\| \leqslant c\, \|x_1 - x_0\|.$$

But $x_2 = T(x_1)$ and $x_1 = T(x_0)$. Thus

$$\|x_2 - x_1\| \leqslant c\, \|x_1 - x_0\|. \tag{1.7}$$

Similarly,

$$\|x_3 - x_2\| = \|T(x_2) - T(x_1)\| \leqslant c\, \|x_2 - x_1\|. \tag{1.8}$$

Combining (1.7) and (1.8) we have

$$\|x_3 - x_2\| \leqslant c^2\, \|x_1 - x_0\|.$$

Continuing in this way we see that

$$\|x_{n+1} - x_n\| \leqslant c^n\, \|x_1 - x_0\|, \qquad n = 1, 2, \ldots.$$

Now if $m > n$, we have

$$\begin{aligned} \|x_m - x_n\| &= \|x_m - x_{m-1} + x_{m-1} - x_{m-2} + \ldots + x_{n+1} - x_n\| \\ &\leqslant \|x_m - x_{m-1}\| + \|x_{m-1} - x_{m-2}\| + \ldots + \|x_{n+1} - x_n\| \\ &\leqslant (c^{m-1} + c^{m-2} + \ldots + c^n)\, \|x_1 - x_0\| \\ &\leqslant \|x_1 - x_0\| \sum_{j=n}^{\infty} c^j = c^n \frac{\|x_1 - x_0\|}{1-c}. \end{aligned} \tag{1.9}$$

For a given positive number ϵ, let

$$N = \frac{\ln \dfrac{\epsilon(1-c)}{\|x_1 - x_0\|}}{\ln c}$$

If $m > n > N$, then

$$c^n < c^N = e^{N \ln c} = \epsilon(1-c)/\|x_1 - x_0\|.$$

Thus

$$\|x_m - x_n\| \leqslant c^n\, \|x_1 - x_0\|/(1-c) < \epsilon.$$

If $n > m > N$, one can reverse the roles of m and n in (1.9) and reach the same conclusion. Thus $\{x_n\}$ is a Cauchy sequence and has a limit.

Let $v = \lim_{n\to\infty} x_n$. Now from Theorem 1.1, v is a fixed point of T. (How do we know that T is continuous?) In order to show that v is the only fixed point, suppose that w is a different fixed point of T. Then

$$\|v - w\| = \|T(v) - T(w)\| \leqslant c\,\|v - w\| < \|v - w\|.$$

This contradiction rules out the possibility of another fixed point.

Frequently the transformation T is a contraction mapping only on a subset of the linear space. The following theorem can often be applied in this case.

THEOREM 1.3. *Suppose X is a Banach space, T is a transformation from X into X, x_0 is a member of X, and r is a positive number. Let $A = \{x \in X: \|x - x_0\| \leqslant r\}$, and let $x_{n+1} = T(x_n)$, $n = 0, 1, 2, \ldots$. If for some $c \in [0,1)$,*

$$\text{(i)} \qquad \|T(x) - T(y)\| \leqslant c\,\|x - y\| \qquad \textit{for all} \quad x,y \in A,$$

$$\text{(ii)} \qquad \|T(x_0) - x_0\| \leqslant (1 - c)r,$$

then $\lim_{n\to\infty} x_n$ exists and belongs to A. In addition, if we let $v = \lim_{n\to\infty} x_n$, then v is the only fixed point of T in A and

$$\|x_n - v\| \leqslant c^n r, \qquad n = 1, 2, \ldots.$$

Proof. The argument of Theorem 1.2 will establish the fact that $\lim_{n\to\infty} x_n$ exists and is a unique fixed point provided that $x_n \in A$ for every n. This will be proved by induction. Now

$$\|x_1 - x_0\| = \|T(x_0) - x_0\| \leqslant (1 - c)r \leqslant r;$$

thus $x_1 \in A$. Suppose that $x_1, x_2, \ldots, x_n \in A$ and that

$$\|x_k - x_{k-1}\| \leqslant c^{k-1}(1 - c)r \qquad \text{for } k = 1, 2, \ldots, n.$$

Then

$$\begin{aligned}\|x_{n+1} - x_n\| &= \|T(x_n) - T(x_{n-1})\| \leqslant c\,\|x_n - x_{n-1}\| \\ &\leqslant c\,c^{n-1}(1 - c)r = c^n(1 - c)r.\end{aligned}$$

Thus

$$\begin{aligned}\|x_{n+1} - x_0\| &= \|x_{n+1} - x_n + x_n - x_{n-1} + \ldots + x_1 - x_0\| \\ &\leqslant \|x_{n+1} - x_n\| + \|x_n - x_{n-1}\| + \ldots + \|x_1 - x_0\| \\ &\leqslant (1 - c)r(c^n + c^{n-1} + \ldots + 1) \qquad (1.10) \\ &\leqslant (1 - c)r(1 + c + c^2 + \ldots + c^n + \ldots) \\ &= r.\end{aligned}$$

Hence $x_{n+1} \in A$ and by induction $x_n \in A$ for all n. Now

$$\|x_n - v\| = \|T(x_{n-1}) - T(v)\| \leqslant c \, \|x_{n-1} - v\|.$$

It follows by induction that

$$\|x_n - v\| \leqslant c^n \, \|x_0 - v\| \leqslant c^n r.$$

(The fact that $\|x_0 - v\| \leqslant r$ follows by taking the limit as $n \to \infty$ on both sides of (1.10).)

EXAMPLE 1.3. Let $T(x) = (x^2 - 11)/10$ for each $x \in R$. Now

$$\|T(x) - T(y)\| = |T(x) - T(y)| = \frac{|x^2 - y^2|}{10}$$
$$= \frac{|x - y| \cdot |x + y|}{10}.$$

Let $x_0 = 0$, $r = 2$, and $A = \{x\colon |x| < r\}$. Then if x and $y \in A$, we have $|T(x) - T(y)| \leqslant 0.4|x - y|$. Thus part (i) of Theorem 1.3 is satisfied with $c = 0.4$. Furthermore,

$$|T(x_0) - x_0| = 11/10 \leqslant (0.6)2 = (1 - c)r.$$

Hence the conditions of Theorem 1.3 are satisfied and T has a unique fixed point in $[-2,2]$. With $x_0 = 0$, the first two Picard iterates are given by

$$x_1 = -1.1,$$
$$x_2 = -.979.$$

Theorem 1.3 gives the error bound

$$|x_2 - v| \leqslant c^2 r = (0.16)\cdot 2 = 0.32.$$

The actual magnitude of the error is 0.021 since $v = -1$. The error bound in this case is not very sharp. (The numerical analyst uses the term *sharp error bound* to mean one which is close to the actual error.)

EXAMPLE 1.4. Define a transformation from E_2 into E_2 by the equation

$$T(x) = \frac{x}{20} + (3,3).$$

Now

$$\|T(x) - T(y)\| = \left\|\frac{x}{20} + (3,3) - \frac{y}{20} - (3,3)\right\|$$
$$= 0.05 \, \|x - y\|.$$

Thus T is a contraction mapping on all of E_2 and Theorem 1.2 can be applied. Choosing $x_0 = (0,0)$, we obtain the following:

$$x_0 = (0,0),$$
$$x_1 = T(0,0) = (3,3),$$
$$x_2 = T(3,3) = (3.15,3.15),$$
$$x_3 = T(3.15,3.15) = (3.1575,3.1575).$$

In this case it is easy to solve the equation and obtain $v = \frac{20}{19}(3,3)$ as the true value of the fixed point.

PROBLEMS

1.1. Let $T(x) = (x + 1)/2$ for every $x \in R$. Let $x_0 = 0.5$ and compute the first five approximations to a fixed point by Picard's method.

1.2. For each $x \in E_3$, let

$$T(x) = Ax + \begin{bmatrix} 2 \\ 1 \\ 2 \end{bmatrix},$$

where

$$A = \begin{bmatrix} 0.1 & 0 & 0.2 \\ 0.3 & 0.1 & 0.2 \\ 0 & 0 & 0.3 \end{bmatrix}.$$

Show that T is a contraction mapping. Hint: Note that

$$Ax = \begin{bmatrix} a_{1\cdot} \cdot x \\ a_{2\cdot} \cdot x \\ a_{3\cdot} \cdot x \end{bmatrix}$$

and use the Schwarz inequality.

1.3. Let $T(x) = ax + b$ for each $x \in R$. For what values of a and b will T be a contraction mapping?

1.4. Let $T(x) = \sqrt{x}$ for each $x > 0$. Let $x_0 = 18$ and $r = 2$. Show that T is a contraction mapping on the set $[16, 20]$, but condition (ii) of Theorem 1.3 is not satisfied.

1.5. Let T be a contraction mapping with contraction factor $c < 1$, and let x^* be the fixed point of T. Suppose that $x_{n+1} = x_n$, but that a round-off error of ϵ_n was incurred in computing $T(x_n)$. Then we have

$$x_n = T(x_n) + \epsilon_n.$$

Show that if $\|\epsilon_n\| < \epsilon$, then

$$\|x^* - x_n\| \leqslant \frac{\epsilon}{1-c}.$$

1.6. Consider the equation

$$x = \frac{x+1}{10}.$$

Let $x_0 = 0$ and use fixed-point arithmetic with four decimal places to compute the approximations (1.4). Stop when two successive values agree. Then use the result of Problem 1.5 to find a bound on the error. Compare the bound with the actual error.

1.7. Consider the initial value problem

$$y'(x) = f(x,y(x)), \qquad y(0) = \alpha. \tag{1.11}$$

The Picard method for approximating a solution is described by the equations

$$y_0(x) = \alpha,$$
$$y_{n+1}(x) = \alpha + \int_0^x f(t,y_n(t))dt, \qquad n = 1, 2, \ldots.$$

Describe a transformation T which shows that this method is really of the type defined by (1.4). Assuming that f and y are continuous, show that $T(y) = y$ if and only if y satisfies the initial-value problem (1.11). This problem will be considered further in Section 6.

2. CONTRACTION MAPPINGS ON THE REAL LINE

In this section we will consider equations of the form $x = g(x)$, where g is a real-valued function of a real variable. The crucial factor governing the convergence of the Picard process is the slope of g.

THEOREM 2.1. *Suppose that r is a positive number, $c \in [0,1)$, $x_0 \in R$, and $x_{n+1} = g(x_n)$, $n = 0, 1, 2, \ldots$. If*

(i) $|g'(x)| \leqslant c$ *whenever* $|x - x_0| \leqslant r$,

(ii) $|g(x_0) - x_0| \leqslant (1 - c)r$,

then there is a unique fixed point v in the interval $[x_0 - r, x_0 + r]$. Furthermore, $v = \lim_{n\to\infty} x_n$ and

$$|x_n - v| \leqslant c^n r, \qquad n = 1, 2, \ldots. \tag{2.1}$$

Proof. Let $A = [x_0 - r, x_0 + r]$. Applying the mean-value theorem, we obtain

$$|g(x) - g(y)| = |g'(z)(x - y)| \leqslant c|x - y|$$

whenever $x,y \in A$, where z is some number between x and y. Thus g is a contraction mapping on A and the result follows from Theorem 1.3.

EXAMPLE 2.1. Consider the equation $x = (x^3 + 1)/4$ used in Example 1.1. Then $g(x) = (x^3 + 1)/4$ and $g'(x) = 3x^2/4$. On the interval $[-1/2,1/2]$ we have $|g'(x)| \leqslant 3/16 = 0.1875$. Let $x_0 = 0$. Then $|g(x_0) - x_0| = 0.25 \leqslant (0.5)(0.8125)$. Thus Theorem 2.1 applies with

$x_0 = 0$, $r = 0.5$, and $c = 0.1875$. Some error bounds (ignoring round-off) are given by

$$\|x_2 - v\| \leqslant c^2 r = (0.1825)^2(0.5) \leqslant (0.2)^2(0.5) = 0.02,$$
$$\|x_4 - v\| \leqslant (0.2)^4(0.5) = 0.0008,$$
$$\|x_6 - v\| \leqslant (0.2)^6(0.5) = 0.000\,032.$$

EXAMPLE 2.2. Let $g(x) = 2 - x^2$ and $x_0 = -1$. Then $x_1 = 1, x_2 = 1$, and the process converges to the fixed point 1. On the other hand, if $x_0 = 0.99$, then to four decimals

$$x_1 = 1.0199,$$
$$x_2 = 0.9598,$$
$$x_3 = 1.0788,$$
$$\vdots$$
$$x_7 = 1.9050,$$
$$x_8 = -1.6291.$$

The derivative of g at 1 is -2. This example shows that the Picard iterates might converge even though the magnitude of the derivative is greater than 1. However, convergence under this condition is not very likely. In fact, the theorem below shows that the sequence converges in this case only if the process stumbles exactly onto the fixed point.

THEOREM 2.2. *Suppose that $v \in (a,b)$, $v = g(v)$, and $|g'(x)| \geqslant 1$ for all $x \in [a,b]$. Let $x_{n+1} = g(x_n)$, $n = 1, 2, \ldots$. Then $\lim_{n\to\infty} x_n = v$ if and only if $x_n = v$ for some n.*

Proof. If $x_N = v$ for some N, then it is clear that $x_n = v$ for all $n > N$ and hence $\lim_{n\to\infty} x_n = v$. Conversely, suppose $\lim_{n\to\infty} x_n = v$ and x_n is never equal to v. There exists a positive integer N so that $x_n \in [a,b]$ whenever $n > N$ (why?). Using the mean-value theorem we have

$$\begin{aligned} |x_{n+1} - v| &= |g(x_n) - g(v)| = |g'(z)(x_n - v)| \\ &\geqslant |x_n - v| \end{aligned}$$

whenever $n > N$. Thus $|x_n - v| \geqslant |x_{N+1} - v|$ for all $n > N + 1$. It follows that

$$0 = \lim_{n\to\infty} |x_n - v| \geqslant |x_{N+1} - v| > 0.$$

This contradiction completes the proof.

For all practical purposes, then, the requirement that g' have a magnitude smaller than 1 in the vicinity of v is a necessary requirement. In fact, the convergence is extremely slow if g' is near 1; so Picard's method is seldom used unless $|g'(x)| \leqslant 1/2$ in the vicinity of v. (Nonbelievers should work Problem 2.1.)

The inequality (2.1) provides a means for obtaining a bound on the error in Picard's method. In actual practice, people frequently dispense with such an error estimate and terminate the process when two successive iterates agree to within the desired accuracy. The next two theorems show that this procedure is entirely justified when g' is between -1 and $1/2$.

THEOREM 2.3. *Suppose the hypotheses of Theorem 2.1 hold, and that*

$$-1 < g'(x) \leqslant 1/2$$

whenever $|x - x_0| \leqslant r$. *Then*

$$|x_n - v| \leqslant |x_n - x_{n-1}|, \qquad n = 1, 2, \ldots. \tag{2.2}$$

Proof. Using the mean-value theorem, we obtain the equation

$$x_n - v = g(x_{n-1}) - g(v) = g'(z)(x_{n-1} - v), \tag{2.3}$$

where z is between x_{n-1} and v. If $g'(z) < 0$, then x_n and x_{n-1} lie on opposite sides of v and (2.2) must hold. From (2.3) we have

$$\begin{aligned} v - x_n &= x_n - x_{n-1} + x_{n-1} - v + 2(v - x_n) \\ &= x_n - x_{n-1} + (2g'(z) - 1)(v - x_{n-1}). \end{aligned} \tag{2.4}$$

If $0 \leqslant g'(z) \leqslant 1/2$ and $x_{n-1} < v$, then (2.3) and (2.4) imply that

$$0 \leqslant v - x_n \leqslant x_n - x_{n-1}$$

and hence (2.2) holds. If $0 \leqslant g'(z) \leqslant 1/2$ and $x_{n-1} \geqslant v$, then (2.3) and (2.4) imply that

$$0 \geqslant v - x_n \geqslant x_n - x_{n-1}$$

and again (2.2) holds. This completes the proof.

We will refer to the error bound (2.2) as the *eyeball error bound*. It does not take into account any truncation or round-off errors incurred in computing $x_1, x_2, \ldots, x_n$. The next theorem takes such errors into account.

THEOREM 2.4. *Suppose the hypotheses of Theorem 2.3 hold except that the equation* $x_{n+1} = g(x_n)$ *is not satisfied exactly. If for a fixed value of* n *and* ϵ

(i) $x_n = g(x_{n-1}) + \epsilon_n$,

(ii) $|\epsilon_n| \leqslant \epsilon$,

(iii) $x_{n-1} \in [x_0 - r, x_0 + r]$,

(iv) $x_n \in [x_0 - r + \epsilon, x_0 + r - \epsilon]$,

then

(2.5) $$|x_n - v| \leqslant |x_n - x_{n-1}| + 2\epsilon.$$

Proof. Let $x_n^* = x_n - \epsilon_n$ and note that $x_n^* = g(x_{n-1})$. Now x_n^* and $x_{n-1} \in [x_0 - r, x_0 + r]$; so the argument used in proving Theorem 2.3 can be used if x_n^* replaces x_n in (2.3), (2.4), and so on. Thus

$$|x_n^* - v| \leqslant |x_n^* - x_{n-1}|.$$

Replacing x_n^* by $x_n - \epsilon_n$, we have

$$|x_n - \epsilon_n - v| \leqslant |x_n - \epsilon_n - x_{n-1}|.$$

But

$$|x_n - \epsilon_n - x_{n-1}| \leqslant |x_n - x_{n-1}| + |\epsilon_n|$$

and

$$|x_n - \epsilon_n - v| \geqslant |x_n - v| - |\epsilon_n|.$$

Thus

$$|x_n - v| - |\epsilon_n| \leqslant |x_n - x_{n-1}| + |\epsilon_n|$$

and the result follows.

The important thing to notice about the error bound (2.5) is this: It doesn't matter how careless we were in computing $x_1, x_2, \ldots, x_{n-1}$. If x_{n-1} is still in the interval and x_n is not too close to an endpoint, then the eyeball error bound is valid provided that we add the term 2ϵ which takes into account the rounding and truncation errors at the last step.

EXAMPLE 2.3. Consider the equation $x = (x^3 + 1)/4$ used in Examples 2.1 and 1.1. There we had

$$x_6 = 0.2541\ 0168,$$

$$x_7 = 0.2541\ 0168.$$

Thus (2.5) gives

$$|x_7 - v| \leqslant |x_7 - x_6| + 2\epsilon = 2\epsilon,$$

where ϵ is a bound on the round-off error incurred in computing x_7 from x_6. The calculations were performed on an electronic desk computer which truncates to eight decimal places. Let $[\cdot]$ and $[\div]$ represent this type of fixed-point multiplication and division. Then we have

$$x_6[\cdot]x_6 = x_6^2 + \eta_1,$$

where $|\eta_1| \leqslant 10^{-8}$;

$$x_6[\cdot]x_6[\cdot]x_6 = [x_6^2 + \eta_1][\cdot]x_6 = x_6^3 + \eta_1 x_6 + \eta_2$$
$$= x_6^3 + \eta_3,$$

where $|\eta_3| \leqslant 1.26 \cdot 10^{-8}$;

$$x_7 = (x_6^3 + \eta_3 + 1)[\div]4 = \frac{(x_6^3 + \eta_3 + 1)}{4} + \eta_4$$
$$= \frac{(x_6^3 + 1)}{4} + \frac{\eta_3}{4} + \eta_4,$$

where $|\eta_4| \leqslant 10^{-8}$. Thus the total error in computing x_7 is at most

$$\epsilon = (1.26/4 + 1) \cdot 10^{-8} \leqslant 1.32 \cdot 10^{-8}.$$

Finally,

$$|x_7 - v| \leqslant 2\epsilon \leqslant 2.64 \cdot 10^{-8}$$

gives a bound on the error.

PROBLEMS

2.1. Consider the equation $x = 0.9x + 1$. Let $x_0 = 9.9$ and use the Picard method to approximate the solution to eight decimal places. This problem is feasible on an electronic desk computer. If one is not available, you may stop this grueling process by admitting that a contraction factor of $c = 0.9$ yields extremely slow convergence.

2.2. Compute x_n using Picard iteration on the following equations for the given values of x_0 and n. Use fixed-point arithmetic to m decimal places.
(a) $x = (x^3 + 1)/10$, $x_0 = 0$, $n = 5$, $m = 8$
(b) $x = (x + 5/x)/2$, $x_0 = 100$, $n = 10$, $m = 6$, (Ans. $\sqrt{5}$)
(c) $x = (x^2 + 1)/2$, $x_0 = 1.2$, $n = 16$, $m = 4$
(d) $x = (x^2 + 1)/2$, $x_0 = 0.99$, $n = 3$, $m = 4$
(e) $x = (x^2 + 1)/2$, $x_0 = 0.999$, $n = 3$, $m = 4$

2.3. Suppose that the hypotheses of Theorem 2.1 are satisfied. If in addition $g'(x) > 0$ whenever $|x - x_0| \leqslant r$, then:
(a) the sequence $\{x_n\}$ is strictly increasing if $x_0 < v$;
(b) the sequence $\{x_n\}$ is strictly decreasing if $x_0 > v$.

2.4. Suppose the hypotheses of Theorem 2.1 are satisfied. Show that the eyeball error bound (2.2) cannot hold if $g'(x) > 1/2$ for all $x \in [x_0 - r, x_0 + r]$.

2.5. Test the conclusion of Problem 2.4 on the equation $x = .99x + 1$, $x_0 = 99$.

2.6. Prove that the Picard sequence in Problem 2.2(d) is convergent even though condition (i) of Theorem 2.1 cannot hold on an interval containing the fixed point 1. Assume no round-off error.

2.7. Draw some pictures showing how Picard's method works (or doesn't work) when $g'(v)$ is:
(a) small and positive
(b) large and positive
(c) small in magnitude and negative
(d) large in magnitude and negative
See Froberg [7].

2.8. Prove the following generalization of the eyeball error bound. Let T be a transformation which satisfies the conditions of Theorem 1.3. If the contraction factor c does not exceed 1/2, then

$$\|x_n - v\| \leqslant \|x_n - x_{n-1}\|.$$

3. NEWTON'S METHOD

In this section we will consider equations of the form

$$f(x) = 0, \tag{3.1}$$

where f is a differentiable real-valued function of a real variable. (Note that an equation of the form $x = g(x)$ can be converted to this form by setting $f(x) = g(x) - x$.) Let x_0 be an initial approximation to a solution of (3.1). In the vicinity of x_0, f can be approximated by its first-degree Taylor polynomial. This gives

$$f(x) \doteq f(x_0) + (x - x_0)f'(x_0).$$

An approximate solution to (3.1) can now be found by solving the equation

$$f(x_0) + (x - x_0)f'(x_0) = 0 \tag{3.2}$$

for x. Call this solution x_1. From (3.2) we have $x_1 = x_0 - f(x_0)/f'(x_0)$. Now repeat the process with x_1 in place of x_0 to obtain a second approximation x_2, and so on. In this way we obtain a sequence $\{x_n\}$ which satisfies the equation

$$x_{n+1} = x_n - \frac{f(x_n)}{f'(x_n)}, \qquad n = 0, 1, 2, \ldots. \tag{3.3}$$

This iterative process is called *Newton's method.*

The graphical interpretation of Newton's method, shown in Figure 3.1, is that at each step the line tangent to f at $(x_n, f(x_n))$ is drawn, and x_{n+1} is the abscissa of the point where this tangent line crosses the x-axis.

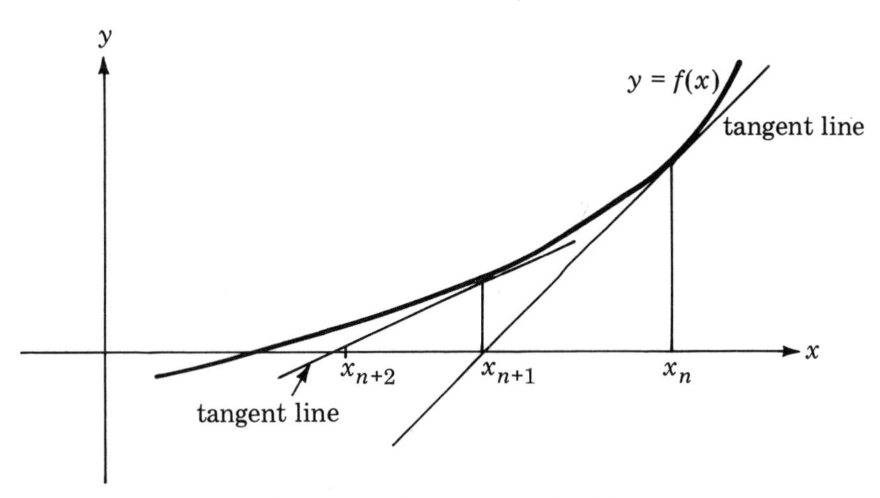

FIGURE 3.1. Graphical Interpretation of Newton's Method.

EXAMPLE 3.1. Let $f(x) = x^3 - 4x + 1$. Now $f(0) = 1$ and $f(0.5) = -0.875$. Thus there is a root to the equation $f(x) = 0$ in the interval $[0, 0.5]$. Letting $x_0 = 0$, we obtain

$$x_1 = x_0 - f(x_0)/f'(x_0) = 0 - 1/(-4) = 0.25.$$

Similarly,

$$x_2 = 0.2540\ 9836,$$

$$x_3 = 0.2541\ 0168,$$

$$x_4 = 0.2541\ 0168.$$

This agrees with the result given in Example 1.1. Note that the convergence is faster than the Picard method on the equivalent equation $x = (x^3 + 1)/4$.

Now let $g(x) = x - f(x)/f'(x)$. The equation $x = g(x)$ is then equivalent to the equation $f(x) = 0$. Thus Newton's method is really a special type of Picard iteration process and the theory of the preceding section can be applied. For example, after noting that $g'(x) = f(x)f''(x)/[f'(x)]^2$, we can translate Theorem 2.1 into the following theorem concerning Newton's method.

THEOREM 3.1. *Suppose that* r *is a positive number,* $c \in [0,1)$, $x_0 \in R$, *and* $x_{n+1} = x_n - f(x_n)/f'(x_n)$, $n = 0, 1, 2, \ldots$. *If*

(i) f' *is never zero on* $[x_0 - r, x_0 + r]$,

(ii) $|f(x)f''(x)/[f'(x)]^2| \leqslant c$ *whenever* $|x - x_0| \leqslant r$,

(iii) $|f(x_0)/f'(x_0)| \leqslant (1 - c)r$,

then there is a unique zero v of f in the interval $[x_0 - r, x_0 + r]$. *Furthermore,* $v = \lim_{n\to\infty} x_n$ *and*

$$|x_n - v| \leqslant c^n r, \qquad n = 1, 2, \ldots. \tag{3.4}$$

Convergence of the Newton process is usually quite rapid. The reason for this is that $g'(v) = f(v)f''(v)/[f'(v)]^2 = 0$ when v is a zero of f (assuming $f'(v) \neq 0$). Thus the convergence factor c in (3.4) tends to zero as x_n converges to v. This produces what is known as *second-order* or *quadratic convergence.* To see this another way, consider the Taylor expansion of g about v:

$$g(x) = g(v) + (x - v)g'(v) + \tfrac{1}{2}(x - v)^2 g''(\eta),$$

where η lies between x and v. Letting $x = x_n$ and using the relationships $v = g(v)$, $x_{n+1} = g(x_n)$, $g'(v) = 0$, we obtain

$$x_{n+1} - v = \tfrac{1}{2}(x_n - v)^2 g''(\eta).$$

If $|g''(\eta)/2| \leqslant M$, then we have

$$|x_{n+1} - v| \leqslant M|x_n - v|^2.$$

If, for example, $M = 1$ and x_n is within 0.01 units of v, then x_{n+1} is within 0.0001 units of v.

Another consequence of the fact that $g'(v) = 0$ is that the eyeball error bound (2.5) will be valid as soon as the approximations are sufficiently close to v.

If $f'(v) = 0$, then Newton's method usually still works in theory, but the convergence is apt to be much slower and the eyeball error bound is not necessarily valid. Further discussion of this problem can be found in [8] or [10].

There are variations of Newton's method of the form

$$x_{n+1} = x_n - f(x_n)/h(x_n),$$

where h is normally a function which is easier to compute than f'. For example, if f' stays nearly constant, it is not necessary to compute it for every x_n. The general idea is that we are sometimes willing to give up the quadratic convergence in order to have a function h which is relatively easy to compute. It should also be mentioned that Newton's method can be used for computing zeros of a function of a complex variable.

Newton's method has been popular for several centuries and there is a large amount of literature devoted to it. The *American Mathematical Monthly* is a good source for convergence theorems and variations of this method [4, 11, 14].

PROBLEMS

3.1. We wish to find a positive solution to the equation

$$x^3 - 2x^2 - 1 = 0.$$

(a) Use Newton's method to find an approximate solution correct to six decimal places.

(b) Compare this with Picard's method applied to the equivalent equation $x = 2 + 1/x^2$.

3.2. Show that the equation $e^{-x} - 0.1x = 0$ has one and only one solution. (Hint: Differentiate.) Approximate this solution correct to three decimal places using Newton's method. If necessary, use linear interpolation in an exponential table.

3.3. Let $f(x) = x^2$. Apply Newton's method with $x_0 = 1/2$. Repeat using $f(x) = x^5$. Why is the convergence relatively slow?

3.4. Use Newton's method to compute the cube root of 10 correct to six decimals. Check by computing $(x^* + 10^{-6})^3$ and $(x^* - 10^{-6})^3$, where x^* is your final approximation.

3.5. What would be the effect of

(a) a small mistake in arithmetic

(b) a large mistake in arithmetic

in computation of x_1 in Newton's or Picard's method?

3.6. Write a flow diagram for a computer program to calculate square roots by Newton's method. This is the method in use on many computer systems.

3.7. Let

$$A = \begin{bmatrix} 1 & 2 & 5 \\ 2 & 3 & 1 \\ 5 & 1 & 2 \end{bmatrix}.$$

(a) Compute the characteristic polynomial of A. Check your polynomial by evaluating it at λ and comparing with $|A - \lambda I|$ for $\lambda = 0, 1, 2, 3$.

(b) Compute the eigenvalues of A correct to five decimals using Newton's method on the characteristic polynomial.

(c) Recompute the largest eigenvalue by the Givens-Householder method.

3.8. Consider the equation $x = \Phi(x)$. Derive a general formula for the sequence of approximations obtained by using Newton's method on the equation $\Phi(x) - x = 0$. Try this formula on the equation $x = 3/x$ (see Example 1.2). Compare this method with Newton's method applied to the equation $x^2 - 3 = 0$.

3.9. We wish to approximate $\sqrt{5}$ by applying Newton's method to the equation $x^2 - 5 = 0$. Show that any positive starting value will work.

4. SYSTEMS OF NON-LINEAR EQUATIONS

Newton's method can also be applied to non-linear systems of equations. For notational simplicity, we will first discuss a system of two equations of the form

$$(4.1) \qquad \begin{aligned} f(x,y) &= 0, \\ g(x,y) &= 0. \end{aligned}$$

Let (x_0, y_0) be an initial approximation to a solution of (4.1). Now approximate f and g by their first-degree Taylor expansions about (x_0, y_0) to obtain the system of equations

$$(4.2) \qquad \begin{aligned} f(x_0,y_0) + (x - x_0)D_1f(x_0,y_0) + (y - y_0)D_2f(x_0,y_0) &= 0, \\ g(x_0,y_0) + (x - x_0)D_1g(x_0,y_0) + (y - y_0)D_2g(x_0,y_0) &= 0, \end{aligned}$$

where D_1f and D_2f represent the partial derivatives of f with respect to x and y, respectively. Then solve (4.2) for x and y and denote the solution by (x_1, y_1). Replace (x_0, y_0) in (4.2) by (x_1, y_1), solve again to obtain (x_2, y_2), and so on. In this way we obtain a sequence (x_n, y_n) which satisfies the matrix equation

$$(4.3) \qquad \begin{bmatrix} x_{n+1} \\ y_{n+1} \end{bmatrix} = \begin{bmatrix} x_n \\ y_n \end{bmatrix} - J^{-1}(x_n,y_n) \begin{bmatrix} f(x_n,y_n) \\ g(x_n,y_n) \end{bmatrix},$$

where

$$J(x,y) = \begin{bmatrix} D_1f(x,y) & D_2f(x,y) \\ D_1g(x,y) & D_2g(x,y) \end{bmatrix}$$

is the so-called Jacobian matrix.

EXAMPLE 4.1. Let $f(x,y) = x^2 + y^2 - 4$, and $g(x,y) = x^2 - 6x + y^2 + 5$. To obtain an approximate solution we plot the graphs of the equations $f(x,y) = 0$ and $g(x,y) = 0$. From the graphs we obtain (1,1) as a rough approximation to the solution in the first quadrant. Now $D_1f(x,y) = 2x$, $D_2f(x,y) = 2y$, $D_1g(x,y) = 2x - 6$, and $D_2g(x,y) = 2y$. Substituting (1,1) for (x_0,y_0) in (4.2) we get

$$\begin{aligned} -2 + 2\Delta x + 2\Delta y &= 0, \\ 1 - 4\Delta x + 2\Delta y &= 0, \end{aligned}$$

where $\Delta x = x - x_0 = x - 1$ and $\Delta y = y - y_0 = y - 1$. Solving this system we get $\Delta x = 1/2 = \Delta y$ and hence $(x_1,y_1) = (3/2,3/2)$. Substitution of (x_1,y_1) for (x_0,y_0) in (4.2) yields the system

$$\begin{aligned} 1/2 + 3\Delta x + 3\Delta y &= 0, \\ 1/2 - 3\Delta x + 3\Delta y &= 0. \end{aligned}$$

Solving this system we get $\Delta x = 0$, $\Delta y = -1/6$ and hence $(x_2,y_2) = (3/2,4/3) = (1.5, 1.333\ldots)$. In this case the original system can be solved algebraically to show that the solution in the first quadrant is actually $(3/2, \sqrt{7}/2) \doteq (1.5, 1.322\,876)$.

Now consider k equations of the form

$$
\begin{aligned}
&f_1(x_1,x_2, \ldots, x_k) = 0,\\
&f_2(x_1,x_2, \ldots, x_k) = 0,\\
&\vdots\\
&f_k(x_1,x_2, \ldots, x_k) = 0.
\end{aligned}
\tag{4.4}
$$

It is convenient to use vector notation here. The equations (4.4) can be expressed in vector form as

$$f(x) = \emptyset,$$

where $x = (x_1,x_2, \ldots, x_k)$, $f = (f_1,f_2, \ldots, f_k)$, and $\emptyset = (0,0,\ldots,0)$. Now we want to discuss the sequence $\{x_n\}$ of approximate solution vectors; so we will use double subscripts for the coordinates of x_n, that is, $x_n = (x_{n1},x_{n2},\ldots,x_{nk}) \in E_k$ is the nth approximation to a solution of the system (4.4). As before, each f_i, $i = 1, 2, \ldots, k$ is approximated by its first-degree Taylor expansion about x_0 to obtain the linear system

$$
\begin{aligned}
&f_1(x_0) + \sum_{j=1}^{k} D_j f_1(x_0)\Delta x_j = 0,\\
&f_2(x_0) + \sum_{j=1}^{k} D_j f_2(x_0)\Delta x_j = 0,\\
&\vdots\\
&f_k(x_0) + \sum_{j=1}^{k} D_j f_k(x_0)\Delta x_j = 0.
\end{aligned}
\tag{4.5}
$$

This system can be written in matrix form as

$$J_f(x_0)\Delta x = -f(x_0),$$

where Δx and $f(x_0)$ are column vectors (k by 1 matrices) and $J_f(x_0)$ is the Jacobian matrix given by

$$
J_f(x_0) = \begin{bmatrix}
D_1 f_1(x_0) & D_2 f_1(x_0) & \ldots & D_k f_1(x_0)\\
D_1 f_2(x_0) & D_2 f_2(x_0) & \ldots & D_k f_2(x_0)\\
\vdots & & & \\
D_1 f_k(x_0) & D_2 f_k(x_0) & \ldots & D_k f_k(x_0)
\end{bmatrix}.
$$

The system (4.5) is solved for Δx, and the next approximation is given by

$$x_1 = x_0 + \Delta x = x_0 - J_f^{-1}(x_0)f(x_0).$$

Continuing in this fashion, we obtain a sequence of vectors $\{x_n\}$ which satisfy the equations

$$x_{n+1} = x_n - J_f^{-1}(x_n)f(x_n), \qquad n = 0, 1, \ldots . \tag{4.6}$$

Note that the vector $\Delta x = J_f^{-1}(x_n)f(x_n)$ is computed by solving (4.5) (with x_n replacing x_0) rather than by actually computing $J_f^{-1}(x_n)$. Also note the similarity between (4.6) and (3.3)—they are the same except that J_f^{-1} replaces $1/f'$.

We will not discuss the theoretical aspects of Newton's method for systems of equations. It can be shown that the process will converge to the solution v if x_0 is sufficiently close to v and if $J_f(v)$ is non-singular (see Henrici [8] or Householder [9]). Error bounds, as well as convergence theorems, have been given by Ostrowski [12]. Using the bounds given by Ostrowski, it can be shown that the eyeball error bound is valid if x_n is sufficiently close to v and $J_f(v)$ is non-singular.

PROBLEMS

4.1. Use at least two iterations of Newton's method to approximate the solution of the system of equations

$$x^2 - 4x + y^2 = 0,$$

$$y - 2x^2 = 0,$$

which lies in the first quadrant. Check your result by solving the system algebraically (with the help of the one-dimensional Newton method, if necessary).

4.2. Consider the system of equations

$$e^{xy} - 3y^2 + yz = 4,$$

$$e^{-xy} + 2x^2 + z^3 = 10,$$

$$e^{zx} + y^2 + 2z = 3.$$

Write a flow diagram describing a computer program to carry out Newton's method on this system. Assume that a subroutine called GAUSS4 is available for solving linear systems of equations.

5. BAIRSTOW'S METHOD FOR POLYNOMIALS

Bairstow's method is a technique for extracting quadratic factors from a polynomial. This is a convenient way to compute the imaginary zeros of a polynomial with real coefficients. Let

$$p(x) = a_n x^n + a_{n-1}x^{n-1} + \ldots + a_1 x + a_0$$

be a polynomial of degree n with real coefficients. We wish to determine real numbers u and v such that $x^2 + ux + v$ is a factor of $p(x)$. Then the roots of the equation

$$x^2 + ux + v = 0$$

will be zeros of p.

Now division of $p(x)$ by $x^2 + ux + v$ results in an equation of the form

$$\frac{p(x)}{x^2 + ux + v} = q(x) + \frac{Ax + B}{x^2 + ux + v},$$

where q is a polynomial of degree $n - 2$ and $Ax + B$ is the remainder. The quadratic $x^2 + ux + v$ is a factor of $p(x)$ if and only if $A = 0$ and $B = 0$. But A and B depend on u and v; say $A = f(u,v)$ and $B = g(u,v)$. Thus quadratic factors of p can be found by solving the system of equations

$$f(u,v) = 0,$$

$$g(u,v) = 0.$$

This can be carried out by using Newton's method as described in the preceding section.

EXAMPLE 5.1. Let $p(x) = x^3 - 6x^2 + 37x - 60$. Dividing $p(x)$ by $x^2 + ux + v$, we obtain

$$p(x) = (x^2 + ux + v)(x - 6 - u) + f(u,v)x + g(u,v), \tag{5.1}$$

where

$$f(u,v) = u^2 + 6u - v + 37,$$

$$g(u,v) = uv + 6v - 60.$$

Now p must have at least one real zero (why?). By trial and error we find that there is a zero near 2, and hence (5.1) implies that $6 + u \doteq 2$ or $u \doteq -4$. Setting $u = -4$ in the equation $g(u,v) = 0$, we get $v = 30$. This gives $(u_0, v_0) = (-4,30)$ for a starting value. The first step in the iteration leads to the system

$$-2\Delta u - \Delta v = 1,$$

$$30\Delta u + 2\Delta v = 0.$$

Solving we get $(\Delta u, \Delta v) \doteq (0.0769, -1.1538)$ and $(u_1, v_1) = (-3.9231, 28.8462)$. The results of three more iterations are

$$u_2 = -3.9200\,1965, \qquad v_2 = 28.8464\,2666;$$
$$u_3 = -3.9200\,2046, \qquad v_3 = 28.8464\,3764;$$
$$u_4 = -3.9200\,2046, \qquad v_4 = 28.8464\,3764.$$

The zeros of p corresponding to the factor $x^2 + u_4 x + v_4$, found by the quadratic equation, are $1.9600\,1023 \pm 5.0004\,7973i$. From (5.1) we see that the other zero of p is approximately $6 + u_4 = 2.0799\,7954$. It should be noted that with a cubic equation it would be easier to find the real root by the one-dimensional Newton method, divide out the corresponding linear factor, and solve the resulting quadratic equation. Bairstow's method will normally be used only on polynomials of degree at least 4.

Example 5.1 illustrates the basic idea of Bairstow's method, but there are several problems which arise with higher-degree polynomials. First, there is the problem of obtaining starting values. One method, which is used on some computer routines, proceeds by trying successively the starting values $(u_0, v_0) = (0,1), (0,10), (0,100)$, and so on. If a starting value is found which produces convergence, then the computed quadratic is factored out of the original polynomial and the process is repeated on the reduced polynomial. The degree of the polynomial is repeatedly reduced in this manner until all factors have been found. The round-off errors tend to accumulate in the reduction process; so as a final step each quadratic factor is refined by using Bairstow's method on the original polynomial.

A more systematic method for obtaining starting values is provided by the *Quotient-Difference algorithm*. This technique, which is described in Henrici [8], produces starting values for the real zeros of p and for quadratic factors corresponding to conjugate imaginary zeros.

The other complication which arises with higher-degree polynomials is that the derivation of explicit formulas for $f(u,v)$ and $g(u,v)$ by long division is quite tedious. In practice, f,g and their partial derivatives are evaluated by recurrence relations which we will now derive.

The result of dividing $p(x)$ by $x^2 + ux + v$ can be written in the form

$$(5.2) \qquad p(x) = (x^2 + ux + v)q(x) + Ax + B,$$

where A, B and the coefficients of q depend on u and v. Let the jth coefficients of p and q be denoted by a_j and b_j, respectively. Equation (5.2) then becomes

$$\sum_{j=0}^{n} a_j x^j = \sum_{j=0}^{n-2} [b_j x^{j+2} + u b_j x^{j+1} + v b_j x^j] + Ax + B$$

$$\begin{aligned} &= b_{n-2} x^n + (b_{n-3} + u b_{n-2}) x^{n-1} \\ &\quad + \sum_{j=2}^{n-2} (b_{j-2} + u b_{j-1} + v b_j) x^j \\ &\quad + (u b_0 + v b_1 + A) x + (v b_0 + B). \end{aligned} \tag{5.3}$$

Equating coefficients of like powers of x on both sides of (5.3), we obtain the following recurrence relations:

$$\begin{aligned} b_{n-2} &= a_n, \\ b_{n-3} &= a_{n-1} - u b_{n-2}, \\ b_{n-j} &= a_{n+2-j} - u b_{n+1-j} - v b_{n+2-j}, \qquad j = 4, 5, \ldots, n, \\ f(u,v) &= A = a_1 - u b_0 - v b_1, \\ g(u,v) &= B = a_0 - v b_0. \end{aligned} \tag{5.4}$$

These formulas enable us to evaluate $f(u,v)$ and $g(u,v)$. We also need formulas for the partial derivatives of g and f. These can be obtained by differentiating the equations (5.4). After noting that A, B, and b_j, $j = 0, 1, \ldots, n-2$ are functions of u and v, we obtain

$$\begin{aligned} &D_1 b_{n-2} = 0, \qquad D_2 b_{n-2} = 0, \\ &D_1 b_{n-3} = -b_{n-2}, \qquad D_2 b_{n-3} = 0, \\ &D_1 b_{n-j} = -b_{n+1-j} - u D_1 b_{n+1-j} - v D_1 b_{n+2-j}, \\ &\qquad j = 4, 5, \ldots, n, \\ &D_2 b_{n-j} = -u D_2 b_{n+1-j} - b_{n+2-j} - v D_2 b_{n+2-j}, \\ &\qquad j = 4, 5, \ldots, n, \\ &D_1 f(u,v) = -b_0 - u D_1 b_0 - v D_1 b_1, \\ &D_2 f(u,v) = -u D_2 b_0 - b_1 - v D_2 b_1, \\ &D_1 g(u,v) = -v D_1 b_0, \\ &D_2 g(u,v) = -b_0 - v D_2 b_0. \end{aligned} \tag{5.5}$$

EXAMPLE 5.2. Let $p(x) = 3x^4 + 2x^3 + 5x^2 + x + 6$. To illustrate the use of formulas (5.4) and (5.5) let us arbitrarily choose a starting value of $(u_0, v_0) = (2,3)$. From (5.4) we obtain

$$\begin{aligned} b_2 &= a_4 = 3, \\ b_1 &= a_3 - u_0 b_2 = 2 - 2 \cdot 3 = -4, \end{aligned}$$

$$b_0 = a_2 - u_0 b_1 - v_0 b_2 = 5 - 2\cdot(-4) - 3\cdot 3 = 4,$$
$$f(u_0,v_0) = a_1 - u_0 b_0 - v_0 b_1 = 1 - 2\cdot 4 - 3(-4) = 5,$$
$$g(u_0,v_0) = a_0 - v_0 b_0 = 6 - 3\cdot 4 = -6.$$

Similarly, from (5.5) we obtain

$$D_1 b_2 = 0, \qquad D_2 b_2 = 0,$$
$$D_1 b_1 = -b_2 = -3, \qquad D_2 b_1 = 0,$$
$$D_1 b_0 = -b_1 - u_0 D_1 b_1 - v_0 D_1 b_2 = 10,$$
$$D_2 b_0 = -u_0 D_2 b_1 - b_2 - v_0 D_2 b_2 = -3,$$
$$D_1 f(u_0,v_0) = -b_0 - u_0 D_1 b_0 - v_0 D_1 b_1 = -15,$$
$$D_2 f(u_0,v_0) = -u_0 D_2 b_0 - b_1 - v_0 D_2 b_1 = 10,$$
$$D_1 g(u_0,v_0) = -v_0 D_1 b_0 = -30,$$
$$D_2 g(u_0,v_0) = -b_0 - v_0 D_2 b_0 = 5.$$

Thus, the first set of equations for Bairstow's method is

$$-15\Delta u + 10\Delta v = -5,$$
$$-30\Delta u + 5\Delta v = 6.$$

Detailed descriptions of several programs for computing the zeros of polynomials are given in Ralston and Wilf [13, Parts II and V].

PROBLEMS

5.1. Let $p(x) = x^3 + x^2 - 4x - 25$. Find an approximate quadratic factor for $p(x)$ using two iterations of Bairstow's method. Use slide rule accuracy for the first iteration.

5.2. Let $p(x) = x^4 + 2x^3 + 6x^2 + 2x + 7$. Find an approximate quadratic factor of p using formulas (5.4) and (5.5) with $(u_0,v_0) = (2,5)$. Also compute the four zeros of p.

6. PICARD ITERATION FOR DIFFERENTIAL EQUATIONS

In this section we will discuss an iterative technique for solving certain differential equations. Let us consider a first-order differential equation, together with an initial condition, of the form

$$y'(x) = f(x,y(x)), \qquad y(a) = b, \tag{6.1}$$

where f is a function of two variables which is continuous on a neighborhood of the point (a,b). The problem is to find a function y, continuous on an interval I containing a, which satisfies the conditions:

$$\text{(i)} \quad y(a) = b$$

$$\text{(ii)} \quad y'(x) = f(x,y(x)) \qquad \text{for every} \quad x \in I.$$

Such a problem is called an *initial-value problem.*

The first step is to convert the differential equation into an integral equation. Suppose that y is a solution to the initial-value problem (6.1). Then both sides of the differential equation can be integrated to obtain

$$y(x) - y(a) = \int_a^x y'(t)dt = \int_a^x f(t,y(t))dt.$$

Using the fact that $y(a) = b$, we have

$$y(x) = b + \int_a^x f(t,y(t))dt. \tag{6.2}$$

Conversely, if y is continuous and satisfies (6.2), then it is a solution to (6.1). Thus (6.1) and (6.2) are equivalent. In Problem 1.7, the reader was asked to show that equation (6.2) is really of the form $y = T(y)$ when T is appropriately defined.

In the Picard method, successive approximations are formed according to the equations

$$(6.3) \quad \begin{aligned} y_0(x) &= b, \\ y_{n+1}(x) &= b + \int_a^x f(t,y_n(t))dt. \end{aligned}$$

Under appropriate conditions, which will be discussed below, this sequence of functions converges to a solution of the initial-value problem.

EXAMPLE 6.1. Let us apply this technique to the problem

$$y'(x) = 2x + 5y(x), \qquad y(0) = 1. \tag{6.4}$$

Now $f(x,y) = 2x + 5y$, $a = 0$, and $b = 1$. From (6.3) we have

$$y_0(x) = 1,$$

$$\begin{aligned} y_1(x) &= 1 + \int_0^x f(t,1)dt = 1 + \int_0^x (2t + 5)dt \\ &= 1 + 5x + x^2, \end{aligned}$$

$$\begin{aligned} y_2(x) &= 1 + \int_0^x f(t,1 + 5t + t^2)dt \\ &= 1 + \int_0^x [2t + 5(1 + 5t + t^2)]\, dt \\ &= 1 + 5x + \tfrac{27}{2}x^2 + \tfrac{5}{3}x^3, \end{aligned}$$

$$y_3(x) = 1 + \int_0^x [2t + 5(1 + 5t + \tfrac{27}{2}t^2 + \tfrac{5}{3}t^3)]\,dt$$
$$= 1 + 5x + \tfrac{27}{2}x^2 + \tfrac{45}{2}x^3 + \tfrac{25}{12}x^4.$$

In practice, Picard's method would be applied only when an explicit solution is difficult to compute by other means. However, in this example we can apply the standard method for a first-order linear equation to obtain the exact solution to (6.4):

$$y(x) = \frac{27e^{5x} - 10x - 2}{25}.$$

Table 6.1 gives a comparison between the approximate solution $y_3(x)$ and the exact solution $y(x)$ for selected values of x. The error bound given in the last column will be explained in Example 6.2.

x	$y_3(x)$	$y(x)$	$y(x) - y_3(x)$	*Error Bound*
0.00	1.00000	1.00000	0.00000	0.0
0.05	1.28658	1.28675	0.00017	$5.02 \cdot 10^{-4}$
0.10	1.65771	1.66062	0.00291	$8.94 \cdot 10^{-3}$
0.50	9.81771	12.87709	3.05938	
1.00	44.08333	159.80621	115.72288	

TABLE 6.1. Comparison of Approximate and Exact Solutions to (6.4).

We will now prove a theorem concerning the convergence of the sequence of approximations defined by (6.3). As an important by-product we will obtain a means for computing error bounds. It should be emphasized again that this is frequently the case—a convergence proof often contains a method of computing error bounds for the corresponding approximations.

THEOREM 6.1. *Let f be continuous on a rectangle R which has the point (a,b) as its center and has sides parallel to the coordinate axes. If there exists a constant K such that*

$$|f(x,y_2) - f(x,y_1)| \leqslant K|y_2 - y_1| \tag{6.5}$$

whenever (x,y_1) and $(x,y_2) \in R$, then the sequence of approximations given by (6.3) converges to a solution of the initial-value problem (6.1) on some closed interval I containing the point a in its interior. Furthermore, this solution is unique on I.

Proof. Let us first specify the interval I on which convergence can be guaranteed. Let

$$M = \max_{(x,y)\in R} |f(x,y)|.$$

Now the line $y - b = M(x - a)$ intersects the rectangle R in two points. Let x_1, x_2, with $x_1 < x_2$, be the x coordinates of these points. The interval I is then taken to be $[x_1, x_2]$. Note that if

$$x_1 \leqslant x \leqslant x_2 \quad \text{and} \quad |y - b| \leqslant M|x - a|, \tag{6.6}$$

then $(x,y) \in R$. (The region defined by (6.6) is the shaded region in Figure 6.2. The figure is placed here for convenience only—it is not considered to be part of the proof.)

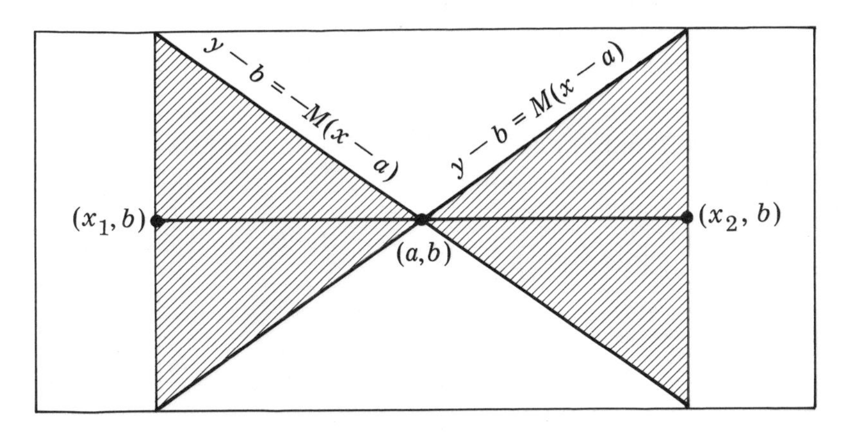

FIGURE 6.2. The Region Defined by (6.6).

Next, we will show that $(x, y_n(x)) \in R$ for all $x \in I$ and all non-negative integers n. This will be proved by induction on n. Now $y_0(x) = b$ for all $x \in I$; thus the inequalities (6.6) are satisfied and $(x, y_0(x)) \in R$. Assuming that $(x, y_n(x)) \in R$ for all $x \in I$, we have

$$|y_{n+1}(x) - b| = \left|\int_a^x f(t, y_n(t))dt\right| \leqslant M|x - a|.$$

Thus the inequalities (6.6) are satisfied and $(x, y_{n+1}(x)) \in R$. This completes the induction proof.

The next step is to show that

$$|y_n(x) - y_{n-1}(x)| \leqslant \frac{MK^{n-1}|x - a|^n}{n!} \tag{6.7}$$

for all $x \in I$ and all n. This will also be proved by induction. We will only prove (6.7) for the case $x \geqslant a$; the proof of the case $x < a$ is left as an exercise (Problem 6.2). For all $x \geqslant a$, we have

$$|y_1(x) - y_0(x)| = |y_1(x) - b| = \left|\int_a^x f(t,b)dt\right| \leqslant M(x - a).$$

Supposing that (6.7) is true for a certain n and all $x \geqslant a$, we can apply (6.5) to obtain

$$\begin{aligned}|y_{n+1}(x) - y_n(x)| &\leqslant \int_a^x |f(t,y_n(t)) - f(t,y_{n-1}(t))|dt \\ &\leqslant \int_a^x K|y_n(t) - y_{n-1}(t)|dt \\ &\leqslant \int_a^x \frac{KMK^{n-1}(t-a)^n}{n!}\,dt \\ &= \frac{MK^n(x-a)^{n+1}}{(n+1)!}.\end{aligned}$$

It follows by induction that (6.7) holds for all n and all $x \in [a,x_2]$.

Now let us show that the series

$$y_0(x) + [y_1(x) - y_0(x)] + \ldots + [y_n(x) - y_{n-1}(x)] + \ldots \tag{6.8}$$

converges uniformly on I. By uniform convergence we mean convergence in the uniform norm as described in Section 4 of Chapter 2. From (6.7) we see that each term after the first in (6.8) is bounded above in absolute value by $MK^{n-1}|x-a|^n/n!$, which in turn is bounded above by $MK^{n-1}(x_2-a)^n/n!$ (making use of the fact that the interval I is symmetric about a). But the series

$$\sum_{n=1}^{\infty} \frac{MK^{n-1}(x_2-a)^n}{n!}$$

converges to

$$M[e^{K(x_2-a)} - 1]/K.$$

It then follows from the Weierstrass M test [1] that the series (6.8) is uniformly convergent on I.

Next, note that the nth partial sum of (6.8) is just $y_n(x)$. Thus the sequence $\{y_n\}$ converges uniformly on I to some function which we will call y. The function y is continuous on I because each y_n is continuous and the convergence is uniform. Furthermore, from (6.7)

$$\begin{aligned}|y(x) - y_n(x)| &= \left|\sum_{j=n+1}^{\infty} [y_j(x) - y_{j-1}(x)]\right| \\ &\leqslant M \sum_{j=n+1}^{\infty} \frac{K^{j-1}|x-a|^j}{j!},\end{aligned} \tag{6.9}$$

which is the promised error bound. It remains to show that y is the unique solution to (6.1) on I.

Note that $(x,y(x)) \in R$ for each $x \in I$ since $(x,y_n(x)) \in R$ for all n. Thus

$$|f(x,y(x)) - f(x,y_n(x))| \leqslant K|y(x) - y_n(x)|.$$

But $|y(x) - y_n(x)|$ tends to zero uniformly on I. Hence $f(x,y_n(x))$ tends to $f(x,y(x))$ uniformly. This enables us to take a limit under the integral sign [1] to obtain

$$y(x) = \lim_{n\to\infty} y_{n+1}(x) = \lim_{n\to\infty} \left[b + \int_a^x f(t,y_n(t))dt\right]$$
$$= b + \int_a^x \lim_{n\to\infty} f(t,y_n(t))dt = b + \int_a^x f(t,y(t))dt.$$

Thus y satisfies (6.2) which is equivalent to (6.1).

In order to show that y is unique, suppose that φ is another function which satisfies (6.1) on I. Let us first show that the graph of φ cannot go outside of R on I. Suppose that the graph of φ is outside of R for some $x \in I$. To be specific, suppose that this happens to the right of a. Now φ is necessarily continuous since it must have a derivative in order to satisfy (6.1). Thus there must be a first point, call it x^*, to the right of a where the graph of φ intersects the boundary of R. Then

$$|\varphi(x^*) - b| > M|x^* - a|$$

since the opposite inequality implies that $(x^*, \varphi(x^*))$ is interior to R. But $(x, \varphi(x)) \in R$ for all $x \in [a,x^*)$. Thus

$$|\varphi(x^*) - b| = \left|\int_a^{x^*} f(t,\varphi(t))dt\right| \leqslant M|x^* - a|.$$

This contradiction establishes the fact that $(x,\varphi(x)) \in R$ whenever $a \leqslant x \leqslant x_2$. A similar argument can be used for the portion of I to the left of a.

Knowing that $(x,\varphi(x)) \in R$, we can apply (6.5) to obtain the inequality

$$|y(x) - \varphi(x)| = \left|\int_a^x [f(t,y(t)) - f(t,\varphi(t))]dt\right|$$
$$\leqslant |x - a|KL,$$

where $L = \max_{t\in I} |y(t) - \varphi(t)|$. Proceeding inductively as in the proof of (6.7), we can establish the inequality

$$(6.10) \qquad |y(x) - \varphi(x)| \leqslant LK^n \frac{|x-a|^n}{n!}$$

for all n and all $x \in I$. Letting $n \to \infty$, we infer that $y(x) = \varphi(x)$ since the right side of (6.10) tends to zero. Thus y is the unique solution to (6.1) on I and the proof of Theorem 6.1 is complete.

The error bound (6.9) can be expressed in a more convenient way.

COROLLARY 6.2. *Suppose that the hypotheses of Theorem* 6.1 *hold and* $M = \max_{(x,y)\in R} |f(x,y)|$. *If* y *is the solution to the initial-value problem* (6.1), *then*

$$(6.11) \qquad |y(x) - y_n(x)| \leqslant \frac{M\, e^{K|x-a|}K^n|x-a|^{n+1}}{(n+1)!}$$

for all $x \in I$.

A proof of this corollary can be based on the remainder formula for Taylor series. It is left as an exercise (Problem 6.3).

EXAMPLE 6.2. Consider again the initial-value problem (6.4). Here $f(x,y) = 2x + 5y$, $a = 0$, and $b = 1$. Let us take for the rectangle R the one with corners at (1,2), (−1,2), (−1,0), and (1,0). Then

$$M = \max_{(x,y)\in R} |2x + 5y| = 12.$$

Also

$$|f(x,y_2) - f(x,y_1)| = |5(y_2 - y_1)| \leqslant 5|y_2 - y_1|;$$

so K can be set equal to 5. The line $y - 1 = 12x$ intersects R at $(-1/12, 0)$ and $(1/12, 2)$; so the interval I is $[-1/12, 1/12]$. Setting $x = 0.05$ and $n = 3$ in (6.11), we have

$$|y(0.05) - y_3(0.05)| \leqslant \frac{12e^{5(0.05)}5^3(0.05)^4}{4!}$$
$$\leqslant 5.02\cdot 10^{-4}.$$

This can be compared with the actual error given in Table 6.1. We would like to complete the last column of Table 6.1, but 0.1, 0.5, and 1.0 are not in I. We can try a different rectangle—the choice of R is somewhat arbitrary. Let R be the rectangle with corners at (0.1, 2.04), (−0.1, 2.04), (−0.1, −0.04), and (0.1, −0.04). Now $M = 10.4$ and $I = [-0.1, 0.1]$. From (6.11) we obtain

$$|y(0.1) - y_3(0.1)| \leqslant \frac{10.4e^{0.5}5^3\cdot 10^{-4}}{4!} \leqslant 8.94\cdot 10^{-3}.$$

An error bound for $x = 0.5$ and 1.0 cannot be obtained using rectangles (see Problem 6.4).

Other techniques for handling initial-value problems are discussed in Chapter 6.

PROBLEMS

6.1. Use n iterations of Picard's method to approximate solutions to the initial-value problems:
(a) $y'(x) = 4xy(x)$, $y(0) = 1, n = 3$
(b) $y'(x) = x + y(x)$, $y(0) = 0, n = 4$
Also find bounds on the error for $x = 0.2$.

6.2. Prove that (6.7) holds for values of $x < a$.

6.3. Prove Corollary 6.2. Hint: Look up the remainder formula for Taylor series. This can be found in many calculus texts or in Chapter 5.

6.4. In Example 6.2, show that it is impossible to use a rectangle to find an error bound at $x = 1$.

6.5. Use three iterations of Picard's method to find an approximate solution to the initial-value problem

$$y'(x) = x^2 + y^2(x), \qquad y(0) = 0.$$

Circular regions are more convenient than rectangles for bounding $f(x,y)$ in this case. Show how Theorem 6.1 can be modified to replace the rectangle R by a circular disc about $(0,0)$. Then compute an error bound for this problem.

7. THE NEUMANN SERIES

Let X be a Banach space and let T be a bounded linear transformation mapping X into X. In this section we will consider equations of the form

$$x = c + \lambda T(x), \tag{7.1}$$

where $c \in X$ and $\lambda \in K$ are constants. Let I be the identity transformation (that is, $I(x) = x$ for all x). Then (7.1) can be written in the form

$$(I - \lambda T)(x) = c. \tag{7.2}$$

The study of equation (7.2) will reveal information concerning the existence of an inverse for $I - \lambda T$, as well as information about the largest eigenvalue of T.

Let us apply the Picard method to equation (7.1) with $x_0 = c$. This gives the following:

$$\begin{aligned} x_0 &= c, \\ x_1 &= c + \lambda T(x_0) = c + \lambda T(c), \\ x_2 &= c + \lambda T(x_1) = c + \lambda T(c + \lambda T(c)) \\ &= c + \lambda T(c) + \lambda^2 T^2(c), \\ &\ \ \vdots \\ x_n &= c + \lambda T(c) + \lambda^2 T^2(c) + \ldots + \lambda^n T^n(c), \end{aligned} \tag{7.3}$$

where T^n represents the composite of T on itself n times (that is, $T^2(x) = T[T(x)]$, $T^3(x) = T[T(T(x))]$, etc.). We now wish to find conditions which will guarantee the convergence of the sequence (7.3) to a solution of (7.1).

Now T is a bounded linear transformation; so there exists a constant M such that

$$\|T(x)\| \leqslant M\|x\|$$

for all $x \in X$. We define a norm for transformations by the equation

$$\|T\| = \sup_{\|x\| \neq 0} \frac{\|T(x)\|}{\|x\|}, \tag{7.4}$$

where "sup" stands for *supremum* or *least upper bound*. From (7.4) it follows that

$$\|T(x)\| \leqslant \|T\| \; \|x\|$$

for all $x \in X$. Furthermore,

$$\|T^2(x)\| = \|T[T(x)]\| \leqslant \|T\| \; \|T(x)\| \leqslant \|T\|^2 \|x\|.$$

In a similar fashion one can show that

$$\|T^n(x)\| \leqslant \|T\|^n \|x\|. \tag{7.5}$$

We are now in a position to prove a theorem concerning the convergence of the sequence (7.3).

THEOREM 7.1. *Let T be a bounded linear transformation mapping the Banach space X into itself. If $\|T\| < 1/|\lambda|$, then the sequence defined by* (7.3) *converges to a solution of* (7.1). *Furthermore, this solution is unique.*

Proof. Let $G(x) = c + \lambda T(x)$. We will show that G is a contraction mapping. In fact,

$$\|G(x) - G(y)\| = \|\lambda T(x-y)\| \leqslant |\lambda| \; \|T\| \; \|x-y\|.$$

But $|\lambda| \; \|T\| < 1$ by hypothesis; thus G is a contraction mapping. The result now follows from Theorem 1.2.

The theory of infinite series involving elements from a Banach space is quite analogous to the corresponding theory for series of complex numbers. In view of Theorem 7.1, we can express the unique solution, x^*, of equation (7.1) in terms of an infinite series:

$$x^* = \sum_{j=0}^{\infty} \lambda^j T^j(c),$$

where T^0 would be interpreted as the identity I. Since equation (7.1) (and hence (7.2)) has a unique solution for every $c \in X$ when $|\lambda| < 1/\|T\|$, the transformation $I - \lambda T$ has an inverse given by

$$(I - \lambda T)^{-1} = I + \lambda T + \lambda^2 T^2 + \ldots + \lambda^n T^n + \ldots. \tag{7.6}$$

The interesting thing about (7.7) is its close resemblence to the equation

$$\frac{1}{1-\lambda T} = 1 + \lambda T + \lambda^2 T^2 + \ldots + \lambda^n T^n + \ldots$$

which would be valid if λ and T were complex numbers with $|\lambda T| < 1$.

Using the idea of an infinite series it is easy to establish an error bound for the approximate solutions to (7.1).

THEOREM 7.2. *Suppose that T is a bounded linear transformation mapping the Banach space X into itself, $\|T\| < 1/|\lambda|$, x^* is the unique solution to* (7.1), *and x_n is the sequence defined by* (7.3). *Then*

$$\|x^* - x_n\| \leqslant \frac{\|c\|(|\lambda|\,\|T\|)^{n+1}}{1-|\lambda|\,\|T\|}. \tag{7.7}$$

Proof. Note that

$$x^* = \sum_{j=0}^{\infty} \lambda^j T^j(c)$$

and

$$x_n = \sum_{j=0}^{n} \lambda^j T^j(c).$$

Applying (7.5) we have

$$\|x^* - x_n\| = \left\| \sum_{j=n+1}^{\infty} \lambda^j T^j(c) \right\| \leqslant \sum_{j=n+1}^{\infty} |\lambda^j|\,\|T\|^j\,\|c\|. \tag{7.8}$$

The result then follows by noting that the last series on the right in (7.8) is a geometric series with ratio $|\lambda|\,\|T\|$.

Another consequence of Theorem 7.1 is that T cannot have any eigenvalues larger than $\|T\|$.

THEOREM 7.3. *Let T be a bounded linear transformation mapping the Banach space X into itself. Then no eigenvalue of T has a magnitude which exceeds $\|T\|$.*

Proof. Suppose μ is an eigenvalue of T and $|\mu| > \|T\|$. Then

$$T(x_0) = \mu x_0$$

for some $x_0 \neq \emptyset$. Letting $\lambda = 1/\mu$, we see that x_0 satisfies the equation

$$x = \emptyset + \lambda T(x). \tag{7.9}$$

Now $\|T\| < |\mu| = 1/|\lambda|$; so Theorem 7.1 implies that (7.9) has a unique solution. But the vectors $\emptyset$ and x_0 both satisfy (7.9). This contradiction establishes the fact that $|\mu| \leqslant \|T\|$ if μ is an eigenvalue of T.

EXAMPLE 7.1. Consider a linear integral equation of the form

$$(7.10) \qquad y(x) = f(x) + \lambda \int_a^b K(x,t)y(t)dt,$$

where f is continuous on $[a,b]$ and $K(x,t)$ is continuous for $a \leqslant x \leqslant b$, $a \leqslant t \leqslant b$. The object is to find a function y which is continuous on $[a,b]$ and satisfies (7.10). Let T be the transformation from $C[a,b]$ into $C[a,b]$ defined by the equation

$$T(y)(x) = \int_a^b K(x,t)y(t)dt$$

for every $x \in [a,b]$. It is easy to see that T is a linear transformation and that (7.10) can be written as

$$y = f + \lambda T(y).$$

Note that this is of the same form as (7.1). Let

$$M = \max_{\substack{a \leqslant x \leqslant b \\ a \leqslant t \leqslant b}} |K(x,t)|.$$

Using the uniform norm on $C[a,b]$ we have

$$\|T(y)\| = \max_{x \in [a,b]} \left| \int_a^b K(x,t)y(t)dt \right| \leqslant M \, \|y\| \, (b-a).$$

Thus T is a bounded transformation and $\|T\| \leqslant M(b-a)$. When $|\lambda| < 1/[M(b-a)]$, the solution to (7.10) is given by

$$y = f + \lambda T(f) + \lambda^2 T^2(f) + \ldots + \lambda^n T^n(f) + \ldots .$$

This series converges uniformly (i.e., in the uniform norm) on $[a,b]$.

PROBLEMS

7.1. Consider the integral equation

$$y(x) = \cos(x) + \lambda \int_0^{\pi/2} \sin(x-t)y(t)dt.$$

(a) Compute the first three terms in the Neumann series of the solution.

(b) For what values of λ can you guarantee that this series converges to a solution of the integral equation?

(c) Approximate $y(\pi/4)$ for $\lambda = 0.1$ and find a bound on the error.

(d) Let T be defined by

$$T(y) = \int_0^{\pi/2} \sin(x-t)y(t)dt.$$

Find an upper bound for the magnitude of the eigenvalues of T.

(e) Compute the eigenvalues of T and compare with part (d) above.

7.2. Let A be an n by n real matrix, I_n the n by n identity matrix. Let matrices be normed according to definition 5.1 in Chapter 3. Show that if $\|A\| < 1$, then

(a) $(I_n - A)^{-1}x = x + Ax + A^2x + \ldots + A^kx + \ldots$ for all $x \in E_n$.
(b) $\lim_{k\to\infty} \|(I_n - A)^{-1} - I_n - A - A^2 - \ldots - A^k\| = 0$.
(c) The (i,j)th element of $I_n + A + \ldots + A^k$ converges to the (i,j)th element of $(I_n - A)^{-1}$ as $k \to \infty$.

7.3. Use the results of Problem 7.2 to find an approximate inverse for the matrix

$$\begin{bmatrix} 0.99 & 0.01 \\ 0.01 & 0.99 \end{bmatrix}.$$

Compare your result with the actual inverse.

8. ITERATIVE IMPROVEMENT IN MATRIX PROBLEMS

Iterative techniques can sometimes be used to obtain a correction for the rounding errors incurred in solving a linear system of equations or in computing the inverse of a matrix. Let us first consider the matrix inverse problem.

Let A be a square matrix and suppose B is an approximation to the inverse of A. The matrix B might have been obtained by the Gauss or Jordan method, but that is not important for the discussion which follows. Now let

$$E = I - BA. \tag{8.1}$$

Note that if $B = A^{-1}$, then $E = \emptyset$ (the zero matrix); so one might consider E to be a rough measure of the inaccuracy of B as an approximation to A^{-1}. Solving (8.1) for A^{-1} we obtain

$$A^{-1} = (I - E)^{-1}B. \tag{8.2}$$

If $\|E\| < 1$ (using the norm given by definition 5.1 in Chapter 3), then it can be shown (Problem 7.2) that

$$(I - E)^{-1} = I + E + E^2 + \ldots + E^n + \ldots \doteq I + E.$$

Replacing $(I - E)^{-1}$ by $I + E$ in (8.2) we obtain the approximation

$$A^{-1} \doteq B + EB. \tag{8.3}$$

If B is a first approximation to A^{-1}, then one might expect that $B + EB$ would be a better approximation. This suggests several possibilities for constructing a sequence of approximations to A^{-1}. One such scheme is to form a sequence according to the equations

$$\begin{aligned} X_0 &= B, \\ X_{k+1} &= B + EX_k. \end{aligned} \tag{8.4}$$

Let us now show that this sequence converges to A^{-1} if B is nonsingular and $\|E\| < 1$.

THEOREM 8.1. *Suppose A and B are n by n matrices and $E = I - BA$. If B is non-singular and $\|E\| < 1$, then A^{-1} exists and the sequence defined by (8.4) converges to A^{-1}. Furthermore,*

$$\|A^{-1} - X_k\| \leqslant \frac{\|B\|\,\|E\|^{k+1}}{1 - \|E\|}. \tag{8.5}$$

Proof. We will outline the proof while leaving several important details as exercises (Problem 8.4). Let $M(n,n)$ be the collection of all n by n matrices, and let T be the transformation defined by

$$T(X) = EX$$

for each $X \in M(n,n)$. Now T is linear and we can show that:

(i) $\|AB\| \leqslant \|A\|\,\|B\|$ whenever A and $B \in M(n,n)$.

(ii) $\|T\| \leqslant \|E\| < 1$.

(iii) The space $M(n,n)$ is complete with respect to the norm under consideration.

By Theorem 7.1, the sequence (8.4) converges to the unique solution of the equation

$$X = B + EX. \tag{8.6}$$

If X satisfies (8.6), then

$$(I - E)X = B.$$

But $I - E = BA$; so $BAX = B$, $AX = I$, and X is the inverse of A. Thus A^{-1} exists and the sequence (8.4) converges to it. The error estimate (8.5) follows from Theorem 7.2.

Note that Theorem 8.1 does not take into account the round-off error incurred in computing X_k from X_{k-1}. Problem 8.6 indicates how one might analyze this round-off error.

EXAMPLE 8.1. Let A be the 3 by 3 matrix

$$\begin{bmatrix} 2 & 3 & 1 \\ 6 & 4 & 2 \\ 5 & 3 & 1 \end{bmatrix}.$$

Then

$$A^{-1} = \begin{bmatrix} -\frac{1}{3} & 0 & \frac{1}{3} \\ \frac{2}{3} & -\frac{1}{2} & \frac{1}{3} \\ -\frac{1}{3} & \frac{3}{2} & -\frac{5}{3} \end{bmatrix}.$$

Let B be given by

$$B = \begin{bmatrix} -0.3 & 0 & 0.3 \\ 0.6 & -0.4 & 0.3 \\ -0.3 & 1.4 & -1.6 \end{bmatrix}.$$

Using fixed-point arithmetic truncated to four decimals, we have

$$BA = \begin{bmatrix} 0.9 & 0 & 0 \\ 0.3 & 1.1 & 0.1 \\ -0.2 & -0.1 & 0.9 \end{bmatrix}$$

and

$$E = I - BA = \begin{bmatrix} 0.1 & 0 & 0 \\ -0.3 & -0.1 & -0.1 \\ 0.2 & 0.1 & 0.1 \end{bmatrix}.$$

Note that

$$Ex = \begin{bmatrix} e_{1.} \cdot \bar{x} \\ e_{2.} \cdot \bar{x} \\ e_{3.} \cdot \bar{x} \end{bmatrix},$$

where $e_{i.}$ represents the ith row of E and $\bar{x}$ is the conjugate of x. Using the Schwarz inequality and the fact that $\|\bar{x}\| = \|x\|$, we have

$$\begin{aligned} \|Ex\| &= \sqrt{|e_{1.} \cdot \bar{x}|^2 + |e_{2.} \cdot \bar{x}|^2 + |e_{3.} \cdot \bar{x}|^2} \\ &\leqslant \sqrt{\|e_{1.}\|^2 \|\bar{x}\|^2 + \|e_{2.}\|^2 \|\bar{x}\|^2 + \|e_{3.}\|^2 \|\bar{x}\|^2} \\ &= \|x\| \sqrt{0.01 + 0.11 + .06} \\ &= 0.4243\ \|x\|. \end{aligned}$$

Thus $\|E\| \leqslant 0.4243$. Note also that $\det(B) = 0.15 \neq 0$; so B is nonsingular. The conditions of Theorem 8.1 are satisfied and we are justified in applying the iterative procedure (8.4). Further calculations give:

$$X_1 = B + EB = \begin{bmatrix} -0.33 & 0 & 0.33 \\ 0.66 & -0.5 & 0.34 \\ -0.33 & 1.5 & -1.67 \end{bmatrix},$$

$$X_2 = B + EX_1 = \begin{bmatrix} -0.333 & 0 & 0.333 \\ 0.666 & -0.5 & 0.334 \\ -0.333 & 1.5 & -1.667 \end{bmatrix},$$

$$X_3 = B + EX_2 = \begin{bmatrix} -0.3333 & 0 & 0.3333 \\ 0.6666 & -0.5 & 0.3334 \\ -0.3333 & 1.5 & -1.6667 \end{bmatrix},$$

$$X_4 = X_3.$$

This is as far as we can go without carrying more decimal places in the calculations.

Let us now describe an iterative procedure which will to some extent correct for the round-off error incurred in solving a linear system of equations. Consider the system of equations

$$(8.7) \qquad Ax = b,$$

where $A \in M(n,n)$, x and $b \in EK_n$. We will first solve this system using Gaussian elimination. In effect, we find a matrix M such that MA is a triangular matrix, and then replace (8.7) by the equivalent system

$$(8.8) \qquad MAx = Mb.$$

The matrix M, which was not discussed earlier, can be obtained by performing on the identity matrix all of the row operations used in reducing A to triangular form. Now proceed to solve the system (8.8) in the usual way and denote the solution by x_0. In the ideal situation $x_0 = A^{-1}b$. However, there is usually some round-off error, so that

$$x_0 = B_{-1}b,$$

where B_{-1} is only approximately equal to A^{-1}. Now let $r_0 = b - Ax_0$, solve the system $Ay = r_0$, and denote the solution by y_0. Let $x_1 = x_0 + y_0$ and note that if there were no round-off in computing r_0 and y_0, then

$$Ax_1 = Ax_0 + Ay_0 = Ax_0 + b - Ax_0 = b;$$

that is, x_1 would be a solution to the original system (8.7). For the moment we will assume that r_0 is computed exactly, but that

$$y_0 = B_0 r_0,$$

where B_0 is only approximately equal to A^{-1}. Now we repeat the process with x_1 to obtain a new approximation x_2, and so on; that is, we form sequences $\{r_k\}$, $\{y_k\}$, and $\{x_k\}$ according to the equations

$$(8.9) \qquad \begin{aligned} x_0 &= B_{-1}b, \\ r_k &= b - Ax_k, \\ y_k &= B_k r_k, \\ x_{k+1} &= x_k + y_k, \end{aligned}$$

where B_k is an approximation to A^{-1}. Normally, the matrices MA and M will be computed and stored during the computation of x_0. Thus the reduction of A to triangular form need only be carried out once, and y_k is computed by calculating Mr_k and carrying out the back substitution on the system

$$MAy = Mr_k.$$

Forsythe [5] recommends that the computation of Ax_k be carried out in double precision.

We will now show that under certain conditions the sequence $\{x_k\}$ defined by (8.9) converges to a solution of the linear system.

THEOREM 8.2. *Let the sequence $\{x_k\}$ satisfy the equations* (8.9). *If A^{-1} exists and there is a constant c such that*

$$\|I - AB_k\| \leqslant c < 1$$

for all $k = -1, 0, 1, 2, \ldots$, then $\{x_k\}$ converges to a solution of the equation $Ax = b$.

Proof. Let us first establish that

$$\|r_k\| \leqslant c^{k+1} \|b\| \tag{8.10}$$

and

$$\|B_k\| \leqslant \|A^{-1}\| (1 + c) \tag{8.11}$$

for all $k = 0, 1, 2, \ldots$. Now

$$r_0 = b - Ax_0 = b - AB_{-1}b = (I - AB_{-1})b.$$

Hence

$$\|r_0\| \leqslant \|I - AB_{-1}\| \|b\| \leqslant c \|b\|.$$

In general,

$$\begin{aligned} r_{k+1} &= b - Ax_{k+1} = b - Ax_k - Ay_k \\ &= r_k - AB_k r_k = (I - AB_k)r_k. \end{aligned}$$

Hence

$$\|r_{k+1}\| \leqslant \|I - AB_k\| \|r_k\| \leqslant c \|r_k\|$$

and (8.10) follows by induction. To establish (8.11), let $E_k = I - AB_k$ and note that

$$B_k = A^{-1}(I - E_k).$$

Thus

$$\begin{aligned} \|B_k\| &\leqslant \|A^{-1}\| \|I - E_k\| \\ &\leqslant \|A^{-1}\| (\|I\| + \|E_k\|). \end{aligned}$$

The inequalities (8.11) then follow from the fact that $\|I\| = 1$ and $\|E_k\| \leqslant c$.

Next, we will show that the series

$$(8.12) \qquad x_0 + y_0 + y_1 + \ldots + y_k + \ldots$$

converges. In fact, letting $m = \|A^{-1}\| \, (1 + c)$ and using (8.10) and (8.11), we have

$$\begin{aligned} \|x_0\| + \|y_0\| + \ldots + \|y_k\| + \ldots \\ &= \|x_0\| + \|B_0 r_0\| + \ldots + \|B_k r_k\| + \ldots \\ &\leqslant \|x_0\| + m\|b\| \, (c + c^2 + \ldots + c^{k+1} + \ldots). \end{aligned}$$

This last series, being geometric with ratio less than 1, is convergent. The convergence of (8.12) then follows from the fact that EK_n is complete.

Let x be the sum of the series (8.12). Note that

$$x_k = x_0 + y_0 + \ldots + y_{k-1}$$

is a partial sum, and hence

$$x = \lim_{k \to \infty} x_k .$$

It remains to show that $Ax = b$. Now

$$\begin{aligned} \|Ax - b\| &= \|Ax - Ax_k + Ax_k - b\| \\ &= \|A(x - x_k) - r_k\| \\ &\leqslant \|A\| \, \|x - x_k\| + \|r_k\| \to 0 \end{aligned}$$

as $k \to \infty$. Thus $Ax = b$ and the proof is complete.

Theorem 8.2 does not take into account the rounding error involved in computing $r_k = b - Ax_k$. However, if this computation is carried out in double precision, then it seems reasonable that this round-off error would be negligible for the first few values of k. An error analysis based on the calculation of the number c in Theorem 8.2 appears to be difficult because the matrices $I - AB_k$ are not actually computed. Problem 8.8 gives some hints on how to relate the error in x_k to the condition number of A.

EXAMPLE 8.2. In this example we will use floating-point arithmetic with three significant digits. Additions and subtractions will be carried out to six significant figures and the result will be rounded back to three. Let us solve the matrix equation

$$Ax = b,$$

where

$$A = \begin{bmatrix} 4.21 & 2.75 \\ 2.85 & 1.03 \end{bmatrix}, \qquad b = \begin{bmatrix} 0.17 \\ 2.61 \end{bmatrix}.$$

In order to compute the matrix M in (8.8), we attach the identity to the augmented matrix

$$\begin{bmatrix} 4.21 & 2.75 & 0.17 & 1 & 0 \\ 2.85 & 1.03 & 2.61 & 0 & 1 \end{bmatrix}.$$

Adding -0.677 times the first row to the second row we have

$$\begin{bmatrix} 4.21 & 2.75 & 0.17 & 1 & 0 \\ 0 & -0.83 & 2.50 & -0.677 & 1 \end{bmatrix}.$$

Thus $x_{02} = -2.50/0.83 = -3.01$, and

$$x_{01} = [0.17 + (2.75)(3.01)]/4.21 = 2.01.$$

Also

$$M = \begin{bmatrix} 1 & 0 \\ -0.677 & 1 \end{bmatrix}.$$

Now, using double precision, we obtain

$$r_0 = b - Ax_0 = \begin{bmatrix} 0.17 \\ 2.61 \end{bmatrix} - \begin{bmatrix} 4.21 & 2.75 \\ 2.85 & 1.03 \end{bmatrix} \begin{bmatrix} 2.01 \\ -3.01 \end{bmatrix}$$

$$= \begin{bmatrix} -0.0146 \\ -0.0182 \end{bmatrix}.$$

Returning to single precision, we have

$$Mr_0 = \begin{bmatrix} -0.0146 \\ -0.00832 \end{bmatrix}.$$

This gives

$$y_0 = \begin{bmatrix} -0.0100 \\ 0.0100 \end{bmatrix}, \qquad x_1 = x_0 + y_0 = \begin{bmatrix} 2.00 \\ -3.00 \end{bmatrix}.$$

It happens that x_1 is the exact solution; but this is not typical.

Further information on iterative procedures for matrices can be found in Varga [15], Faddeev and Faddeeva [3], and Fox [6].

PROBLEMS

8.1. Let

$$A = \begin{bmatrix} 4 & 0.1 & 0.2 \\ 0.1 & 5 & 0 \\ 0.2 & 0 & 6 \end{bmatrix}.$$

(a) Let

$$B = \begin{bmatrix} 0.25 & 0 & 0 \\ 0 & 0.20 & 0 \\ 0 & 0 & 0.16 \end{bmatrix}$$

be an approximation to A^{-1}. Use two iterations of (8.4) to find a (hopefully) better approximation to A^{-1}.

(b) Find a bound on $\|A^{-1} - X_2\|$.

(c) Compute A^{-1} by the adjoint method and compare with part (a) above.

8.2. Find a series expansion for $(A - \Delta)^{-1}$ in terms of A^{-1} and Δ. Explain how this could be used to approximate the inverse of the matrix in Problem 8.1. Would you get the same result as that obtained in part (a) of Problem 8.1?

8.3. Assuming the initial approximation B is available, draw a flow diagram for the iterative scheme (8.4). To make it easy, assume that subroutines MATADD and MATMLT are available for carrying out addition and multiplication of matrices.

8.4. Supply the missing details labeled (i), (ii), and (iii) in the proof of Theorem 8.1.

8.5. An iterative scheme similar to (8.4) is defined by

$$X_0 = B,$$
$$X_{k+1} = X_k(2I - AX_k).$$

Show that this method converges roughly twice as fast as (8.4). Also show that (8.4) takes roughly half as much work per iteration so that the two methods are essentially equivalent. See Froberg [7].

8.6. Suppose that the sequence $\{X_k\}$ defined by (8.4) has "numerically converged". By this we mean that for some k, $X_{k+1} = X_k$, but that various quantities have been rounded in the calculation of X_{k+1}. It follows that

$$X_k = B + EX_k + \Delta,$$

where Δ represents the round-off error. Show that in this case

$$\|A^{-1} - X_k\| \leqslant \frac{\|\Delta\|}{1 - \|E\|}.$$

8.7. Draw a flow diagram for the iterative procedure (8.9). Assume that the subroutines mentioned in Problem 8.3 are available.

8.8. Suppose that the sequence $\{x_k\}$ defined by (8.9) has numerically converged. Show how to relate the relative error in the solution to the condition number of A. Hint: See Problem 8.6 above and Problem 5.5 in Chapter 3.

9. THE POWER METHOD FOR COMPUTING EIGENVALUES

In this section we will describe an iterative technique, called the *power method*, for computing some of the eigenvalues of a matrix. This technique can also be used on certain linear transformations which are not directly related to a matrix, but a discussion in terms of matrices will illustrate the basic ideas of the method.

Let A be an n by n matrix with eigenvalues $\lambda_1, \lambda_2, \ldots, \lambda_n$. We will assume that

$$(9.1) \qquad |\lambda_1| > |\lambda_2| \geqslant |\lambda_3| \geqslant \ldots \geqslant |\lambda_n|,$$

and that there is a linearly independent set of eigenvectors u_1, $u_2, \ldots, u_n$ corresponding to the above eigenvalues.

Now the power method proceeds as follows: Let x_0 be an arbitrary vector in EK_n. Then form a sequence $\{x_k\}$ of vectors according to the recurrence relation.

$$x_{k+1} = Ax_k.$$

Let x_{kj} represent the jth coordinate of x_k, and define a sequence of numbers y_k by setting

$$y_k = \frac{x_{kj}}{x_{k-1,j}},$$

where j, which can depend on k, is chosen so that $x_{k-1,j} \neq 0$. Then, under suitable conditions,

$$\lim_{k\to\infty} y_k = \lambda_1$$

and

$$\lim_{k\to\infty} \frac{x_k}{y_k^k} = u,$$

where u is an eigenvector corresponding to λ_1.

We will not give a formal convergence theorem for the power method, but we hope the discussion which follows will show why the method works. Since $u_1, \ldots, u_n$ are linearly independent, there exist constants $c_1, \ldots, c_n$ such that

$$x_0 = c_1 u_1 + \ldots + c_n u_n.$$

Now $u_1, \ldots, u_n$ are eigenvectors; so

$$Au_1 = \lambda_1 u_1, \ldots, Au_n = \lambda_n u_n.$$

Hence

$$\begin{aligned} x_1 &= Ax_0 = A(c_1 u_1 + \ldots + c_n u_n) \\ &= c_1 \lambda_1 u_1 + \ldots + c_n \lambda_n u_n. \end{aligned}$$

Similarly,

$$x_2 = Ax_1 = A(c_1\lambda_1 u_1 + \ldots + c_n\lambda_n u_n)$$
$$= c_1\lambda_1^2 u_1 + \ldots + c_n\lambda_n^2 u_n.$$

Continuing in this fashion, we see that

$$x_k = c_1\lambda_1^k u_1 + \ldots + c_n\lambda_n^k u_n.$$

It follows that

$$y_k = \frac{x_{kj}}{x_{k-1,j}} = \frac{c_1\lambda_1^k u_{1j} + c_2\lambda_2^k u_{2j} + \ldots + c_n\lambda_n^k u_{nj}}{c_1\lambda_1^{k-1} u_{1j} + c_2\lambda_2^{k-1} u_{2j} + \ldots + c_n\lambda_n^{k-1} u_{nj}}.$$

Assuming that $c_1 u_{1j} \neq 0$, we can factor the numerator and denominator to obtain

$$y_k = \frac{\lambda_1\left(1 + \frac{c_2}{c_1}\left(\frac{\lambda_2}{\lambda_1}\right)^k \frac{u_{2j}}{u_{1j}} + \ldots + \frac{c_n}{c_1}\left(\frac{\lambda_n}{\lambda_1}\right)^k \frac{u_{nj}}{u_{1j}}\right)}{1 + \frac{c_2}{c_1}\left(\frac{\lambda_2}{\lambda_1}\right)^{k-1} \frac{u_{2j}}{u_{1j}} + \ldots + \frac{c_n}{c_1}\left(\frac{\lambda_n}{\lambda_1}\right)^{k-1} \frac{u_{nj}}{u_{1j}}}.$$

But from (9.1),

$$\lim_{k\to\infty}\left(\frac{\lambda_i}{\lambda_1}\right)^k = 0, \qquad i = 2, 3, \ldots, n.$$

Hence $\lim_{k\to\infty} y_k = \lambda_1$.

In a similar manner, we see that

$$\frac{x_k}{y_k^k} \doteq \frac{x_k}{\lambda_1^k} = c_1 u_1 + c_2\left(\frac{\lambda_2}{\lambda_1}\right)^k u_2 + \ldots + c_n\left(\frac{\lambda_n}{\lambda_1}\right)^k u_n \doteq c_1 u_1.$$

But $c_1 u_1$ is an eigenvector of A corresponding to λ_1 (see Problem 3.3 in Chapter 2); so x_k/y_k^k can be used as an approximate eigenvector.

EXAMPLE 9.1. Let us apply the power method to the matrix

$$A = \begin{bmatrix} 1 & 2 \\ 1 & 1 \end{bmatrix}.$$

Let us arbitrarily take

$$x_0 = \begin{bmatrix} 1 \\ 1 \end{bmatrix}.$$

Then

$$x_1 = Ax_0 = \begin{bmatrix} 3 \\ 2 \end{bmatrix},$$

$$x_2 = Ax_1 = \begin{bmatrix} 7 \\ 5 \end{bmatrix},$$

$$x_3 = Ax_2 = \begin{bmatrix} 17 \\ 12 \end{bmatrix},$$

$$x_4 = Ax_3 = \begin{bmatrix} 41 \\ 29 \end{bmatrix}.$$

Hence

$$\lambda_1 \doteq \tfrac{41}{17} = 2.4118 \qquad \text{or} \qquad \lambda_1 \doteq \tfrac{29}{12} = 2.4167.$$

In fact, $\lambda_1 = 1 + \sqrt{2} = 2.4142$. An approximate eigenvector is given by

$$u = \frac{x_4}{(2.4118)^4} = \begin{bmatrix} 1.2118 \\ 0.8571 \end{bmatrix}.$$

The ratio $u_1/u_2 = 1.2118/0.8571 = 1.4138$, while in fact we should have $u_1/u_2 = \sqrt{2} = 1.4142$.

The power method is not in general a good method for obtaining all of the eigenvalues of a moderately large matrix. However, there are tricks available for obtaining the smallest eigenvalue, the next to largest eigenvalue, and so on. The exploration of these ideas is left for an exercise (Problem 9.2).

The power method can also be used in case

$$|\lambda_1| = |\lambda_2| > |\lambda_3| \geqslant |\lambda_4| \geqslant \ldots \geqslant |\lambda_n|.$$

This might happen, for example, if A has conjugate imaginary roots for its largest eigenvalues. For further details on this aspect, as well as the power method in general, see Faddeev and Faddeeva [3].

PROBLEMS

9.1. Apply the power method to approximate the largest eigenvalue of the matrix

$$A = \begin{bmatrix} 1 & 1 \\ 1 & 2 \end{bmatrix}.$$

Also find an approximate eigenvector. Compare your results with exact eigenvalues and eigenvectors.

9.2. Let A be an n by n matrix with eigenvalues $\lambda_1, \ldots, \lambda_n$. Suppose that

$$|\lambda_1| > |\lambda_2| > \ldots > |\lambda_n| > 0.$$

(a) Show that the eigenvalues of $A - \lambda_1 I$ are $0, \lambda_2 - \lambda_1, \ldots, \lambda_n - \lambda_1$. Hence the power method applied to $A - \lambda_1 I$ will yield $\lambda_j - \lambda_1$, where λ_j is the furthest eigenvalue from λ_1.

(b) Show that the eigenvalues of A^{-1} are

$$\frac{1}{\lambda_1}, \frac{1}{\lambda_2}, \ldots, \frac{1}{\lambda_n}.$$

Hence $1/\lambda_n$ is the largest eigenvalue of A^{-1}.

(c) Use (a) to compute the second eigenvalue in Problem 9.1.

(d) Use (b) to compute the second eigenvalue in Problem 9.1.

9.3. Let A be the matrix given in Problem 3.7. Compute an eigenvector corresponding to the largest eigenvalue by means of formula (9.11) in Chapter 3. Explain why this is called the *inverse power method.*

REFERENCES

1. T. M. Apostol, *Mathematical Analysis*, Addison-Wesley, 1957.
2. E. W. Cheney, *Introduction to Approximation Theory*, McGraw-Hill, 1966.
3. D. K. Faddeev and V. N. Faddeeva, *Computational Methods of Linear Algebra*, Freeman, 1963.
4. G. E. Forsythe, "Singularity and near singularity in numerical analysis," *Amer. Math. Monthly*, 65 (1958), 229-240.
5. G. E. Forsythe, "Today's computational methods of linear algebra," *SIAM Review*, 9 (1967), 489-515.
6. L. Fox, *An Introduction to Numerical Linear Algebra*, Oxford, 1965.
7. C.-E. Froberg, *Introduction to Numerical Analysis*, Addison-Wesley, 1965.
8. P. Henrici, *Elements of Numerical Analysis*, Wiley, 1964.
9. A. S. Householder, *Principles of Numerical Analysis*, McGraw-Hill, 1953.
10. E. Isaacson and H. B. Keller, *Analysis of Numerical Methods*, Wiley, 1966.
11. T. E. Mott, "Newton's method and multiple roots," *Amer. Math. Monthly*, 64 (1957), 635–638.
12. A. Ostrowski, "Konvergenzdiskussion und Fehlerabschätzung für die Newton'sche Methode bei Gleichungssystemen," *Comm. Math. Helvetici*, 9 (1936), 79–103.
13. A. Ralston and H. S. Wilf, *Mathematical Methods for Digital Computers*, Volume 2, Wiley, 1967.
14. J. K. Stewart, "Another variation of Newton's method," *Amer. Math. Monthly*, 58 (1951), 331-334.
15. R. S. Varga, *Matrix Iterative Analysis*, Prentice-Hall, 1962.

CHAPTER 5

Interpolation

1. THE LINEAR INTERPOLATION PROBLEM

The word *interpolation* has several meanings in mathematics. Historically, it referred to the process of approximating the value of a function between known values. In modern usage, the word also refers to the process of constructing an approximating function (frequently a polynomial) which agrees with a given function or its derivatives at prescribed points. Of course, the reason for constructing such an approximating function is frequently to interpolate in the first sense of the word.

We will first describe a general linear interpolation problem, and then we will consider several examples of this problem. Let $g_0, g_1, \ldots, g_n$ be linearly independent elements of a linear space X, and let $L_0, L_1, \ldots, L_n$ be linear functionals defined on X. The general linear interpolation problem is to find an element $p \in \text{span}\{g_0, g_1, \ldots, g_n\}$ satisfying the equations

$$
\begin{aligned}
&L_0(p) = y_0,\\
&L_1(p) = y_1,\\
(1.1)\qquad &\vdots\\
&L_n(p) = y_n,
\end{aligned}
$$

where $y_0, y_1, \ldots, y_n$ are prescribed numbers.

EXAMPLE 1.1. Let X be the set of functions whose nth derivative exists at a fixed number x_0, and let $g_0(x) = 1, g_1(x) = x, g_2(x) = x^2, \ldots, g_n(x) = x^n$. Then span $\{g_0, g_1, \ldots, g_n\} = \mathcal{P}_n$, the set of all polynomials of degree $\leqslant n$. Now let $L_0(f) = f(x_0), L_1(f) = f'(x_0), \ldots, L_n(f) = f^{(n)}(x_0)$, where x_0 is a fixed number. In this case the interpolation problem (1.1) is to find a polynomial $p \in \mathcal{P}_n$ such that its value at x_0 is y_0, and its first n derivatives evaluated at x_0 are $y_1, y_2, \ldots, y_n$, respectively. We will call this the *Taylor interpolation problem*. This type of interpolation will be discussed in more detail in Section 6.

EXAMPLE 1.2. Let X be the set of complex-valued functions defined at each of the points $x_0, x_1, \ldots, x_n$, where $x_0, x_1, \ldots, x_n$ are distinct numbers, and let $g_0, g_1, \ldots, g_n$ be as in Example 1.1. Define $L_0(f) = f(x_0), L_1(f) = f(x_1), \ldots, L_n(f) = f(x_n)$. In this case the interpolation problem is to find a polynomial $p \in \mathcal{P}_n$ such that $p(x_0) = y_0$, $p(x_1) = y_1, \ldots, p(x_n) = y_n$. This will be called the *Lagrange interpolation problem*. Sections 2–5 will be concerned with this type of interpolation.

Let us now consider conditions under which the interpolation problem (1.1) has a unique solution. Let $p = a_0 g_0 + a_1 g_1 + \ldots + a_n g_n$ be a typical element of span $\{g_0, g_1, \ldots, g_n\}$. It follows from the linearity of $L_0, L_1, \ldots, L_n$ that the interpolation equations (1.1) are equivalent to the system of equations

$$\begin{aligned} a_0 L_0(g_0) + a_1 L_0(g_1) + \ldots + a_n L_0(g_n) &= y_0, \\ a_0 L_1(g_0) + a_1 L_1(g_1) + \ldots + a_n L_1(g_n) &= y_1, \\ &\vdots \\ a_0 L_n(g_0) + a_1 L_n(g_1) + \ldots + a_n L_n(g_n) &= y_n. \end{aligned} \tag{1.2}$$

Let B represent the matrix whose (i,j)th element is given by

$$b_{ij} = L_i(g_j), \qquad i,j = 0, 1, \ldots, n. \tag{1.3}$$

Then (1.2) can be written in matrix form as

$$Ba = y,$$

where

$$a = \begin{bmatrix} a_0 \\ \vdots \\ a_n \end{bmatrix} \quad \text{and} \quad y = \begin{bmatrix} y_0 \\ \vdots \\ y_n \end{bmatrix}.$$

Now a system of equations such as (1.2) has a unique solution if and only if B is non-singular. Thus we have the following theorem.

THEOREM 1.1. *The interpolation problem* (1.1) *has a unique solution if and only if the matrix B defined by* (1.3) *is non-singular.*

If B is singular, then the interpolation problem either has no solutions or an infinite number of solutions, depending on the nature of the vector y.

The discussion above shows that one way of constructing a solution to the interpolation problem is to solve the linear system of equations (1.2). Another technique is based on the following theorem.

THEOREM 1.2. *Suppose that* $p_0, p_1, \ldots, p_n$ *are members of* span $\{g_0, g_1, \ldots, g_n\}$ *which satisfy the equations*

$$(1.4) \qquad L_i(p_j) = \delta_{ij}, \qquad i,j = 0, 1, \ldots, n,$$

where δ_{ij} *is* 0 *if* $i \neq j$ *and* 1 *if* $i = j$. *Then a solution to the interpolation problem* (1.1) *is given by the equation*

$$p = y_0 p_0 + y_1 p_1 + \ldots + y_n p_n.$$

Proof. Using the linearity of L_i, we obtain

$$L_i(p) = y_0 L_i(p_0) + y_1 L_i(p_1) + \ldots + y_n L_i(p_n).$$

But $L_i(p_j) = \delta_{ij}$. Hence

$$L_i(p) = y_0 \delta_{i0} + y_1 \delta_{i1} + \ldots + y_n \delta_{in} = y_i,$$

and p satisfies the system (1.1).

As we will see in Sections 2 and 6, it is sometimes fairly easy to construct elements $p_0, p_1, \ldots, p_n$ which satisfy (1.4). The next theorem guarantees that such elements exist if the matrix B is nonsingular.

THEOREM 1.3. *Suppose the matrix B defined by* (1.3) *is nonsingular. Then there exists a unique set of elements* $p_0, p_1, \ldots, p_n$ *from* span $\{g_0, g_1, \ldots, g_n\}$ *which satisfy equations* (1.4).

Proof. Let $C = B^{-1}$ and

$$p_j = c_{0j} g_0 + c_{1j} g_1 + \ldots + c_{nj} g_n, \qquad j = 0, 1, \ldots, n.$$

Using the linearity of L_i and the fact that $L_i(g_k) = b_{ik}$, we obtain

$$(1.5) \qquad L_i(p_j) = c_{0j} b_{i0} + c_{1j} b_{i1} + \ldots + c_{nj} b_{in}.$$

Now the right side of (1.5) is just the (i,j)th element of the matrix BC. But $C = B^{-1}$; hence $L_i(p_j) = \delta_{ij}$. The proof of uniqueness is left as an exercise (Problem 1.3).

EXAMPLE 1.3. Let $n = 2$, $x_0 = 0$, $x_1 = 1$, and $x_2 = 2$ in Example 1.2. Let $q_0(x) = (x-1)(x-2)$, $q_1(x) = x(x-2)$, and $q_2(x) = x(x-1)$. It is easy to see that $L_i(q_j) = 0$ if $i \neq j$, and $L_i(q_j) \neq 0$ if $i = j$. Let

$$p_0(x) = \frac{q_0(x)}{q_0(0)},$$

$$p_1(x) = \frac{q_1(x)}{q_1(1)},$$

$$p_2(x) = \frac{q_2(x)}{q_2(2)} \, .$$

Now $L_i(p_j) = \delta_{ij}$, $i,j = 0, 1, 2$. Note that the elements $p_0, p_1, \ldots, p_n$ which satisfy (1.4) can be constructed for this type of problem without computing B^{-1}.

PROBLEMS

1.1. In Example 1.2 let $n = 2$, $x_0 = -1$, $x_1 = 0$, and $x_2 = 1$.
   (a) Show that the matrix B defined by (1.3) is non-singular and compute B^{-1}.
   (b) Find elements p_0, p_1, and p_2 which satisfy the equations $L_i(p_j) = \delta_{ij}$. Use the method employed in the proof of Theorem 1.3.
   (c) Repeat (b) using the method employed in Example 1.3.
   (d) Find a second-degree polynomial p satisfying the equations $p(-1) = 4$, $p(0) = 7$, $p(1) = 5$.

1.2. In Example 1.1, let $n = 3$ and $x_0 = 0$.
   (a) Show that the matrix B defined by (1.3) is non-singular and compute B^{-1}.
   (b) Find elements p_0, p_1, p_2, and p_3 of $\mathcal{P}_3$ which satisfy the equations $L_i(p_j) = \delta_{ij}$.
   (c) Find a third-degree polynomial p which satisfies the equations $p(0) = 3$, $p'(0) = 8$, $p''(0) = 10$, and $p'''(0) = 1$.

1.3. Prove the uniqueness part of Theorem 1.3.

2. LAGRANGE INTERPOLATION

We will now consider in more detail the problem given in Example 1.2. Let $x_0, x_1, \ldots, x_n$ be distinct numbers, and let $y_0, y_1, \ldots, y_n$ be numbers which are not necessarily distinct. The Lagrange interpolation problem is to find a polynomial p of degree at most n such that

$$\begin{aligned} p(x_0) &= y_0, \\ p(x_1) &= y_1, \\ &\vdots \\ p(x_n) &= y_n. \end{aligned} \tag{2.1}$$

Let $p(x) = a_n x^n + a_{n-1} x^{n-1} + \ldots + a_0$ be a typical polynomial in $\mathcal{P}_n$. Then equations (2.1) can be written as

$$
\begin{aligned}
&a_0 + a_1 x_0 + \ldots + a_n x_0^n = y_0, \\
&a_0 + a_1 x_1 + \ldots + a_n x_1^n = y_1, \\
&\vdots \\
&a_0 + a_1 x_n + \ldots + a_n x_n^n = y_n.
\end{aligned} \tag{2.2}
$$

Now the interpolation problem has a unique solution if the matrix of coefficients in (2.2) is non-singular.

THEOREM 2.1. *If $x_0, x_1, \ldots, x_n$ are distinct numbers (real or imaginary), then the interpolation problem* (2.1) *has a unique solution.*

Proof. (Davis [4]*). For each integer j, $1 \leqslant j \leqslant n$, let

$$
v_j(x_0, x_1, \ldots, x_j) = \begin{vmatrix} 1 & x_0 & x_0^2 & \ldots & x_0^j \\ 1 & x_1 & x_1^2 & \ldots & x_1^j \\ \vdots & & & & \\ 1 & x_j & x_j^2 & \ldots & x_j^j \end{vmatrix}. \tag{2.3}
$$

We will show by induction that $v_j(x_0, x_1, \ldots, x_j) \neq 0$ for each j. First note that

$$
v_1(x_0, x_1) = \begin{vmatrix} 1 & x_0 \\ 1 & x_1 \end{vmatrix} = x_1 - x_0 \neq 0,
$$

since x_0 and x_1 are distinct. Next, let

$$
q(x) = v_j(x_0, x_1, \ldots, x_{j-1}, x)
$$

and note that $q \in \mathcal{P}_j$. Now

$$
q(x_0) = v_j(x_0, x_1, \ldots, x_{j-1}, x_0) = 0,
$$

since the determinant in (2.3) will have two equal rows when x_j is replaced by x_0. Similarly, $q(x_1) = 0 = q(x_2) = \ldots = q(x_{j-1})$. It follows from the factorization theorem for polynomials [1] that

$$
q(x) = \alpha(x - x_0)(x - x_1) \ldots (x - x_{j-1}),
$$

where α is the coefficient of x^j in $q(x)$. Expanding the determinant in (2.3) (with x_j replaced by x) about the last row, we see that α must equal $v_{j-1}(x_0, x_1, \ldots, x_{j-1})$. Thus

*A shorter proof is outlined in Problem 2.8.

$$
\begin{aligned}
q(x_j) &= v_j(x_0, x_1, \ldots, x_j) \\
&= v_{j-1}(x_0, x_1, \ldots, x_{j-1})(x_j - x_0) \prod_{k=0}^{j=1} (x_j - x_k)
\end{aligned} \tag{2.4}
$$

From the fact that $x_0, x_1, \ldots, x_j$ are distinct numbers, it follows that $v_j(x_0, x_1, \ldots, x_j) \neq 0$ if $v_{j-1}(x_0, x_1, \ldots, x_{j-1}) \neq 0$. By induction, $v_j(x_0, x_1, \ldots, x_j) \neq 0$ for each j.

Now let B represent the matrix of coefficients in the system (2.2). Then $|B| = v_n(x_0, x_1, \ldots, x_n) \neq 0$ and hence the system (2.2) has a unique solution. This completes the proof.

The determinant $v_n(x_0, x_1, \ldots, x_n)$ is known as the *Vandermonde determinant*. It follows readily from (2.4) that

$$
v_n(x_0, x_1, \ldots, x_n) = \prod_{i=1}^{n} \prod_{j=0}^{i-1} (x_i - x_j).
$$

The coefficients of the interpolation polynomial can be found by solving the linear system of equations (2.2) for $a_0, a_1, \ldots, a_n$. The interpolation polynomial can also be found by means of Theorem 1.2. To see this, let

$$
\begin{aligned}
q_0(x) &= (x - x_1)(x - x_2)\ldots(x - x_n), \\
q_1(x) &= (x - x_0)(x - x_2)\ldots(x - x_n), \\
&\vdots \\
q_n(x) &= (x - x_0)(x - x_1)\ldots(x - x_{n-1}).
\end{aligned}
$$

It is easy to see that $q_i(x_j)$ is zero if $i \neq j$ and not zero if $i = j$. Now let $l_j(x) = q_j(x)/q_j(x_j)$ and note that $l_j(x_i) = \delta_{ij}$. Thus the solution to the interpolation problem is given by

$$
p(x) = \sum_{j=0}^{n} y_j l_j(x). \tag{2.5}
$$

EXAMPLE 2.1. Let $x_0 = 5$, $x_1 = 6$, $x_2 = 7$, $y_0 = 15$, $y_1 = 10$, and $y_2 = 20$. The interpolation polynomial can be found by solving the system of equations

$$
\begin{aligned}
a_0 + 5a_1 + 25a_2 &= 15, \\
a_0 + 6a_1 + 36a_2 &= 10, \\
a_0 + 7a_1 + 49a_2 &= 20.
\end{aligned}
$$

Another way to obtain the interpolation polynomial is to use (2.5). This gives

$$p(x) = \frac{15(x-6)(x-7)}{(5-6)(5-7)} + \frac{10(x-5)(x-7)}{(6-5)(6-7)} + \frac{20(x-5)(x-6)}{(7-5)(7-6)}$$

$$= \frac{15}{2}x^2 - \frac{175}{2}x + 265.$$

Let f be a real or complex-valued function defined at the distinct points $x_0, x_1, \ldots, x_n$. By choosing $y_j = f(x_j)$, $j = 0, 1, \ldots, n$, we can see from Theorem 2.1 that there is a unique polynomial $p \in \mathcal{P}_n$ such that

$$p(x_i) = f(x_i), \qquad i = 0, 1, \ldots, n.$$

The polynomial p is called the *Lagrange interpolation polynomial for f at $x_0, x_1, \ldots, x_n$*, and p is said to interpolate f at these points. From (2.5) we have

$$p(x) = \sum_{j=0}^{n} f(x_j) l_j(x).$$

PROBLEMS

2.1. We wish to find a polynomial $p \in \mathcal{P}_2$ such that $p(2) = 8$, $p(3) = 5$, and $p(4) = 10$.
   (a) Compute p by solving the system of equations (2.2).
   (b) Compute p using (2.5).

2.2. Let $\pi(x) = (x - x_0)(x - x_1)\ldots(x - x_n)$. Show that the polynomials l_j used in (2.5) are given by

$$l_j(x) = \frac{\pi(x)}{(x - x_j)\pi'(x_j)} .$$

2.3. Let $f(x) = e^x$. Write out (but do not simplify) an expression for the polynomial $p \in \mathcal{P}_2$ which interpolates f at 0, 0.5, and 1.

2.4. Let $q \in \mathcal{P}_n$. Show that if p interpolates q at the distinct points $x_0, x_1, \ldots, x_n$, then $p = q$.

2.5. Let X be the collection of all complex-valued functions which are defined at the distinct points $x_0, x_1, \ldots, x_n$. For each $f \in X$, let $L(f)$ be the Lagrange interpolation polynomial for f at $x_0, x_1, \ldots, x_n$. Show that L is a linear operator.

2.6. (a) Write a Fortran program to calculate the coefficients of the interpolation polynomial satisfying (2.1). Assume $x_0, \ldots, x_n, y_0, \ldots, y_n$ are real, and that a subprogram called GAUSS4 is available for solving linear systems of equations.
   (b) Use the result of Problem 2.4 to make up several test problems for checking out your program.

2.7. The polynomial p from Example 2.1 can be expressed in at least three ways:
   (i) $p(x) = 7.5(x-6)(x-7)$
   $\quad + 10[(x-5)(x-6) - (x-5)(x-7)]$

(ii) $p(x) = 7.5x^2 - 87.5x + 265$
(iii) $p(x) = x(7.5x - 87.5) + 265$
Which formula will require the shortest computing time if multiplications require 10 μ sec each, and additions (or subtractions) require 5 μ sec each?

2.8. Prove Theorem 2.1 by showing that if the matrix of coefficients in (2.2) were singular, then there would be a non-zero polynomial in $\mathcal{P}_n$ with more than n zeros.

3. ERROR BOUNDS

Let $x_0, x_1, \ldots, x_n$ be real numbers, let f be a real-valued function defined on an interval containing $x_0, x_1, \ldots, x_n$, and let p be the unique polynomial in $\mathcal{P}_n$ such that

$$p(x_i) = f(x_i), \qquad i = 0, 1, 2, \ldots, n.$$

It would seem reasonable to use p to approximate f. We will now derive a formula which can be used to find a bound on the error incurred by this approximation.

THEOREM 3.1. *Suppose that f is a real-valued continuous function on the interval $[a,b]$, that $f^{(n+1)}$ exists on (a,b), and that $x_0, x_1, \ldots, x_n$ are distinct numbers in $[a,b]$. Let p be the unique polynomial in $\mathcal{P}_n$ which interpolates f at these points. If $x \in [a,b]$, then*

$$(3.1) \qquad f(x) - p(x) = \frac{f^{(n+1)}(\xi)}{(n+1)!} \prod_{j=0}^{n} (x - x_j)$$

for some number ξ lying between $\min(x, x_0, x_1, \ldots, x_n)$ *and* $\max(x, x_0, x_1, \ldots, x_n)$. (In general, ξ depends on x.)

Proof. Let x be a fixed but arbitrary number in $[a,b]$. If x is one of the interpolation points, then (3.1) is satisfied since both sides are zero. Otherwise, let $\pi(t) = (t - x_0)(t - x_1)\ldots(t - x_n)$ and

$$(3.2) \qquad \varphi(t) = f(t) - p(t) - \frac{\pi(t)}{\pi(x)}\,[f(x) - p(x)].$$

Now $\varphi(x) = 0 = \varphi(x_0) = \varphi(x_1) = \ldots = \varphi(x_n)$. According to Rolle's theorem, φ' must have a zero between each two zeros of φ. Thus φ' has at least $n+1$ distinct zeros on (a,b). Applying Rolle's theorem to φ', we see that φ'' has at least n distinct zeros. Continuing in this manner, we infer that $\varphi^{(n+1)}$ has at least one zero which is between $\min(x, x_0, \ldots, x_n)$ and $\max(x, x_0, \ldots, x_n)$. Call this zero ξ. Now $p^{(n+1)}(\xi) = 0$ and $\pi^{(n+1)}(\xi) = (n+1)!$. By differentiating (3.2) we obtain the equation

$$0 = \varphi^{(n+1)}(\xi) = f^{(n+1)}(\xi) - \frac{(n+1)!\,[f(x) - p(x)]}{\pi(x)}$$

and (3.1) follows.

EXAMPLE 3.1. Let $f(x) = ln\ x$ and let p be the polynomial in $\mathcal{P}_2$ which interpolates f at the points 1.25, 1.50, and 1.75. Now $f^{(3)}(x) = 2/x^3$ and thus (3.1) yields the equation

$$ln\ x - p(x) = \frac{(x - 1.25)(x - 1.50)(x - 1.75)}{3\xi^3}. \tag{3.3}$$

Suppose, for example, we wish to approximate $ln(1.35)$. Then ξ lies between 1.25 and 1.75. From (3.3) we have

$$ln(1.35) - p(1.35) = \frac{(0.1)(-0.15)(-0.40)}{3\xi^3} = \frac{0.002}{\xi^3}.$$

It follows that

$$\frac{0.002}{(1.75)^3} < ln(1.35) - p(1.35) < \frac{0.002}{(1.25)^3}.$$

Thus the error is at least 0.000373 but not larger than 0.001024.

Frequently we want to find an upper bound on the magnitude of the error which will be valid over an entire interval. For example, we might want a bound on $|ln\ x - p(x)|$ which would be valid for all $x \in [1,2]$. In this case ξ in (3.3) would be between 1 and 2. Hence, if $x \in [1,2]$, then

$$|ln\ x - p(x)| \leqslant \frac{1}{3} \max_{x \in [1,2]} |\pi(x)|,$$

where $\pi(x) = (x - 1.25)(x - 1.50)(x - 1.75)$. A crude bound on $|\pi(x)|$, valid for $x \in [1,2]$, is $(3/4)^2 \cdot 1/2$. A better bound can be obtained by computing the maximum of $|\pi(x)|$. To do this, let z_1 and z_2 be the two zeros of π' which lie in $[1,2]$. (Why are these two zeros in $[1,2]$?) Then

$$\max_{x \in [1,2]} |\pi(x)| = \max\,(|\pi(1)|, |\pi(z_1)|, |\pi(z_2)|, |\pi(2)|).$$

The maximum value of $\prod_{j=0}^{n} (x - x_j)$ for equally spaced interpolation points is tabulated in [6]. In the next section we will discuss some non-equally spaced interpolation points which give the smallest possible maximum value for this term.

PROBLEMS

A complete reference for the CRC Tables mentioned in Problems 3.1 and 3.2 is given at the end of this chapter [7]. Any edition should be satisfactory.

3.1. The value of $e^{1.0143}$ is to be approximated from the CRC Exponential Functions Table by linear (first-degree) interpolation between the two tabulated values which bracket 1.0143. Find a bound on the error assuming:
(a) Exact table entries and no round-off error in the computation.
(b) No round-off error in the computation, but taking into account the rounding of the table entries.
(c) The intermediate results in the computation are rounded to four decimal places, and taking into account the rounding of the table entries.

3.2. The value of $\log_{10}x$ is to be computed using linear interpolation between the values which bracket x in the CRC Five-Place Logarithms Table. Find a bound on the error in this approximation which will be valid for all x between 1 and 10. Check to see if your result is reasonable by computing the error for several examples with the aid of the Seven-Place Logarithms Table.

3.3. (a) Let $Q(x) = (x-1)(x-2)(x-3)$. Compute:

$$\max_{x \in [0,4]} |Q(x)|.$$

(b) Let $Q_0(x) = (x-0.25)(x-0.50)(x-0.75)$. Compute:

$$\max_{x \in [0,1]} |Q_0(x)|.$$

Hint: Let $x = 4z$ in part (a).

3.4. Let p be the polynomial in $\mathcal{P}_2$ which interpolates the exponential function at 0.25, 0.50, and 0.75. Use Problem 3.3 to find a bound on $|e^x - p(x)|$ valid for all $x \in [0,1]$.

3.5. Let p be the polynomial in $\mathcal{P}_2$ which interpolates the sine function at $\pi/8$, $\pi/4$, and $3\pi/8$. Find a bound on $|\sin x - p(x)|$ valid for all $x \in [0,\pi/2]$.

3.6. Let $P(x) = (x-1)(x-2)(x-3)(x-4)$.
(a) Compute:

$$\max_{x \in [0,5]} |P(x)|.$$

Hint: Use Newton's method to compute the zeros of P'.
(b) Suppose p interpolates f at 0.2, 0.4, 0.6, and 0.8, and that $|f^{(4)}(x)| \leqslant M$ for all $x \in [0,1]$. Find a bound on $|f(x) - p(x)|$ valid for $x \in [0,1]$.

4. INTERPOLATION AT THE ZEROS OF THE CHEBYSHEV POLYNOMIAL

Suppose that we wish to approximate the continuous function f on the interval $[a,b]$ by an interpolation polynomial p from $\mathcal{P}_n$. For each continuous function φ, let

$$\|\varphi\| = \max_{x \in [a,b]} |\varphi(x)|.$$

In this discussion let us agree that p is a better approximation to f

than q iff

$$\|f-p\| < \|f-q\|.$$

In other words, we prefer the polynomial which yields the smaller maximum error over $[a,b]$. Now consider the question: Might one set of interpolation points provide a better approximation to f than another set? The answer is yes. In fact, it will be shown in Chapter 7 that there is an interpolation polynomial $p^* \in \mathcal{P}_n$ such that

$$\|f-p^*\| = \min_{p\in\mathcal{P}_n} \|f-p\|.$$

The optimal interpolation points—those which yield p^*—depend on f and are somewhat difficult to compute. For the moment we will only describe what might be termed a good set of interpolation points.

Suppose $p \in \mathcal{P}_n$ interpolates f at the distinct points $x_0, x_1, \ldots, x_n$. Let

$$\varphi(x) = (x-x_0)(x-x_1)\ldots(x-x_n).$$

From (3.1) we have

$$f(x) - p(x) = \frac{f^{(n+1)}(\xi(x))\varphi(x)}{(n+1)!}.$$

Now the term $f^{(n+1)}(\xi(x))$ is rather difficult to analyze because little is known about the behavior of $\xi(x)$. What we can do without too much trouble is to minimize the maximum value of $|\varphi(x)|$ over the interval $[a,b]$. It should be emphasized that this will not in general minimize the maximum value of the product

$$|f^{(n+1)}(\xi(x))\varphi(x)|,$$

but only the one factor $|\varphi(x)|$. Let us now establish the fact that $\|\varphi\|$ is smallest when

$$x_j = \tfrac{1}{2}(b+a) + \tfrac{1}{2}(b-a)\cos\frac{(n+\frac{1}{2}-j)\pi}{n+1}, \qquad j=0,1,2,\ldots,n. \tag{4.1}$$

DEFINITION 4.1. *Let* $T_n(x) = \cos(n \arccos x)$ *for each* $x \in [-1,1]$. *The function* T_n *is called the nth-degree Chebyshev polynomial of the first kind.*

THEOREM 4.1. *For each* $x \in [-1,1]$ *and* $n = 1, 2, \ldots$

$$T_{n+1}(x) = 2xT_n(x) - T_{n-1}(x).$$

Proof. Recall the trigonometric identity

$$\cos(A+B) + \cos(A-B) = 2\cos A\cos B. \tag{4.2}$$

Let $\theta = \text{arc cos } x$, $A = n\theta$, and $B = \theta$. Then (4.2) gives

$$\cos (n + 1)\theta + \cos (n - 1)\theta = 2 \cos n\theta \cos \theta.$$

Now use the definition of T_n and the fact that $x = \cos \theta$ to obtain

$$T_{n+1}(x) + T_{n-1}(x) = 2xT_n(x),$$

and the result follows.

THEOREM 4.2. *The Chebyshev polynomial $T_n(x)$ is indeed a polynomial of degree n, and the coefficient of x^n is 2^{n-1}.*

Proof. This follows readily from Theorem 4.1 by using mathematical induction.

THEOREM 4.3. *Let*

$$x_j = \cos \frac{[n + \frac{1}{2} - j]\pi}{n + 1}, \qquad j = 0, 1, \ldots, n;$$

$$y_j = \cos \frac{(n + 1 - j)\pi}{n + 1}, \qquad j = 0, 1, \ldots, n+1.$$

Then

(i) $T_{n+1}(x_j) = 0, \qquad j = 0, 1, \ldots, n;$

(ii) $T_{n+1}(y_j) = (-1)^{n+1-j}, \qquad j = 0, 1, \ldots, n+1;$

(iii) $\|T_{n+1}\| = 1$ *when* $[a,b] = [-1,1]$.

Proof. Note that $(n + 1) \text{ arc cos } x_j = [2(n-j) + 1]\pi/2$ and $(n + 1) \text{ arc cos } y_j = (n + 1 - j)\pi$. Parts (i) and (ii) now follow from the definition of T_{n+1}. Obviously,

$$\|T_{n+1}\| = \max_{x \in [-1,1]} |\cos [(n+1) \text{ arc cos } x]| \leqslant 1.$$

But from (ii) it follows that $\|T_{n+1}\| = 1$.

THEOREM 4.4. *Of all polynomials in $\mathcal{P}_{n+1}$ whose coefficient of x^{n+1} is 1, $T_{n+1}/2^n$ has the smallest maximum absolute value on $[-1,1]$.*

Proof. Suppose $q \in \mathcal{P}_{n+1}$ has a smaller maximum absolute value. Let y_j be as in Theorem 4.3. Then

$$2^n q(y_{n+1}) < T_{n+1}(y_{n+1}) = 1;$$

otherwise we would have $\|q\| \geqslant 2^{-n} = \|2^{-n} T_{n+1}\|$. Similarly,

$$2^n q(y_n) > T_{n+1}(y_n) = -1.$$

In general, we have

$$(-1)^{n+1-j}[2^n q(y_j) - T_{n+1}(y_j)] < 0, \qquad j = 0, 1, \ldots, n+1.$$

Thus the polynomial $2^n q - T_{n+1}$ alternates sign at the $n+2$ points, $y_0, y_1, \ldots, y_{n+1}$. It follows that $2^n q - T_{n+1}$ has at least $n+1$ zeros on $[-1,1]$. But $2^n q - T_{n+1} \in \mathcal{P}_n$, since the coefficient of x^{n+1} is 2^n for both polynomials. Now a polynomial in $\mathcal{P}_n$ can have at most n zeros unless it is the zero polynomial. Hence $2^n q = T_{n+1}$. This contradicts the assumption that q has a smaller norm than $T_{n+1}/2^n$, and the result follows.

Theorem 4.4 can be extended to an arbitrary interval $[a,b]$ by using the change of variable

$$z = \frac{2x - b - a}{b - a}, \tag{4.3}$$

which is equivalent to

$$x = [b + a + z(b - a)]/2. \tag{4.4}$$

Note that as x increases from a to b, z increases continuously from -1 to 1, and vice versa.

THEOREM 4.5. *Let $T^*_{n+1}(x) = (x - x_0)(x - x_1)\ldots(x - x_n)$, where $x_0, \ldots, x_n$ are given by (4.1). Of all polynomials in $\mathcal{P}_{n+1}$ whose leading coefficient is 1, T^*_{n+1} has the smallest maximum absolute value on $[a,b]$. This smallest maximum is given by*

$$\|T^*_{n+1}\| = \frac{(b-a)^{n+1}}{2^{2n+1}}.$$

Proof. Let $\varphi \in \mathcal{P}_{n+1}$ be a polynomial with leading coefficient 1. Let x and z satisfy (4.3) and define

$$q(z) = \varphi(x) = \varphi\left[\frac{b + a + z(b-a)}{2}\right].$$

Now q is a polynomial of degree $n+1$ with a leading coefficient of $[(b-a)/2]^{n+1}$. From Theorem 4.4 it follows that

$$\max_{z \in [-1,1]} |q(z)| \geqslant \frac{(b-a)^{n+1}}{2^{2n+1}}.$$

But, as z increases from -1 to 1, x increases continuously from a to b. Thus

$$\max_{x \in [a,b]} |\varphi(x)| = \max_{z \in [-1,1]} |q(z)| \geqslant \frac{(b-a)^{n+1}}{2^{2n+1}}.$$

Let

$$z_j = \cos \frac{[n + \frac{1}{2} - j]\pi}{n + 1}$$

be the jth zero of T_{n+1}. Then

$$\begin{aligned} x - x_j &= x - \tfrac{1}{2}(b + a) - \tfrac{1}{2}(b - a)z_j \\ &= \frac{b-a}{2}\left[\frac{2x - b - a}{b - a} - z_j\right] \\ &= \frac{b-a}{2}[z - z_j]. \end{aligned}$$

Thus

$$\begin{aligned} T^*_{n+1}(x) &= (x - x_0)(x - x_1)\ldots(x - x_n) \\ &= \left(\frac{b-a}{2}\right)^{n+1}(z - z_0)(z - z_1)\ldots(z - z_n) \\ &= \frac{(b-a)^{n+1}}{2^{2n+1}}\, T_{n+1}(z). \end{aligned}$$

Thus

$$\begin{aligned} \max_{x\in[-1,1]} |T^*_{n+1}(x)| &= \frac{(b-a)^{n+1}}{2^{2n+1}} \max_{z\in[-1,1]} |T_{n+1}(z)| \\ &= \frac{(b-a)^{n+1}}{2^{2n+1}}, \end{aligned}$$

and the result follows.

It follows from Theorem 4.5 that if the interpolation points satisfy (4.1), then the part of the error bound corresponding to

$$\|\varphi\| = \max_{x\in[a,b]} |(x - x_0)(x - x_1)\ldots(x - x_n)|$$

will have its smallest possible value. The following corollary summarizes the information needed to apply these results to interpolation problems.

COROLLARY 4.6. *Let $x_0, x_1, \ldots, x_n$ be given by (4.1) and suppose p interpolates f at these points. If f is continuous on $[a,b]$, $f^{(n+1)}(x)$ exists, and $|f^{(n+1)}(x)| \leqslant M$ for all $x \in (a,b)$, then*

$$(4.5) \qquad |f(x) - p(x)| \leqslant \frac{M(b-a)^{n+1}}{(n+1)!\, 2^{2n+1}}$$

for every $x \in [a,b]$.

EXAMPLE 4.1. Let us approximate e^x on the interval $[0,1]$ using a third-degree interpolation polynomial. From (4.1) the "good" interpolation points are given by

$$x_j = \frac{1}{2} + \frac{1}{2} \cos \frac{(7-2j)\pi}{8}.$$

This gives approximately

$$x_0 = 0.03806,$$
$$x_1 = 0.30866,$$
$$x_2 = 0.69134,$$
$$x_3 = 0.96194.$$

The interpolation polynomial can now be computed either by using (2.5) or by solving the system (2.2). A bound on the error, obtained from (4.5), is

$$|e^x - p(x)| \leqslant \frac{e}{4!2^7} = 8.85 \cdot 10^{-4}.$$

This bound is valid for all $x \in [0,1]$.

PROBLEMS

4.1. We wish to approximate $ln\ x$ on the interval $[2,4]$ by an nth-degree interpolation polynomial.
   (a) Find the interpolation points given by (4.1) if $n = 5$.
   (b) Find a bound on the error if the points found in (a) are used.
   (c) How large should n be in order to have a maximum error of at most 10^{-8}?

4.2. How large should n be so that the nth-degree interpolation polynomial for $\sin x$ (at the points given by (4.1)) will yield a maximum error of 10^{-6} on the interval $[0,\pi/2]$? Find the corresponding interpolation points.

5. DIFFERENCE FORMULAS

There are a number of interpolation formulas based on finite difference operators which can be applied with equally spaced interpolation points. The formulas are convenient for computations carried out by hand or with desk calculators. As such, they are not as popular today as they were before the advent of the electronic computer. We will derive only one such formula—the *Newton forward interpolation formula*. Other formulas can be found in various texts on numerical analysis, for example, Henrici [5].

We will first describe the *forward difference operator*. Let $\{y_j\}$ be a sequence of numbers, and let

$$\Delta y_j = y_{j+1} - y_j.$$

Then Δy_j is called the *first difference of* y. The second difference is defined by

$$\Delta^2 y_j = \Delta(\Delta y)_j = \Delta y_{j+1} - \Delta y_j.$$

Now $\Delta y_{j+1} = y_{j+2} - y_{j+1}$; so $\Delta^2 y_j = y_{j+2} - 2y_{j+1} + y_j$. Similarly, for any positive integer k, the kth difference is defined by

$$\Delta^k y_j = \Delta(\Delta^{k-1} y)_j = \Delta^{k-1} y_{j+1} - \Delta^{k-1} y_j.$$

EXAMPLE 5.1. Let $y_j = j^3$, $j = 0, 1, 2, \ldots$. It is convenient to display some of the differences of y in a difference table as shown below.

j	y_j	Δy_j	$\Delta^2 y_j$	$\Delta^3 y_j$	$\Delta^4 y_j$
0	0	1	6	6	0
1	1	7	12	6	0
2	8	19	18	6	0
3	27	37	24	6	
4	64	61	30		
5	125	91			
6	216				

TABLE 5.1. Difference Table for $y_j = j^3$.

We also want to define the forward difference operator for functions defined over an interval. Let h be a number which will normally remain constant in any one discussion. Then the first difference of a function f is defined by

$$\Delta f(x) = f(x + h) - f(x).$$

The kth difference is defined by

$$\Delta^k f(x) = \Delta(\Delta^{k-1} f)(x).$$

This difference operator can be related to the difference operator for sequences by setting $y_j = f(x + jh)$, $j = 0, 1, 2, \ldots$. Then

$$\Delta^k f(x) = \Delta^k y_0.$$

We will have need for the generalized binomial coefficient. For any real number z and positive integer n, this is defined by the equation

$$\binom{z}{n} = \frac{z(z-1)\ldots(z-n+1)}{n!}.$$

By definition, $\binom{z}{0} = 1$ for all z. In case z is a positive integer,

$$\binom{z}{n} = \frac{z!}{n!(z-n)!}.$$

We will now show that $\Delta^k y_m$ can be expressed in terms of $y_m, y_{m+1}, \ldots, y_{m+k}$, and the binomial coefficients.

THEOREM 5.1. *Let $\{y_j\}$ be a sequence. Then*

$$\Delta^k y_m = \sum_{j=0}^{k} (-1)^j \binom{k}{j} y_{m+k-j}.$$

Proof. We will prove this by induction. Note that

$$\sum_{j=0}^{1} (-1)^j \binom{1}{j} y_{m+1-j} = y_{m+1} - y_m = \Delta y_m.$$

Thus the formula holds for $k = 1$. Now suppose the formula holds for some k. Then

$$\begin{aligned} \Delta^{k+1} y_m &= \Delta^k y_{m+1} - \Delta^k y_m \\ &= \sum_{j=0}^{k} (-1)^j \binom{k}{j} y_{m+1+k-j} - \sum_{j=0}^{k} (-1)^j \binom{k}{j} y_{m+k-j}. \end{aligned} \tag{5.1}$$

Let $j = \mu - 1$ to obtain

$$\begin{aligned} -\sum_{j=0}^{k} (-1)^j \binom{k}{j} y_{m+k-j} &= -\sum_{\mu=1}^{k+1} (-1)^{\mu-1} \binom{k}{\mu-1} y_{m+k+1-\mu} \\ &= \sum_{j=1}^{k} (-1)^j \binom{k}{j-1} y_{m+k+1-j} \\ &\quad + (-1)^{k+1} y_m. \end{aligned} \tag{5.2}$$

Now separate the $j = 0$ term from the first sum in (5.1), and replace the second sum by (5.2) to obtain

$$\begin{aligned} \Delta^{k+1} y_m &= y_{m+k+1} + \sum_{j=1}^{k} (-1)^j \left[\binom{k}{j} + \binom{k}{j-1}\right] y_{m+k+1-j} \\ &\quad + (-1)^{k+1} y_m \\ &= \sum_{j=0}^{k+1} (-1)^j \binom{k+1}{j} y_{m+k+1-j}. \end{aligned}$$

In the last step we used the identity

$$\binom{k}{j} + \binom{k}{j-1} = \binom{k+1}{j}. \tag{5.3}$$

The proof of (5.3) is left as an exercise (Problem 5.4). Thus the formula holds for $k + 1$ and the proof is complete.

THEOREM 5.2. *If* $\Delta^k y_m = \Delta^k z_m$, $k = 0, 1, \ldots, n$, *then* $y_{m+k} = z_{m+k}$, $k = 0, 1, \ldots, n$. (By $\Delta^0 y_m$ and $\Delta^0 z_m$, we mean y_m and z_m, respectively.)

Proof. This theorem will be proved by induction on n. If $n = 1$, then $y_0 = z_0$ and $y_1 - y_0 = z_1 - z_0$. Hence $y_1 = z_1$, and the theorem holds for $n = 1$. Suppose that it holds for n and that $\Delta^k y_m = \Delta^k z_m$, $k = 0, 1, \ldots, n+1$. By the inductive hypothesis, $y_{m+k} = z_{m+k}$, $k = 0, 1, \ldots, n$. Letting $k = n + 1$ in Theorem 5.1, we obtain the equations

$$\begin{aligned} y_{m+n+1} &= \Delta^{n+1} y_m - \sum_{j=1}^{n+1} (-1)^j \binom{n+1}{j} y_{m+n+1-j} \\ &= \Delta^{n+1} z_m - \sum_{j=1}^{n+1} (-1)^j \binom{n+1}{j} z_{m+n+1-j} = z_{m+n+1}. \end{aligned}$$

Hence $y_{m+n+1} = z_{m+n+1}$ and the proof is complete.

The differences of the binomial coefficients will be needed in order to establish Newton's forward interpolation formula.

THEOREM 5.3. *Let* $f(z) = \binom{z}{n}$ *for a fixed value of* n, *and let* $h = 1$. *Then*

$$\Delta^k f(z) = \binom{z}{n-k}, \qquad k = 0, 1, \ldots, n.$$

The proof of this result is left for an exercise (Problem 5.6). It should be noted that the differences in this theorem are with respect to z rather than n; that is,

$$\Delta f(z) = f(z+h) - f(z) = f(z+1) - f(z).$$

THEOREM 5.4. *Let* f *be defined at* $0, 1, 2, \ldots, n$, *and suppose* $p \in \mathcal{P}_n$ *interpolates* f *at these points. Then*

$$p(z) = \sum_{k=0}^{n} \binom{z}{k} \Delta^k f(0),$$

where the step size h *in the difference operations is taken to be* 1.

Proof. Suppose p interpolates f at $0, 1, \ldots, n$, and

$$q(z) = \sum_{k=0}^{n} \binom{z}{k} \Delta^k f(0) = f(0) + \binom{z}{1} \Delta f(0) + \ldots + \binom{z}{n} \Delta^n f(0). \tag{5.4}$$

Now $q(0) = f(0) = p(0)$, since p interpolates f at 0 and $\binom{0}{k} = 0$ for $k = 1, 2, \ldots, n$. Using Theorem 5.3 on (5.4), we obtain

$$\Delta q(z) = \Delta f(0) + \binom{z}{1} \Delta^2 f(0) + \ldots + \binom{z}{n-1} \Delta^n f(0).$$

Evaluating at $z = 0$, we have

$$\Delta q(0) = \Delta f(0) = \Delta p(0).$$

Continuing in this fashion, one can show that

$$\Delta^k q(0) = \Delta^k p(0), \qquad k = 0, 1, \ldots, n.$$

Then Theorem 5.2 implies that $q(k) = p(k)$, $k = 0, 1, 2, \ldots, n$. But the interpolation polynomial is unique; thus $p = q$ and the proof is complete.

EXAMPLE 5.2. Let $f(z) = z^3$ and suppose we wish to compute the polynomial $p_2 \in \mathcal{P}_2$ which interpolates f at 0, 1, and 2. The differences of f are given in the difference table of Example 5.1. Thus

$$\begin{aligned} p_2(z) &= f(0) + \binom{z}{1}\Delta f(0) + \binom{z}{2}\Delta^2 f(0) \\ &= 0 + z \cdot 1 + \frac{z(z-1)\cdot 6}{2} \\ &= 3z^2 - 2z. \end{aligned}$$

If we wish to compute the polynomial $p_3 \in \mathcal{P}_3$ which interpolates f at 0, 1, 2, 3, we note that the first three terms of p_3 are the same as those for p_2. Hence

$$\begin{aligned} p_3(z) &= p_2(z) + \binom{z}{3}\Delta^3 f(0) \\ &= 3z^2 - 2z + \frac{z(z-1)(z-2)}{6} \cdot 6 \\ &= 3z^2 - 2z + z^3 - 3z^2 + 2z = z^3. \end{aligned}$$

A formula for an arbitrary set of equally spaced interpolation points can be obtained from Theorem 5.4 by the usual change-of-variable trick.

THEOREM 5.5. (Newton's Forward Difference Formula). *Suppose $x_0, x_1, \ldots, x_n$ are equally spaced and $p \in \mathcal{P}_n$ interpolates f at these points. Then*

$$p(x) = \sum_{k=0}^{n} \binom{z}{k} \Delta^k f(x_0),$$

where the step size h in the difference operations is $x_1 - x_0$, and $z = (x - x_0)/h$.

Proof. Let

$$q(z) = \sum_{k=0}^{n} \binom{z}{k} \Delta^k f(x_0).$$

It follows from Theorem 5.4 that $q(j) = f(x_j)$, $j = 0, 1, \ldots, n$. Now let

$$p(x) = q\left(\frac{x - x_0}{h}\right).$$

Note that $(x_j - x_0)/h = j$, since the interpolation points are equally spaced with spacing constant h. Hence

$$p(x_j) = q(j) = f(x_j), \qquad j = 0, 1, \ldots, n,$$

and the proof is complete.

EXAMPLE 5.3. Suppose f has the values 1, 5, 7, 4, −2 at 5.0, 5.5, 6.0, 6.5, 7.0, respectively. We wish to compute the polynomial $p \in \mathcal{P}_4$ which interpolates f at these points. A difference table for f is shown below.

j	x_j	$f(x_j)$	$\Delta f(x_j)$	$\Delta^2 f(x_j)$	$\Delta^3 f(x_j)$	$\Delta^4 f(x_j)$
0	5.0	1	4	−2	−3	5
1	5.5	5	2	−5	2	
2	6.0	7	−3	−3		
3	6.5	4	−6			
4	7.0	−2				

TABLE 5.2. A Difference Table for f.

Let $z = (x - 5)/0.5 = 2x - 10$. Then

$$p(x) = 1 + 4z - 2\frac{z(z-1)}{2} - \frac{3z(z-1)(z-2)}{6} + \frac{5z(z-1)(z-2)(z-3)}{24}. \tag{5.5}$$

Replacing z by $2x - 10$, we obtain

$$p(x) = 1 + 4(2x - 10) - (2x - 10)(2x - 11) - \frac{(2x-10)(2x-11)(2x-12)}{2} + \frac{5(2x-10)(2x-11)(2x-12)(2x-13)}{24}.$$

For many computational purposes it is more convenient to work with the expression (5.5) which gives $p(x)$ in terms of z.

PROBLEMS

5.1. Let $y_j = j^4$, $j = 0, 1, 2, \ldots, 6$. Construct a difference table for this sequence. As a check, $\Delta^5 y_j$ should be zero for any j.

5.2. Let $f(x) = e^x$. Show that $\Delta^n f(x) = e^x(e^h - 1)^n$.

5.3. Compute the following binomial coefficients:

(a) $\binom{20}{2}$ (b) $\binom{10.5}{3}$ (c) $\binom{8}{8}$ (d) $\binom{0.5}{4}$

5.4. Prove equation (5.3).

5.5. Use Theorem 5.1 to compute $\Delta^3 y_0$ if $y_j = j^4$. Check your result with that obtained in Problem 5.1.

5.6. Prove Theorem 5.3.

5.7. Use Theorem 5.4 to construct a polynomial $p \in \mathcal{P}_3$ such that $p(0) = 3$, $p(1) = 8$, $p(2) = 2$, and $p(3) = 12$. How can you check your result?

5.8. Using difference methods, construct a polynomial $p \in \mathcal{P}_2$ such that $p(2) = 7$, $p(2.25) = 1$, and $p(2.5) = 3$. Check your result.

5.9. Let X be the collection of real-valued functions defined on $[0,\infty)$. For a fixed $h > 0$, show that the forward difference operator Δ is a linear mapping from X into X.

6. TAYLOR INTERPOLATION

Let $x_0, y_0, \ldots, y_n$ be real or complex numbers. The Taylor interpolation problem is to find a polynomial $p \in \mathcal{P}_n$ such that

$$p^{(k)}(x_0) = y_k, \qquad k = 0, 1, \ldots, n, \tag{6.1}$$

where $p^{(0)}(x_0)$ is defined to be $p(x_0)$. Normally the y_k's are chosen so that

$$y_k = f^{(k)}(x_0), \qquad k = 0, 1, \ldots, n,$$

where f is a function whose first n derivatives exist at x_0. In this case the solution p to the Taylor interpolation problem is called the *nth-degree Taylor polynomial of f about* x_0. It is easy to establish the fact that a unique solution to the Taylor problem exists.

THEOREM 6.1. *Let $x_0, y_0, \ldots, y_n$ be given numbers. Then there is a unique polynomial $p \in \mathcal{P}_n$ such that $p^{(k)}(x_0) = y_k$, $k = 0, 1, \ldots, n$.*

Proof. Let $p(x) = a_n x^n + a_{n-1}x^{n-1} + \ldots + a_0$ be a typical member of $\mathcal{P}_n$. The system of equations (6.1) can then be written in the form:

$$\begin{aligned}
&a_0 + a_1 x_0 + a_2 x_0^2 + \ldots + a_n x_0^n = y_0, \\
&a_1 + 2a_2 x_0 + \ldots + n a_n x_0^{n-1} = y_1, \\
&2a_2 + \ldots + n(n-1) a_n x_0^{n-2} = y_2, \\
&\vdots \\
&n! a_n = y_n.
\end{aligned}$$

In the matrix notation this system can be written as

$$Ba = y,$$

where

$$B = \begin{bmatrix} 1 & x_0 & x_0^2 & \dots & x_0^n \\ 0 & 1 & 2x_0 & \dots & nx_0^{n-1} \\ 0 & 0 & 2 & \dots & n(n-1)x_0^{n-2} \\ \vdots & & & & \\ 0 & \dots & \dots & 0 & n! \end{bmatrix}.$$

But B is a triangular matrix and hence $|B| = 1 \cdot 1 \cdot 2 \cdot \ldots \cdot n! \neq 0$. Thus B is non-singular and the interpolation problem has a unique solution.

Now let $p_k(x) = (x-x_0)^k/k!$ and note that $p_k^{(i)}(x_0) = \delta_{ik}$. Thus we can apply Theorem 1.2 to obtain the following theorem.

THEOREM 6.2. *Let p be the unique solution to the Taylor interpolation problem* (6.1). *Then*

$$p(x) = \sum_{k=0}^{n} y_k \frac{(x-x_0)^k}{k!}.$$

If p is the nth-degree Taylor polynomial for f about x_0, then

$$p(x) = \sum_{k=0}^{n} f^{(k)}(x_0) \frac{(x-x_0)^k}{k!}. \tag{6.2}$$

EXAMPLE 6.1. Let $f(x) = ln\ x$ and $x_0 = 1$. The nth-degree Taylor polynomial for f about x_0 is given by

$$p(x) = \sum_{k=0}^{n} f^{(k)}(1) \frac{(x-1)^k}{k!}.$$

Now $f^{(k)}(x) = (-1)^{k-1}(k-1)!x^{-k}$ for $k = 1, 2, \ldots$. Hence

$$p(x) = 0 + (x-1) - \frac{(x-1)^2}{2!} + \frac{2(x-1)^3}{3!}$$
$$+ \ldots + \frac{(-1)^{n-1}(n-1)!(x-1)^n}{n!}$$
$$= (x-1) - \frac{(x-1)^2}{2} + \frac{(x-1)^3}{3} + \ldots + \frac{(-1)^{n-1}(x-1)^n}{n}.$$

The next theorem gives a formula which can be used to find a bound on the error in the Taylor approximation to a real-valued function.

THEOREM 6.3. *Suppose $f^{(n+1)}$ exists on the interval $[a,b]$, $x_0 \in [a,b]$, and $p \in \mathcal{P}_n$ is the nth-degree Taylor polynomial for f about x_0. Then for each $x \in [a,b]$*

$$(6.3) \qquad f(x) - p(x) = \frac{f^{(n+1)}(\xi)(x-x_0)^{n+1}}{(n+1)!},$$

where ξ lies between x and x_0. (In general ξ depends on x.)

Proof. Let $\varphi(x) = f(x) - p(x)$ and $g(x) = (x - x_0)^{n+1}$. Note that

$$\begin{aligned} \varphi(x_0) &= g(x_0) = \varphi'(x_0) \\ &= g'(x_0) = \ldots = \varphi^{(n)}(x_0) \\ &= g^{(n)}(x_0) = 0. \end{aligned}$$

Using the second mean-value theorem [2], we have

$$\begin{aligned} \frac{\varphi(x)}{g(x)} &= \frac{\varphi(x) - \varphi(x_0)}{g(x) - g(x_0)} = \frac{\varphi'(\xi_1)}{g'(\xi_1)} = \frac{\varphi'(\xi_1) - \varphi'(x_0)}{g'(\xi_1) - g'(x_0)} \\ &= \frac{\varphi''(\xi_2)}{g''(\xi_2)} = \ldots = \frac{\varphi^{(n+1)}(\xi)}{g^{(n+1)}(\xi)}. \end{aligned}$$

But $g^{(n+1)}(\xi) = (n+1)!$ and $\varphi^{(n+1)}(\xi) = f^{(n+1)}(\xi)$. Hence

$$\frac{\varphi(x)}{g(x)} = \frac{f^{(n+1)}(\xi)}{(n+1)!}$$

and the result follows.

Note the similarity between (6.3) and (3.1). In fact, if we set $x_0 = x_1 = \ldots = x_n$ in (3.1), we obtain (6.3).

EXAMPLE 6.2. Let us approximate $ln(1.01)$ by setting $n = 3$ and $x = 1.01$ in the polynomial obtained in Example 6.1. Then

$$\begin{aligned} ln(1.01) \doteq p(1.01) &= 0.01 - \frac{(0.01)^2}{2} + \frac{(0.01)^3}{3} \\ &= 0.0099\ 5033\ 3. \end{aligned}$$

From (6.3) we have

$$\begin{aligned} |ln(1.01) - p(1.01)| &= \left| \frac{ln^{(4)}(\xi)(.01)^4}{4!} \right| \\ &= \frac{(.01)^4}{4\xi^4} \leqslant \frac{10^{-8}}{4 \cdot 1^4} \\ &= 2.5 \cdot 10^{-9}. \end{aligned}$$

The total error due to round-off and the polynomial approximation is at most $3 \cdot 10^{-9}$.

Suppose we wish to approximate f over the interval $[a,b]$ by its nth-degree Taylor polynomial. Let

$$\varphi(x) = (x - x_0)^{n+1}.$$

The choice $x_0 = (b+a)/2$ will give φ its smallest maximum absolute value over $[a,b]$. This maximum will be $(b-a)^{n+1}/2^{n+1}$; whereas interpolation at the zeros of the Chebyshev polynomial will give a maximum for $|\varphi|$ of $(b-a)^{n+1}/2^{2n+1}$. This indicates that a Lagrange interpolation polynomial would usually provide a better approximation than the Taylor polynomial, and experience has shown that this is the case. Nevertheless the Taylor polynomial is extremely useful for the following reasons:

(i) It is often easier to construct than the Lagrange interpolation polynomial.
(ii) The construction of the Lagrange polynomial requires the values of f at $x_0, x_1, \ldots, x_n$. These are often computed with the aid of a Taylor polynomial.

PROBLEMS

6.1. Approximate $e^{0.2}$ using the fourth-degree Taylor polynomial about zero. Compare your approximation with an exponential table and find a bound on the error.

6.2. Compute cos 3° correct to five decimals using a Taylor polynomial for the cosine about zero.

6.3. Let $f(x) = \sqrt{9+x}$.
(a) Find the third-degree Taylor polynomial for f about zero.
(b) Use this polynomial to approximate $\sqrt{9.1}$.
(c) Find a bound on the error in (b).

6.4. Show that $\sin x < x$ for all $x > 0$.

6.5. Find a Taylor approximation for e^x which will be correct to five decimals on the interval $[-0.5,0.5]$.

6.6. The so-called differential approximation is described by the formula

$$\Delta y \doteq f'(x)\Delta x.$$

Explain how one might obtain a bound on the error for this approximation.

6.7. Use (6.3) and (6.2) to prove the binomial theorem.

6.8. Consider the differential equation

$$\frac{d^2y}{dt^2} + \sin y = 0, \qquad y(0) = 0.05\pi, \qquad y'(0) = 0.$$

Find an approximate solution by replacing $\sin y$ by y. Then compute a bound on the error of this approximation valid for $|y| \leqslant 0.5\pi$. Hint: In order to find an error bound, first show that the initial value problem

$$y'' + y = y - \sin y, \qquad y(0) = 0.05\pi, \qquad y'(0) = 0$$

is equivalent to the integral equation

$$y = 0.05\pi \cos t + \int_0^t [y(u) - \sin y(u)] \sin (t-u)\, du.$$

7. PADÉ APPROXIMATION

A *rational function* is defined as the ratio of two polynomials. These functions are used for approximations because they can normally be evaluated with less time than a polynomial having the same number of parameters, and because they provide better approximations than polynomials for certain types of functions. On the other hand, rational functions are more difficult to differentiate and integrate than polynomials; so they are seldom used for numerical differentiation or integration.

Let $\mathcal{R}(m,n)$ be the class of all rational functions p/q such that $p \in \mathcal{P}_m$ and $q \in \mathcal{P}_n$, and let $x_0, y_0, y_1, \ldots, y_{m+n}$ be given real or complex numbers. The Padé interpolation problem is to find a rational function $r \in \mathcal{R}(m,n)$ such that

$$(7.1) \qquad r^{(k)}(x_0) = y_k, \qquad k = 0, 1, \ldots, m+n.$$

As before, we will usually pick $y_k = f^{(k)}(x_0)$. In this case r will be called the *type* (m,n) *Padé approximation to* f *about* x_0. We should perhaps note that this type of interpolation does not fall under the general linear interpolation problem described in Section 1.

Let us now develop a method for computing the coefficients of a rational function which satisfies (7.1). It simplifies the notation if we first consider the case where $x_0 = 0$. Let

$$\varphi(x) = \sum_{k=0}^{m+n} y_k \frac{x^k}{k!}$$

be the Taylor polynomial which satisfies the equations

$$\varphi^{(k)}(0) = y_k, \qquad k = 0, 1, \ldots, m+n.$$

The next theorem shows how to convert the problem from one involving the derivatives of a quotient to one involving the derivatives of a product.

THEOREM 7.1. *Let $g(x) = \varphi(x)q(x) - p(x)$, $r(x) = p(x)/q(x)$, and suppose $q(0) \neq 0$. Then $r^{(k)}(0) = y_k$, $k = 0, 1, \ldots, m+n$ if and only if $g^{(k)}(0) = 0$, $k = 0, 1, \ldots, m+n$.*

Proof. Let $h(x) = 1/q(x)$ and note that

(7.2) $$\varphi(x) - r(x) = g(x)h(x).$$

Using the Leibnitz rule for differentiating a product [2], we obtain from (7.2)

(7.3) $$\varphi^{(k)}(0) - r^{(k)}(0) = \sum_{j=0}^{k} \binom{k}{j} g^{(j)}(0)h^{(k-j)}(0).$$

Now if $g^{(j)}(0) = 0, j = 0, 1, \ldots, m+n$, then (7.3) implies that $y_k = \varphi^{(k)}(0) = r^{(k)}(0), k = 0, 1, \ldots, m+n$. Conversely, if $y_k = r^{(k)}(0)$, $k = 0, 1, \ldots, m+n$, then the right side of (7.3) will always be zero. Letting $k = 0$ in (7.3) we obtain the equation $0 = g(0)h(0)$. But $h(0) = 1/q(0)$ is not zero by hypothesis. Hence $g(0) = 0$. Similarly, setting $k = 1$ we have

$$0 = g(0)\, h'(0) + g'(0)h(0).$$

But the fact that $g(0) = 0$ and $h(0) \neq 0$ implies that $g'(0) = 0$. One can now readily show by induction that $g^{(k)}(0) = 0, k = 0, 1, \ldots, m+n$, and the proof is complete.

Let

$$p(x) = \sum_{j=0}^{m} a_j x^j, \qquad q(x) = \sum_{j=0}^{n} b_j x^j,$$

and suppose p/q is written in lowest terms (that is, p and q have no common factors). If $p(x) = 0$ for all x we will take $q(x) = 1$ for all x. Otherwise, in order for $p(0)/q(0)$ to be well-defined, we must have $q(0) \neq 0$. The numerator and denominator of p/q can be divided by $b_0 = q(0)$ to obtain an equivalent rational function with $b_0 = 1$. Henceforth, it will always be assumed that $b_0 = 1$.

We will now obtain a linear system of equations which is satisfied by the coefficients of p and q.

THEOREM 7.2. *Let*

$$p(x) = \sum_{j=0}^{m} a_j x^j, \qquad q(x) = \sum_{j=0}^{n} b_j x^j,$$

$b_0 = 1, r = p/q$, *and* $c_j = y_j/j!$. *Also let* $a_j = 0$ *for* $j > m$ *and* $b_j = 0$ *for* $j > n$. *Then*

$$r^{(k)}(0) = y_k, \qquad k = 0, 1, \ldots, m+n$$

if and only if

(7.4) $$a_k - \sum_{j=0}^{k-1} c_j b_{k-j} = c_k, \qquad k = 0, 1, \ldots, m+n.$$

Proof. Let

$$\varphi(x) = \sum_{j=0}^{m+n} c_j x^j = \sum_{j=0}^{m+n} y_j x^j/j!.$$

From Theorem 7.1 the interpolation equations are satisfied if and only if

$$(\varphi q - p)^{(k)}(0) = 0, \qquad k = 0, 1, \ldots, m+n.$$

Now $p^{(k)}(0) = k!a_k$, and from the Leibnitz rule,

$$(\varphi q)^{(k)}(0) = k! \sum_{j=0}^{k} \frac{\varphi^{(j)}(0)}{j!} \frac{q^{(k-j)}(0)}{(k-j)!}$$

$$= k! \sum_{j=0}^{k} c_j b_{k-j}.$$

Thus the interpolation equations are satisfied if and only if for $k = 0$, $1, \ldots, m+n$,

$$(7.5) \qquad 0 = (\varphi q - p)^{(k)}(0) = k! \sum_{j=0}^{k} c_j b_{k-j} - k!a_k.$$

Now divide both sides of (7.5) by $k!$ and use the fact that $b_0 = 1$ to obtain (7.4).

EXAMPLE 7.1. Let $f(x) = (1 + x)^{10}$ and $m = n = 2$. Note that $c_0 = 1$, $c_1 = 10$, $c_2 = 45$, $c_3 = 120$, and $c_4 = 210$. The system (7.4) can be written in matrix form as

$$\begin{bmatrix} 1 & 0 & 0 & 0 & 0 \\ 0 & 1 & 0 & 0 & -1 \\ 0 & 0 & 1 & -1 & -10 \\ 0 & 0 & 0 & -10 & -45 \\ 0 & 0 & 0 & -45 & -120 \end{bmatrix} \begin{bmatrix} a_0 \\ a_1 \\ a_2 \\ b_2 \\ b_1 \end{bmatrix} = \begin{bmatrix} 1 \\ 10 \\ 45 \\ 120 \\ 210 \end{bmatrix}.$$

Solving this system we get $b_1 = -4$, $b_2 = 6$, $a_2 = 11$, $a_1 = 6$, $a_0 = 1$. Thus the type (2,2) Padé approximation to f about 0 is given by

$$r(x) = \frac{11x^2 + 6x + 1}{6x^2 - 4x + 1}.$$

The Padé approximation problem does not always have a solution. To see this, let $m = 1$, $n = 2$, $y_0 = 1$, $y_1 = 1$, $y_2 = 2$, and $y_3 = 18$. Now suppose r is a solution to the system of equations

$$(7.6) \qquad r^{(k)}(0) = y_k, \qquad k = 0, 1, 2, 3.$$

Then r must have a representation of the form

$$(7.7) \qquad r(x) = \frac{a_0 + a_1 x}{1 + b_1 x + b_2 x^2}.$$

By Theorem 7.2 the coefficients must satisfy the system of equations

$$\begin{aligned} a_0 &= 1, \\ a_1 - b_1 &= 1, \\ -b_2 - b_1 &= 1, \\ -b_2 - b_1 &= 3. \end{aligned}$$

But this system of equations is clearly inconsistent. Hence the given approximation problem has no solution.

It turns out that the existence of a solution to the system (7.4) depends only on the last n equations. This will be demonstrated in the following theorem.

THEOREM 7.3. *Let* $c_j = y_j/j!$, $j = 0, 1, \ldots, m+n$, $c_j = 0$ *if* $j < 0$, *and*

$$H = \begin{vmatrix} c_{m-n+1} & c_{m-n+2} & \cdots & c_m \\ c_{m-n+2} & c_{m-n+3} & \cdots & c_{m+1} \\ \vdots & & & \\ c_m & c_{m+1} & \cdots & c_{m+n-1} \end{vmatrix}.$$

If $H \neq 0$, *then the Padé interpolation problem has a unique solution.*

Proof. The system of equations (7.4) can be written in matrix form as follows:

$$\begin{bmatrix} 1 & 0 & 0 & \ldots & 0 & 0 & \ldots & 0 \\ 0 & 1 & 0 & \ldots & 0 & 0 & \ldots & -c_0 \\ \vdots & & & & & & & \vdots \\ 0 & 0 & \ldots & \ldots & 1 & -c_{m-n} & \ldots & -c_{m-1} \\ 0 & 0 & \ldots & \ldots & 0 & -c_{m-n+1} & \ldots & -c_m \\ 0 & 0 & \ldots & \ldots & 0 & -c_{m-n+2} & \ldots & -c_{m+1} \\ \vdots & & & & & \vdots & & \\ 0 & 0 & \ldots & \ldots & 0 & -c_m & \ldots & -c_{m+n-1} \end{bmatrix} \begin{bmatrix} a_0 \\ a_1 \\ \vdots \\ a_m \\ b_n \\ b_{n-1} \\ \vdots \\ b_1 \end{bmatrix} = \begin{bmatrix} c_0 \\ c_1 \\ \vdots \\ c_m \\ c_{m+1} \\ c_{m+2} \\ \vdots \\ c_{m+n} \end{bmatrix}.$$

Let A represent the above coefficient matrix. Expanding the determinant about the first $m + 1$ columns we get

$$|A| = 1 \cdot 1 \cdot \ldots \cdot 1 \cdot (-1)^n H = (-1)^n H.$$

Thus if $H \neq 0$, then $|A| \neq 0$ and the system (7.4) has a unique solution. By Theorem 7.2 the Padé approximation problem then has a unique solution and the proof is complete.

If the determinant H is zero, then the interpolation problem might or might not have a solution. This depends entirely on the consistency of the last n equations in the system (7.4).

There does not appear to be a simple means of finding a bound on the error such as formula (6.3) for the Taylor polynomial. A technique for estimating the error is discussed in Cheney [3].

If $x_0 \neq 0$, then the Padé approximation can be obtained by letting

$$F(z) = f(x_0 + z).$$

Then compute the rational function R associated with F about $z_0 = 0$. The Padé approximation for f about x_0 is then given by $r(x) = R(x - x_0)$. The net result is that the procedure is exactly the same as for $x_0 = 0$ except that p and q are given by

$$p(x) = \sum_{j=0}^{m} a_j(x - x_0)^j$$

and

$$q(x) = \sum_{j=0}^{n} b_j(x - x_0)^j.$$

One of the reasons that rational functions are important in numerical work is that they normally take less time to evaluate on a computer than a polynomial with the same number of parameters. We will illustrate this fact by an example. A more complete discussion can be found in Cheney [3].

EXAMPLE 7.2. Let $r(x) = (11x^2 + 6x + 1)/(6x^2 - 4x + 1)$ be the rational function obtained in Example 7.1. By successive long divisions we have

$$r(x) = \frac{11}{6} + \frac{\frac{40}{3}x - \frac{5}{6}}{6x^2 - 4x + 1} = \frac{11}{6} + \cfrac{\frac{20}{9}}{\cfrac{6x^2 - 4x + 1}{6x - \frac{3}{8}}}$$

$$= \frac{11}{6} + \cfrac{\frac{20}{9}}{x - \frac{29}{48} + \cfrac{\frac{99}{128}}{6x - \frac{3}{8}}}$$

$$= \alpha_0 + \cfrac{\alpha_1}{x - \alpha_2 + \cfrac{\alpha_3}{x - \alpha_4}},$$

where

$$\alpha_0 = 11/6 = 1.8333\ 3333,$$

$$\alpha_1 = 20/9 = 2.2222\ 2222,$$

$$\alpha_2 = 29/48 = 0.6041\ 6667,$$

$$\alpha_3 = 33/256 = 0.1289\ 0625,$$

$$\alpha_4 = 1/16 = 0.0625\ 0000.$$

To evaluate $r(x)$ by (7.8) takes two divisions and four additions (a subtraction is considered to be an addition since the two operations take the same amount of time on most computers). On the other hand, let p be a polynomial of degree four. The most efficient way to evaluate p is by the formula

$$p(x) = x(x(x(a_4 x + a_3) + a_2) + a_1) + a_0. \tag{7.9}$$

But (7.9) requires four multiplications and four additions. Thus the given rational function can be evaluated more quickly than a polynomial of degree four on any machine whose division time is less than twice the multiplication time.

PROBLEMS

7.1. Let $f(x) = ln(1 + x)$. Compute the type (3,2) Padé approximation to f about 0.

7.2. Let r be the Padé approximation obtained in Example 7.1 and let p be the corresponding fourth-degree Taylor polynomial. Evaluate p and r at 0.01, 0.02, and 0.05. Compare these results with the values of $(1 + x)^{10}$ given in the CRC Compound Interest Tables [7]. Also compare the time required to compute r and p by (7.8) and (7.9) on a computer requiring 5μ sec for each addition or subtraction and 10μ sec for each multiplication or division.

7.3. Explain why r must have a representation of the form (7.7) if it satisfies (7.6).

7.4. Let $r = p/q$ where p and q are both polynomials of degree n.
   (a) Show that r can be expressed as a continued fraction similar to (7.8).
   (b) Using the operation times given in Problem 7.2, find the ratio of the time required to evaluate r by its continued fraction expansion to the time required for evaluating a polynomial of degree $2n$.
   (c) Try to find recurrence formulas for the coefficients of the continued fraction in part (a) above.

8. HERMITE INTERPOLATION

The Hermite interpolation polynomial is obtained by requiring the polynomial and some of its derivatives to assume prescribed values on a given set of points. The Lagrange and Taylor interpolation polynomials are special cases of the Hermite type.

We will consider only what is known as the *simple Hermite* or *osculatory problem*. Let $x_1, x_2, \ldots, x_n$ be distinct numbers and suppose $y_1, y_2, \ldots, y_n, y_1', y_2', \ldots, y_n'$ are prescribed values. The osculatory problem is to find a polynomial $p \in \mathcal{P}_{2n-1}$ which satisfies the equations

$$p(x_k) = y_k, \qquad p'(x_k) = y_k', \qquad k = 1, 2, \ldots, n.$$

Let us now show that this problem always has a unique solution.

THEOREM 8.1. *Let $x_1, x_2, \ldots, x_n$ be distinct numbers. Then for each set of values $y_1, y_2, \ldots, y_n, y_1', y_2', \ldots, y_n'$, there exists a unique polynomial $p \in \mathcal{P}_{2n-1}$ such that*

$$p(x_k) = y_k, \qquad p'(x_k) = y_k', \qquad k = 1, 2, \ldots, n. \tag{8.1}$$

Proof. Let

$$l_k(x) = \prod_{\substack{j=1 \\ j \neq k}}^{n} [(x - x_j)/(x_k - x_j)], \qquad k = 1, 2, \ldots, n,$$

be the polynomials in $\mathcal{P}_{n-1}$ which satisfy the equations $l_k(x_i) = \delta_{ki}$. Let

$$p(x) = \sum_{j=1}^{n} [y_j + (x - x_j)(y_j' - 2y_j l_j'(x_j))] l_j^2(x). \tag{8.2}$$

It is left as an exercise (Problem 8.1) to show that $p \in \mathcal{P}_{2n-1}$ and p satisfies (8.1).

In order to show uniqueness, suppose that both p and $q \in \mathcal{P}_{2n-1}$ and satisfy (8.1). Let $\varphi = p - q$. Now φ has a double zero at each point x_k, $k = 1, 2, \ldots, n$. Thus φ has at least $2n$ zeros. But the only polynomial in $\mathcal{P}_{2n-1}$ which is allowed to have $2n$ zeros is the zero polynomial. It follows that $p = q$ and the proof is complete.

An error formula can be obtained which is very similar to that given for Lagrange interpolation.

THEOREM 8.2. *Suppose f' is continuous on $[a,b]$, $f^{(2n)}$ exists on (a,b), and $x_1, x_2, \ldots, x_n$ are distinct numbers in $[a,b]$. Let p be the unique polynomial in $\mathcal{P}_{2n-1}$ which satisfies the equations*

$$p(x_k) = f(x_k), \qquad p'(x_k) = f'(x_k), \qquad k = 1, 2, \ldots, n.$$

If $x \in [a,b]$, then

$$f(x) - p(x) = \frac{f^{(2n)}(\xi)}{(2n)!} [(x - x_1)(x - x_2) \ldots (x - x_n)]^2,$$

where ξ lies between the largest and smallest of the numbers x, x_1, $x_2, \ldots, x_n$.

The proof of this theorem is left as an exercise (Problem 8.2).

The use of Hermite interpolation in numerical integration will be discussed in Chapter 6.

PROBLEMS

8.1. Show that the polynomial p defined by (8.2) is in $\mathcal{P}_{2n-1}$ and satisfies the equations (8.1).

8.2. Prove Theorem 8.2.

8.3. Let $f(x) = (1 + x)^{10}$, $x_1 = 0$, and $x_2 = 1$.
  (a) Find the osculatory interpolation polynomial for f corresponding to the points x_1 and x_2.
  (b) Use this polynomial to approximate $(1.02)^{10}$. Compare this value with the CRC Compound Interest Table [7].
  (c) Find a bound on the error in (b).

8.4. Suppose p is the osculatory interpolation polynomial for f at the points $x_0, x_1, \ldots, x_n$ given by (4.1). Find a bound on $|f(x) - p(x)|$, valid for all $x \in [a, b]$, in terms of M, where M is a bound on $|f^{(?)}|$ over the interval $[a, b]$.

8.5. Show that simple Hermite interpolation is a special case of the general linear interpolation problem described in Section 1. Then solve the Hermite problem by the technique described in Theorem 1.2 and compare the result with (8.2).

9. TRIGONOMETRIC INTERPOLATION

In the previous sections of this chapter we have considered examples of polynomial interpolation. If the function f is periodic, it is sometimes more desirable to approximate f by a linear combination of sines and cosines. We say that s is a *trigonometric polynomial of degree n* iff s is defined by an equation of the form

$$s(x) = a_0 + \sum_{j=1}^{n} (a_j \cos jx + b_j \sin jx),$$

where $a_0, \ldots, a_n, b_1, \ldots, b_n$ are constants and $a_n^2 + b_n^2 \neq 0$. Let $\mathfrak{T}_n$ be the set of all trigonometric polynomials whose degree does not exceed n, and suppose $x_0, \ldots, x_{2n}$ are numbers which satisfy the inequalities

$$(9.1) \qquad -\pi < x_0 < x_1 < \ldots < x_{2n} \leqslant \pi.$$

The trigonometric interpolation problem is to find a member s of $\mathfrak{J}_n$ which satisfies the equations

$$s(x_k) = y_k, \qquad k = 0, 1, \ldots, 2n, \tag{9.2}$$

where $y_0, \ldots, y_{2n}$ are given numbers. Let us now show that a solution to this problem always exists.

THEOREM 9.1. *Let $x_0, x_1, \ldots, x_{2n}$ satisfy (9.1) and let $y_0, \ldots, y_{2n}$ be given. Then there exists a unique trigonometric polynomial $s \in \mathfrak{J}_n$ which satisfies the equations (9.2).*

Proof. Consider the system of equations

$$a_0 + \sum_{j=1}^{n} (a_j \cos jx_k + b_j \sin jx_k) = y_k, \qquad k = 0, 1, \ldots, 2n. \tag{9.3}$$

Our approach will be to show that this interpolation problem can be transformed into a Lagrange problem. Recall that

$$\begin{aligned} \cos x &= \frac{e^{ix} + e^{-ix}}{2}, \\ \sin x &= \frac{e^{ix} - e^{-ix}}{2i}, \end{aligned} \tag{9.4}$$

where i is the imaginary unit. Let

$$z_k = e^{ix_k} \tag{9.5}$$

and note that $z_k = z_j$ implies that $e^{i(x_k - x_j)} = 1$. But this implies that $x_k - x_j$ is a multiple of 2π; so in view of (9.1), the z_k's are distinct. Using (9.4) and (9.5), we see that (9.3) is equivalent to the system

$$a_0 + \sum_{j=1}^{n} (\alpha_j z_k^j + \beta_j z_k^{-j}) = y_k, \qquad k = 0, 1, \ldots, 2n, \tag{9.6}$$

where

$$\alpha_j = \frac{1}{2}(a_j - ib_j), \qquad \beta_j = \frac{1}{2}(a_j + ib_j).$$

Multiplying both sides of (9.6) by z_k^n, we obtain

$$\sum_{j=0}^{2n} \gamma_j z_k^j = z_k^n y_k, \qquad k = 0, 1, \ldots, 2n, \tag{9.7}$$

where

$$\gamma_j = \begin{cases} \beta_{n-j}, & j = 0, 1, \ldots, n-1 \\ a_0, & j = n \\ \alpha_{j-n}, & j = n+1, n+2, \ldots, 2n \end{cases}$$

Since the z_k's are distinct, the Lagrange theory can be applied to conclude that (9.7) has a unique solution $\gamma_0, \gamma_1, \ldots, \gamma_{2n}$. This implies that (9.6) has a unique solution $a_0, \alpha_1, \ldots, \alpha_n, \beta_1, \ldots, \beta_n$, which in turn implies that (9.3) has a unique solution $a_0, a_1, \ldots, a_n$, $b_1, \ldots, b_n$.

One way to obtain the solution to the trigonometric interpolation problem is to solve the linear system of equations (9.3). Another way is to use the Gauss formula which is given in the next theorem.

THEOREM 9.2. *The solution s of the trigonometric interpolation problem is given by*

$$s(x) = \sum_{k=0}^{2n} y_k s_k(x),$$

where

$$s_k(x) = \prod_{\substack{j=0 \\ j\neq k}}^{2n} \frac{\sin \frac{1}{2}(x - x_j)}{\sin \frac{1}{2}(x_k - x_j)}, \qquad k = 0, 1, \ldots, 2n.$$

The proof of this theorem is left as an exercise (Problem 9.1.).

PROBLEMS

9.1. Prove Theorem 9.2. Hint: Use (9.4) to show that $s_k \in \mathfrak{J}_n$.

9.2. Let $f(x) = \cos^4 x$. Find the trigonometric polynomial from $\mathfrak{J}_1$ which interpolates f at $0, \pi/4, \pi/2$. Express the solution in the form $a_0 + a_1 \cos x + b_1 \sin x$. How can you check your result?

9.3. Show that trigonometric interpolation is a special case of the general linear interpolation problem described in Section 1.

REFERENCES

1. R. A. Beaumont and R. S. Pierce, *The Algebraic Foundations of Mathematics*, Addison-Wesley, 1963.
2. L. Brand, *Advanced Calculus*, Wiley, 1955.
3. E. W. Cheney, *Introduction to Approximation Theory*, McGraw-Hill, 1966.
4. P. J. Davis, *Interpolation and Approximation*, Blaisdell, 1963.
5. P. Henrici, *Elements of Numerical Analysis*, Wiley, 1964.
6. *National Bureau of Standards, Tables of Lagrangian Interpolation Coefficients*, Columbia University Press, 1944.
7. S. M. Selby, (Ed.), *C.R.C. Standard Mathematical Tables*, 19th Ed., Chemical Rubber Publishing Co., Cleveland, 1971.

CHAPTER 6

Applications of Interpolation

1. INTRODUCTION

Most numerical techniques for integration, differentiation, and the solution of differential equations involve some type of interpolation. In order to discuss the general approach, consider a linear transformation T defined on a space of functions. It frequently turns out that $T(p)$ is relatively easy to compute if p is a polynomial. Then we can approximate f by a polynomial p and use $T(p)$ as an approximation to $T(f)$.

In most problems it is not necessary to actually compute the coefficients of the interpolation polynomial p. For example, the Lagrange, Taylor, and Hermite interpolation polynomials can all be written in the form

$$p(x) = \sum_{k=0}^{n} y_k p_k(x),$$

where each of the polynomials p_k is independent of the y_k's, and $y_0, \ldots, y_n$ are values of f or its derivatives. Using the linearity of T, we have

$$T(f) \doteq T(p) = \sum_{k=0}^{n} y_k T(p_k) = \sum_{k=0}^{n} w_k y_k,$$

where $w_k = T(p_k)$, $k = 0, 1, \ldots, n$. The important fact here is that the values of the w_k's are independent of f, and hence they can be computed and stored for a given type of problem. Then to approximate $T(f)$, one needs only to compute a weighted sum of the y_k values.

EXAMPLE 1.1. Suppose $T(f) = f'(0)$. Let us approximate $T(f)$ using a second-degree Lagrange polynomial at the points -0.1, 0, and 0.1. This polynomial is given by

$$p(x) = y_0 p_0(x) + y_1 p_1(x) + y_2 p_2(x),$$

where $y_0 = f(-0.1)$, $y_1 = f(0)$, $y_2 = f(0.1)$, $p_0(x) = 50x^2 - 5x$, $p_1(x) = 1 - 100x^2$, and $p_2(x) = 50x^2 + 5x$. Now

$$w_0 = T(p_0) = p_0'(0) = -5,$$
$$w_1 = T(p_1) = p_1'(0) = 0,$$
$$w_2 = T(p_2) = p_2'(0) = 5.$$

Thus

$$f'(0) = T(f) \doteq T(p) = \sum_{k=0}^{2} w_k y_k$$
$$= -5y_0 + 5y_2 = 5[f(0.1) - f(-0.1)].$$

If $f(x) = e^x$, then this formula gives

$$f'(0) \doteq 5[1.1052 - 0.9048] = 1.002.$$

What is the error in this approximation?

PROBLEMS

1.1. Find a general formula for approximating $\int_{-0.1}^{0.1} f(x)dx$ based on the polynomial which interpolates f at -0.1, 0, and 0.1. Find a similar formula based on the second-degree Taylor polynomial about zero. Approximate $\int_{-0.1}^{0.1} e^x dx$ using these formulas and compare your results with the actual value of the integral.

1.2. Use the polynomial in Example 1.1 to approximate $f''(0)$. Try this formula on e^x.

1.3. Approximate $\int_{-h}^{h} f(x)dx$ using an interpolation polynomial at $-h$, 0, and h.

2. NUMERICAL INTEGRATION USING THE TAYLOR POLYNOMIAL

The Taylor polynomial is practical for approximating integrals over small intervals and certain infinite intervals. For integration over intervals of moderate length, the Taylor polynomial usually cannot compete with the Lagrange polynomial.

Suppose we wish to compute $\int_a^b f(x)dx$. Let p be the nth-degree Taylor polynomial for f about the point x_0, which will normally be in $[a,b]$. We then have

$$\int_a^b f(x)dx \doteq \int_a^b p(x)dx = P(b) - P(a),$$

where

$$P(x) = f(x_0)(x - x_0) + \frac{f'(x_0)}{2!}(x - x_0)^2 + \ldots + \frac{f^{(n)}(x_0)(x - x_0)^{n+1}}{(n+1)!}.$$

The crucial question is: How accurate is this approximation? An error bound is given in the following theorem.

THEOREM 2.1. *Suppose $f^{(n+1)}(x)$ exists, $|f^{(n+1)}(x)| \leqslant M$ for all $x \in [a,b]$, $x_0 \in [a,b]$, and p is the nth-degree Taylor polynomial for f about x_0. Then*

$$\left| \int_a^b f(x)dx - \int_a^b p(x)dx \right| \leqslant M \frac{(b - x_0)^{n+2} + (x_0 - a)^{n+2}}{(n+2)!}. \tag{2.1}$$

Proof. Using formula (6.3) from Chapter 5, we have

$$\begin{aligned} \left| \int_a^b f(x)dx - \int_a^b p(x)dx \right| &= \left| \int_a^b [f(x) - p(x)]\,dx \right| \\ &\leqslant \int_a^b \left| \frac{f^{(n+1)}(\xi)(x - x_0)^{n+1}}{(n+1)!} \right| dx \\ &\leqslant \frac{M}{(n+1)!} \int_a^b |x - x_0|^{n+1}\,dx \\ &= M \frac{(b - x_0)^{n+2} + (x_0 - a)^{n+2}}{(n+2)!}. \end{aligned}$$

Verification of the formula

$$\int_a^b |x - x_0|^{n+1} dx = \frac{(b - x_0)^{n+2} + (x_0 - a)^{n+2}}{n + 2}$$

is left as an exercise (Problem 2.2).

EXAMPLE 2.1. Let us approximate $\int_0^{0.1} (1 + x)^{-1} dx$ using a fourth-degree Taylor polynomial about zero. In this case $f(x) = (1 + x)^{-1}$, $x_0 = 0$, and

$$p(x) = 1 - x + x^2 - x^3 + x^4.$$

Hence

$$\int_0^{0.1} f(x)dx \doteq \int_0^{0.1} p(x)dx = \left[x - \frac{x^2}{2} + \frac{x^3}{3} - \frac{x^4}{4} + \frac{x^5}{5} \right]_0^{0.1}$$

$$\doteq 0.0953\,1033.$$

Now $|f^{(5)}(x)| = |-5!(1 + x)^{-6}| \leqslant 5!$ for all $x \in [0,0.1]$. Thus with

$M = 5!$, (2.1) gives a bound on the error of

$$\frac{5!}{6!}(0.1)^6 = 1.67 \cdot 10^{-7}.$$

The actual error is about $1.5 \cdot 10^{-7}$.

EXAMPLE 2.2. In order to approximate $\int_0^{\pi/4} \sin(x^2)dx$, it is probably more convenient to use the Taylor expansion for the sine about zero, and to compute an error bound directly, rather than using (2.1). Now for $z > 0$,

$$\sin z = z - \frac{z^3}{6} + \frac{z^5}{120} + \frac{\sin^{(7)}(\xi) z^7}{7!},$$

where $0 < \xi < z$. But $\sin^{(7)}(\xi) = -\cos(\xi)$. Thus, we can let $z = x^2$ to obtain

$$\sin(x^2) = x^2 - \frac{x^6}{6} + \frac{x^{10}}{120} - \frac{\cos(\xi) x^{14}}{7!},$$

where $0 < \xi < x^2$. Thus

$$\int_0^{\pi/4} \sin(x^2)dx = \left[\frac{x^3}{3} - \frac{x^7}{42} + \frac{x^{11}}{1320}\right]_0^{\pi/4} + E,$$

where

$$|E| \leqslant \int_0^{\pi/4} \frac{x^{14}}{7!}\,dx = \frac{(\pi/4)^{15}}{15 \cdot 7!}.$$

It is perhaps worth noting that this is an example of a simple function whose antiderivative is not an elementary function (sine, cosine, exponential, polynomial, square root, etc.). Thus the Fundamental Theorem of Integral Calculus is of no help in computing the integral, and some type of numerical technique is really necessary.

In order to discuss integration over infinite intervals, consider $\int_a^\infty f(x)dx$, where $a > 0$. Letting $x = 1/z$, $g(z) = f(1/z)$, we obtain

$$\int_a^\infty f(x)dx = \int_{1/a}^0 f\left(\frac{1}{z}\right)(-z^{-2})dz = \int_0^{1/a} \frac{g(z)dz}{z^2}.$$

If $g^{(n)}(0)$ exists and $g(0) = g'(0) = 0$, then we may approximate g by its nth-degree Taylor polynomial p about zero to obtain

$$(2.2) \qquad \int_a^\infty f(x)dx = \int_0^{1/a} \frac{g(z)dz}{z^2} \doteq \int_0^{1/a} \frac{p(z)}{z^2}\,dz.$$

The error will be given by

$$\int_0^{1/a} \frac{g^{(n+1)}(\xi(z))}{(n+1)!} z^{n-1} dz.$$

An error bound may be obtained by finding a bound on this integral.

EXAMPLE 2.3. Let us approximate

$$\int_4^\infty \frac{dx}{1+x^3}.$$

Now $f(x) = (1 + x^3)^{-1}$ and

$$g(z) = f\left(\frac{1}{z}\right) = \frac{1}{1+\dfrac{1}{z^3}} = \frac{z^3}{z^3+1}.$$

One could obtain the Taylor expansion for g about zero by computing derivatives, but it is easier to use the formula

$$\frac{1}{1+w} = 1 - w + w^2 - \frac{w^3}{(1+\xi)^4},$$

where $0 < \xi < w$. Letting $w = z^3$, we obtain the equation

$$\frac{z^3}{1+z^3} = z^3\left(1 - z^3 + z^6 - \frac{z^9}{(1+\xi)^4}\right)$$

$$= z^3 - z^6 + z^9 - \frac{z^{12}}{(1+\xi)^4},$$

where $0 < \xi < z^3$. Hence from (2.2)

$$\int_4^\infty \frac{dx}{1+x^3} = \int_0^{0.25} \frac{g(z)dz}{z^2} = \int_0^{0.25} (z - z^4 + z^7)dz + E$$

$$= \frac{(0.25)^2}{2} - \frac{(0.25)^5}{5} + \frac{(0.25)^8}{8} + E,$$

where

$$|E| \leqslant \int_0^{0.25} z^{10}dz = \frac{(0.25)^{11}}{11} \doteq 2.2\cdot 10^{-8}.$$

Let us also note that integrals of the form $\int_a^\infty f(x)dx$ can sometimes be approximated by a so-called *asymptotic expansion* in terms of a. Some asymptotic expansions produce series which are divergent but nevertheless yield useful approximations (see, for example, Feller [8, Chap. 7, Problem 1]). The general theory is discussed by Erdélyi [7].

PROBLEMS

2.1. Use a Taylor polynomial of degree n to approximate the following integrals. In each case also find a bound on the error.

(a) $\int_0^{0.2} e^x dx, \quad n = 4$

(b) $\int_0^{0.1} \cos(x^2)dx, \quad n = 4$

(c) $\int_0^{0.01} e^{-x^2/2}dx, \quad n = 2$

(d) $\int_2^{\infty} \frac{dx}{1+x^2}, \quad n = 6$

As a check, the integrals in (a) and (d) can be computed by the usual methods of calculus, and the value of

$$\frac{1}{\sqrt{2\pi}} \int_0^{0.01} e^{-x^2/2}dx$$

can be found in the CRC Normal Curve of Error Table [15].

2.2. Show that if $x_0 \in [a, b]$, then

$$\int_a^b |x - x_0|^{n+1} dx = \frac{(b - x_0)^{n+2} + (x_0 - a)^{n+2}}{n+2}.$$

2.3. Approximate $\int_0^2 (1+x)^{-1}dx$ using the second-degree Taylor polynomial for $(1+x)^{-1}$ about zero. Then repeat using the fourth-degree polynomial. Compare your results with those obtained from a table of natural logarithms. What's going on here?

2.4. We wish to approximate $\int_0^1 (1+x)^{-1}dx$ using the nth-degree Taylor polynomial for $(1+x)^{-1}$ about zero. How large should n be in order to have the magnitude of the error be less than:

(a) 0.4
(b) 0.1
(c) 0.01
(d) 0.001
(e) ϵ, where ϵ is some positive number

Prove directly from the definition of a limit that

$$\lim_{n\to\infty} \int_0^1 p_n(x)dx = \int_0^1 \frac{dx}{1+x}$$

2.5. Let p_n be the nth-degree Taylor polynomial for f about zero. How does the formula for $\int_a^b p_n(x)dx$ simplify in case $a = -b$, $b > 0$. Figure out the most efficient way to evaluate this formula. How many multiplications and additions are required? Show that if n is even, then

$$\left| \int_{-b}^b f(x)dx - \int_{-b}^b p_n(x)dx \right| \leqslant \frac{2Mb^{n+3}}{(n+3)!},$$

where M is a bound on $|f^{(n+2)}|$ over $[-b, b]$.

3. NEWTON-COTES FORMULAS

The Newton-Cotes integration formulas are based on Lagrange interpolation at equally spaced points. Suppose $b > a$ and we wish to approximate $\int_a^b f(x)dx$ using a polynomial from $\mathcal{P}_n$. Let $h = (b-a)/n$ and choose the interpolation points according to the formula

(3.1) $$x_0 = a, \; x_1 = a + h, \; \ldots, \; x_n = a + nh = b.$$

We will now show that the integration formula obtained in this way is of the form

$$(3.2) \qquad \int_a^b f(x)dx \doteq \frac{b-a}{n} \sum_{j=0}^{n} w_j f(x_j),$$

where the weighting factors $w_0, \ldots, w_n$, which are independent of a, b, and f, are given by (3.3) below.

THEOREM 3.1. *Let $x_0, x_1, \ldots, x_n$ be given by* (3.1), *and*

$$l_j^*(z) = \prod_{\substack{k=0\\k\neq j}}^{n} \frac{z-k}{j-k}, \quad j = 0, 1, \ldots, n.$$

If $p \in \mathcal{P}_n$ is the polynomial which interpolates f at $x_0, \ldots, x_n$, then

$$\int_a^b p(x)dx = \frac{b-a}{n} \sum_{j=0}^{n} w_j f(x_j),$$

where

$$(3.3) \qquad w_j = \int_0^n l_j^*(z)dz, \qquad j = 0, 1, \ldots, n.$$

Proof. Let

$$l_j(x) = \prod_{\substack{k=0\\k\neq j}}^{n} \frac{x-x_k}{x_j-x_k}, \qquad j = 0, 1, \ldots, n.$$

Then the interpolation polynomial p is given by

$$p(x) = \sum_{j=0}^{n} f(x_j) l_j(x).$$

Hence

$$\int_a^b p(x)dx = \sum_{j=0}^{n} f(x_j) \int_a^b l_j(x)dx.$$

Now make the change of variable $x = a + zh$ to obtain

$$\int_a^b l_j(x)dx = h \int_0^n l_j(a+zh)dz.$$

Using the fact that $x_k = a + kh$, we get

$$l_j(a+zh) = \prod_{\substack{k=0\\k\neq j}}^{n} \frac{a+zh-(a+kh)}{a+jh-(a+kh)} = \prod_{\substack{k=0\\k\neq j}}^{n} \frac{z-k}{j-k} = l_j^*(z).$$

Thus

$$\int_a^b l_j(x)dx = h \int_0^n l_j^*(z)dz,$$

and the result follows from the fact that $h = (b-a)/n$.

Let us now compute the weighting factors w_0, w_1, and w_2 corresponding to $n = 2$. First we have

$$l_0^*(z) = \frac{(z-1)(z-2)}{(0-1)(0-2)} = \frac{1}{2}(z^2 - 3z + 2).$$

Hence

$$w_0 = \int_0^2 l_1^*(z)dz = \frac{1}{2}\int_0^2 (z^2 - 3z + 2)dz = \frac{1}{3}.$$

Similarly,

$$l_1^*(z) = \frac{(z-0)(z-2)}{(1-0)(1-2)} = 2z - z^2,$$

$$l_2^*(z) = \frac{(z-0)(z-1)}{(2-0)(2-1)} = \frac{1}{2}(z^2 - z).$$

Then

$$w_1 = \int_0^2 l_1^*(z)dz = \frac{4}{3} \quad \text{and} \quad w_2 = \int_0^2 l_2^*(z)dz = \frac{1}{3}.$$

From (3.2) we obtain the integration formula

$$\int_a^b f(x)dx \doteq \frac{(b-a)}{6}\,[f(x_0) + 4f(x_1) + f(x_2)].$$

This is a famous integration formula known as *Simpson's rule.*

The weighting factors for other values of n can be computed in a similar fashion (see also Problem 3.1). Table 3.1 gives these factors for $n = 1, 2, 3,$ and 4. Additional values can be found in many numerical analysis texts and the NBS Handbook [1].

n	w_0	w_1	w_2	w_3	w_4	*Remainder Term*
1	$\frac{1}{2}$	$\frac{1}{2}$				$\frac{-h^3 f''(\xi)}{12}$
2	$\frac{1}{3}$	$\frac{4}{3}$	$\frac{1}{3}$			$\frac{-h^5 f^{(4)}(\xi)}{90}$
3	$\frac{3}{8}$	$\frac{9}{8}$	$\frac{9}{8}$	$\frac{3}{8}$		$\frac{-3h^5 f^{(4)}(\xi)}{80}$
4	$\frac{14}{45}$	$\frac{64}{45}$	$\frac{24}{45}$	$\frac{64}{45}$	$\frac{14}{45}$	$\frac{-8h^7 f^{(6)}(\xi)}{945}$

TABLE 3.1. Weights and Remainder Terms for Formula (3.2).

EXAMPLE 3.1. Let us approximate $\int_0^{0.5} \sqrt{x}\,dx$. Using (3.2) with $n = 2$, we have

$$\int_0^{0.5} \sqrt{x}\,dx \doteq \frac{0.5}{2\cdot 3}(\sqrt{0} + 4\sqrt{0.25} + \sqrt{0.5}) = 0.2255\,9223.$$

Using (3.2) with $n = 4$, we get

$$\int_0^{0.5} \sqrt{x}\, dx \doteq \frac{0.5}{4 \cdot 45} (14\sqrt{0} + 64\sqrt{0.125} + 24\sqrt{0.25} + 64\sqrt{0.375} + 14\sqrt{0.5}) \doteq 0.2325\,5208.$$

This can be compared with the integral computed in the usual way:

$$\int_0^{0.5} \sqrt{x}\, dx = \frac{2}{3} x^{3/2} \Big|_0^{0.5} \doteq 0.2357\,0226.$$

Error Bounds. Error formulas in terms of higher derivatives are available for the Newton-Cotes integration rules. These are listed in the last column of Table 3.1. In each case ξ is an unknown point somewhere in the open interval (a,b). We will justify below the error expression for Simpson's rule. A general discussion of these error formulas is given in Isaacson and Keller [13].

THEOREM 3.2. *Suppose f is continuous on $[a,b]$ and $f^{(4)}$ exists on (a,b). Let $h = (b-a)/2$, $x_0 = a$, $x_1 = a + h$, and $x_2 = a + 2h$. Then for some $\xi \in (a,b)$,*

$$\int_a^b f(x)dx - \frac{h}{3}[f(x_0) + 4f(x_1) + f(x_2)] = \frac{-h^5}{90} f^{(4)}(\xi).$$

Proof. (Brand [2]). Let

$$E(t) = \int_{x_1-t}^{x_1+t} f(x)dx - \frac{t}{3}[f(x_1 - t) + 4f(x_1) + f(x_1 + t)].$$

Now

$$\begin{aligned} E'(t) &= f(x_1 + t) + f(x_1 - t) \\ &\quad - \frac{1}{3}[f(x_1 - t) + 4f(x_1) + f(x_1 + t)] \\ &\quad - \frac{t}{3}[f'(x_1 + t) - f'(x_1 - t)] \\ &= \frac{2}{3}[f(x_1 + t) + f(x_1 - t) - 2f(x_1)] \\ &\quad - \frac{t}{3}[f'(x_1 + t) - f'(x_1 - t)]. \end{aligned}$$

Similarly,

$$\begin{aligned} E''(t) &= \frac{1}{3}[f'(x_1 + t) - f'(x_1 - t)] \\ &\quad - \frac{t}{3}[f''(x_1 + t) + f''(x_1 - t)], \end{aligned}$$

and

$$E'''(t) = -\frac{t}{3}[f'''(x_1 + t) - f'''(x_1 - t)].$$

Note that $E(0) = E'(0) = E''(0) = E'''(0) = 0$. Next, we reach into our magic hat and pull out a function φ defined by

$$\varphi(t) = E(t) - \frac{E(h)t^5}{h^5}.$$

Note that $\varphi(h) = 0 = \varphi(0) = \varphi'(0) = \varphi''(0)$. Now by Rolle's theorem we have

$$\begin{aligned} \varphi'(\xi_1) &= 0 \quad \text{for some} \quad \xi_1 \in (0,h), \\ \varphi''(\xi_2) &= 0 \quad \text{for some} \quad \xi_2 \in (0,\xi_1), \\ \varphi'''(\xi_3) &= 0 \quad \text{for some} \quad \xi_3 \in (0,\xi_2). \end{aligned}$$

Hence

$$E'''(\xi_3) = \frac{60\xi_3^2\, E(h)}{h^5} \qquad \text{and} \qquad E(h) = \frac{E'''(\xi_3)h^5}{60\xi_3^2}.$$

Using the mean-value theorem, we have

$$\begin{aligned} E'''(\xi_3) &= \frac{-\xi_3}{3}[f'''(x_1 + \xi_3) - f'''(x_1 - \xi_3)] \\ &= \frac{-2\xi_3^2}{3} f^{(4)}(x_1 + \xi_4) \end{aligned}$$

for some $\xi_4 \in (-\xi_3,\xi_3)$. Letting $\xi = x_1 + \xi_4$, we obtain

$$E(h) = \frac{-2}{3}\xi_3^2 f^{(4)}(\xi)\left[\frac{h^5}{60\xi_3^2}\right] = \frac{-h^5 f^{(4)}(\xi)}{90}$$

and the proof is complete.

EXAMPLE 3.2. In Example 3.1 we obtained the approximation $\int_0^{0.5} \sqrt{x}\,dx \doteq 0.2255\ 9223$. In this case $f^{(4)}(x) = (-15/16)x^{-7/2}$. Using Theorem 3.2, we have

$$\begin{aligned} E &= \int_0^{0.5} \sqrt{x}\,dx - \frac{h}{3}[f(x_0) + 4f(x_1) + f(x_2)] \\ &= \frac{15(0.25)^5\xi^{-7/2}}{16(90)} = \frac{\xi^{-7/2}}{96(4^5)}, \end{aligned}$$

where $\xi \in (0,0.5)$. Now $(0.5)^{-7/2} < \xi^{-7/2} < \infty$; so

$$\frac{2^{7/2}}{96(2^{10})} = 0.0001\ 1509 < E < \infty$$

This can be compared with the actual value to about eight decimals,

$$E = 0.2357\,0226 - 0.2255\,9223 = 0.0101\,1003.$$

Note that in this case we do not obtain a meaningful upper bound on the error because $|f^{(4)}|$ is unbounded on (0, 0.5). This does not imply that the error must be extremely large, but rather that our usual method for obtaining error bounds cannot be applied to this example.

Composite Formulas. A composite formula is obtained by applying an integration rule to a number of subintervals and adding the results. For example, suppose we wish to approximate $\int_a^b f(x)dx$ using a composite formula obtained from Simpson's rule. Let $x_0 = a$, $x_1 = a + h$, $x_2 = a + 2h, \ldots, x_{2n} = a + 2nh = b$, where $h = (b-a)/(2n)$. Now

$$\int_a^b f(x)dx = \sum_{j=1}^{n} \int_{x_{2j-2}}^{x_{2j}} f(x)dx.$$

Applying Simpson's rule to each integral of the form

$$\int_{x_{2j-2}}^{x_{2j}} f(x)dx,$$

we obtain

$$\begin{aligned} \int_a^b f(x)dx &\doteq \sum_{j=1}^{n} \frac{h}{3}\,[f(x_{2j-2}) + 4f(x_{2j-1}) + f(x_{2j})] \\ &= \frac{h}{3}\left[4 \sum_{j=1}^{n} f(x_{2j-1}) + \sum_{j=1}^{n} f(x_{2j-2}) + \sum_{j=1}^{n} f(x_{2j})\right]. \end{aligned} \tag{3.4}$$

Letting $k = j - 1$, we have

$$\sum_{j=1}^{n} f(x_{2j-2}) = \sum_{k=0}^{n-1} f(x_{2k}) = f(x_0) + \sum_{j=1}^{n-1} f(x_{2j}).$$

The last two sums in (3.4) can now be combined to give

$$\begin{aligned} \int_a^b f(x)dx \doteq \frac{h}{3}\Bigg[f(x_0) &+ 4 \sum_{j=1}^{n} f(x_{2j-1}) \\ &+ 2 \sum_{j=1}^{n-1} f(x_{2j}) + f(x_{2n})\Bigg]. \end{aligned} \tag{3.5}$$

We will call this the *composite Simpson's rule*—it is frequently referred to in the literature simply as Simpson's rule or the compound Simpson's rule. Similar composite formulas for other values of n can be found in many numerical analysis texts. These can be easily derived from the weight factors such as those given in Table 3.1 (see Problem 3.6).

Let us illustrate the error analysis for composite rules by again using Simpson's rule. Suppose f is continuous on $[a,b]$ and $f^{(4)}$ exists on (a,b). Let E be the difference between the left- and right-hand sides of (3.5). Applying Theorem 3.2 to each subinterval $[x_{2j-2}, x_{2j}]$, we obtain

$$(3.6) \qquad E = \frac{-h^5}{90} \sum_{j=1}^{n} f^{(4)}(\xi_j),$$

where $\xi_j \in (x_{2j-2}, x_{2j})$ for each j. If $m_j \leqslant f^{(4)}(x) \leqslant M_j$ for $x \in (x_{2j-2}, x_{2j})$, then from (3.6) we have

$$(3.7) \qquad \frac{-h^5}{90} \sum_{j=1}^{n} M_j \leqslant E \leqslant \frac{-h^5}{90} \sum_{j=1}^{n} m_j.$$

A more convenient, but not as sharp, error expression is obtained as follows. Let

$$m = \min_{1 \leqslant j \leqslant n} f^{(4)}(\xi_j) \qquad \text{and} \qquad M = \max_{1 \leqslant j \leqslant n} f^{(4)}(\xi_j).$$

Then

$$nm \leqslant \sum_{j=1}^{n} f^{(4)}(\xi_j) \leqslant Mn.$$

It follows from the intermediate-value theorem for derivatives (see Brand [2], theorem of Darboux) that there exists a number $\xi \in (a,b)$ such that

$$\frac{1}{n} \sum_{j=1}^{n} f^{(4)}(\xi_j) = f^{(4)}(\xi).$$

Substituting this in (3.6), we have

$$(3.8) \qquad E = \frac{-nh^5}{90} f^{(4)}(\xi) = \frac{-(b-a)}{180} h^4 f^{(4)}(\xi)$$

for some $\xi \in (a,b)$.

An interesting error estimate can be obtained from (3.6) by noting that $\Sigma_{j=1}^{n} f^{(4)}(\xi_j)(2h)$ is a Riemann sum for $\int_a^b f^{(4)}(x)dx$. Thus for h sufficiently small,

$$(3.9) \qquad E \doteq \frac{-h^4}{180} \int_a^b f^{(4)}(x)dx = \frac{-h^4}{180} [f^{(3)}(b) - f^{(3)}(a)].$$

Of course, for (3.9) we need the additional assumptions that $\int_a^b f^{(4)}(x)dx$, $f^{(3)}(b)$, and $f^{(3)}(a)$ exist.

EXAMPLE 3.3. Let us approximate $\int_0^1 x^5 dx$ using the composite Simpson's rule at the points 0, 0.1, 0.2, . . . , 1. Substitution in (3.5) yields the approximation

$$\int_0^1 x^5 dx \doteq \frac{(0.1)}{3}\Big[0 + 4[.1^5 + .3^5 + .5^5 + .7^5 + .9^5]$$

$$+ 2[.2^5 + .4^5 + .6^5 + .8^5] + 1\Big]$$

$$= \frac{10^{-6}}{3}[4(1 + 243 + 3125 + 16807 + 59049)$$

$$+ 2(32 + 1024 + 7776 + 32768) + 100{,}000]$$

$$= 0.1667.$$

The error is given by

$$E = 0.1666\,666\ldots - 0.1667 = -3.33\ldots 10^{-5}.$$

Now $f^{(4)}(x) = 120x$. From (3.7) we get

$$\frac{-10^{-5}}{90}(24 + 48 + 72 + 96 + 120) \leqslant E$$

$$\leqslant \frac{-10^{-5}}{90}(0 + 24 + 48 + 72 + 96).$$

This simplifies to

$$-4 \cdot 10^{-5} \leqslant E \leqslant -2.67 \cdot 10^{-5}.$$

From (3.8) we conclude that E is negative and

$$|E| \leqslant \frac{10^{-4}(120)}{180} \leqslant 6.67 \cdot 10^{-5}.$$

Note that (3.7) gave a slightly sharper bound on $|E|$; but the extra sharpness was probably not worth the effort. Finally, (3.9) gives

$$E \doteq \frac{-10^{-4}}{180}[60 - 0] = -3.33\ldots 10^{-5},$$

which happens to be the exact error for this example.

PROBLEMS

3.1. Verify that the weighting factors given in Table 3.1 for $n = 1$ and $n = 3$ are correct.

3.2. Approximate $\int_1^{1.5} (3x + 1)^{-1} dx$ using Simpson's rule with $h = 0.25$. Find a bound on the error.

3.3. Prove that the remainder term given in Table 3.1 for $n = 1$ is correct.

3.4. Use the composite Simpson's rule with $h = 0.25$ to approximate $\int_1^3 x^{-1} dx$. Find a bound on the error. Also estimate the error using (3.9). Note: The CRC Table of Reciprocals [15] can be used to avoid divisions.

3.5. We wish to approximate $\int_0^1 e^{2x} dx$ using the composite Simpson's rule. How small would h be to insure that the error does not exceed:
(a) 10^{-4} (b) ϵ, where ϵ is a positive number

3.6. Derive formulas for the composite Newton-Cotes integration rules corresponding to:
(a) $n = 1$ (b) $n = 3$
The result of (a) is known as the *trapezoidal rule*, while that of (b) is known as *Simpson's three-eighths rule*.

3.7. Approximate π by using a Newton-Cotes rule on

$$\int_0^1 \frac{dx}{1+x^2}.$$

4. GAUSSIAN QUADRATURE

The term *quadrature* means integration in this context. Historically, quadrature referred to the process of constructing a square having the same area as a given plane surface.

The Newton-Cotes integration rule obtained from an nth-degree interpolation polynomial gives the exact value for the integral of a polynomial whose degree does not exceed n if n is odd, or $n + 1$ if n is even. A Gaussian integration rule of the form

$$\int_a^b f(x)dx \doteq \sum_{j=1}^{n} w_j f(x_j)$$

is exact for polynomials of degree $2n - 1$ or less, while the Newton-Cotes rule using n points would be exact at most for polynomials of degree n. The Gaussian rules frequently provide better accuracy than the corresponding Newton-Cotes formulas, although they are more cumbersome to work with.

The theory of Gaussian quadrature can be based on Hermite interpolation. For convenience we consider integrals of the form $\int_{-1}^{1} f(x)dx$; any integral over a finite closed interval can be converted to this form by a change of variable of the type $x = \alpha z + \beta$. Let x_1, $x_2, \ldots, x_n$ be in $[-1,1]$, and let $p \in \mathcal{P}_{2n-1}$ be the simple Hermite interpolation polynomial for f at these points. From (8.2) in Chapter 5 we have

$$p(x) = \sum_{j=1}^{n} f(x_j)p_j(x) + \sum_{j=1}^{n} f'(x_j)q_j(x),$$

where

$$p_j(x) = [1 - 2(x - x_j)l_j'(x_j)]\,l_j^2(x)$$

and

$$q_j(x) = (x - x_j)l_j^2(x). \tag{4.1}$$

Thus

$$\int_{-1}^{1} p(x)dx = \sum_{j=1}^{n} w_j f(x_j) + \sum_{j=1}^{n} \alpha_j f'(x_j),$$

where

$$w_j = \int_{-1}^{1} p_j(x)dx \qquad \text{and} \qquad \alpha_j = \int_{-1}^{1} q_j(x)dx. \tag{4.2}$$

The strategy is to choose $x_1, \ldots, x_n$ so that all of the α_j's are zero.

Let us define an inner product by the equation

$$f \cdot g = \int_{-1}^{1} f(x)g(x)dx.$$

Recall that f and g are said to be orthogonal iff $f \cdot g = 0$. The following theorem gives a condition which will insure that all of the α_j's are zero.

THEOREM 4.1. *Let $x_1, x_2, \ldots, x_n$ be in $[-1,1]$, $\pi(x) = (x - x_1)(x - x_2)\ldots(x - x_n)$, $\varphi_0(x) = 1$, $\varphi_1(x) = x, \ldots, \varphi_{n-1}(x) = x^{n-1}$, and let w_j and α_j be given by (4.2). If π is orthogonal to each of the functions $\varphi_0, \varphi_1, \ldots, \varphi_{n-1}$, then $\alpha_j = 0$ for each $j = 1, 2, \ldots, n$.*

Proof. Note that

$$(x - x_j)l_j(x) = c_j\pi(x), \tag{4.3}$$

where

$$c_j = \prod_{\substack{k=1 \\ k \neq j}}^{n} \frac{1}{x_j - x_k}.$$

Now l_j is a polynomial of degree $n - 1$; so

$$l_j(x) = \sum_{i=0}^{n-1} a_{ji}x^i = \sum_{i=0}^{n-1} a_{ji}\varphi_i(x),$$

where the a_{ji}'s are the coefficients of l_j. From (4.1), (4.2), and (4.3) we obtain

$$\alpha_j = \int_{-1}^{1} q_j(x)dx = c_j \int_{-1}^{1} \pi(x)l_j(x) = c_j\pi \cdot l_j.$$

But $l_j = \sum_{i=0}^{n-1} a_{ji}\varphi_i$. Hence

$$\alpha_j = c_j\pi \cdot \sum_{i=0}^{n-1} a_{ji}\varphi_i = c_j \sum_{i=0}^{n-1} a_{ji}(\pi \cdot \varphi_i) = 0$$

and the proof is complete.

COROLLARY 4.2. *Suppose f' is continuous on $[-1,1]$ and $f^{(2n)}$ exists on $(-1,1)$. Then, under the hypotheses of Theorem 4.1,*

$$\int_{-1}^{1} f(x)dx = \sum_{j=1}^{n} w_jf(x_j) + \beta_n f^{(2n)}(\xi),$$

where $\xi \in (-1,1)$, and β_n is a constant which is independent of f.

Proof. Let p be the simple Hermite interpolation polynomial for f at $x_1, x_2, \ldots, x_n$. The remainder formula for Hermite interpolation implies that

$$f(x) - p(x) = \frac{f^{(2n)}(\xi(x))\pi^2(x)}{(2n)!},$$

where $\xi(x) \in (-1,1)$ for each x. Hence

$$\int_{-1}^{1} f(x)dx - \sum_{j=1}^{n} w_j f(x_j) = \int_{-1}^{1} [f(x) - p(x)]\,dx$$

$$= \int_{-1}^{1} \frac{f^{(2n)}(\xi(x))\pi^2(x)dx}{(2n)!}.$$

Now let m and M be numbers such that

$$m \leqslant f^{(2n)}(x) \leqslant M$$

for all $x \in (-1,1)$. (Choose $m = -\infty$ or $M = \infty$ if $f^{(2n)}$ is unbounded.) From the fact that $\pi^2(x)/(2n)!$ is non-negative, it follows that

$$m\beta_n \leqslant \int_{-1}^{1} \frac{f^{(2n)}(\xi(x))\pi^2(x)dx}{(2n)!} \leqslant M\beta_n,$$

where

$$\beta_n = \int_{-1}^{1} \frac{\pi^2(x)dx}{(2n)!}.$$

Then the intermediate-value theorem for derivatives (see the derivation of (3.8)) implies that

$$\int_{-1}^{1} \frac{f^{(2n)}(\xi(x))\pi^2(x)dx}{(2n)!} = \beta_n f^{(2n)}(\xi_0)$$

for some $\xi_0 \in (-1,1)$. Now β_n depends only on n; so the proof is complete.

COROLLARY 4.3. *If the hypotheses of Theorem 4.1 hold, then the integration rule*

$$\int_{-1}^{1} f(x)dx \doteq \sum_{j=1}^{n} w_j f(x_j)$$

is exact for all polynomials in $\mathcal{P}_{2n-1}$.

Proof. If $f \in \mathcal{P}_{2n-1}$, then $f^{(2n)}(x) = 0$ for all x.

From Theorem 4.1 it is clear that the construction of a Gaussian quadrature formula can be accomplished if one can construct a polynomial $\pi \in \mathcal{P}_n$ with the following properties:

(i) Each of the functions $\varphi_0, \varphi_1, \ldots, \varphi_{n-1}$ is orthogonal to π.
(ii) The zeros of π are distinct and lie in $[-1,1]$.

We will not discuss the theory of orthogonal polynomials which plays an important role in the construction of such a polynomial. (See Davis and Rabinowitz [6], or Szegö [18].) Instead, we will just show that the nth-degree *Legendre polynomial* satisfies the two requirements. This polynomial is defined by

$$P_n(x) = \frac{1}{2^n n!} g^{(n)}(x), \tag{4.4}$$

where $g(x) = (1 - x^2)^n$.

THEOREM 4.4. *Let P_n be defined by* (4.4). *Then*

(i) $P_n \in \mathcal{P}_n$.
(ii) $\int_{-1}^{1} x^m P_n(x)dx = 0, \quad m = 0, 1, 2, \ldots, n-1$.
(iii) The zeros of P_n are distinct and lie in $(-1,1)$.

Proof. Applying the binomial theorem, we get

$$g(x) = \sum_{j=0}^{n} \binom{n}{j} (-1)^j x^{2j}. \tag{4.5}$$

Clearly g is a polynomial of degree $2n$, and hence $g^{(n)}$ is a polynomial of degree n. Thus $P_n \in \mathcal{P}_n$.

Next, note that $g(x) = (1-x)^n(1+x)^n$. It follows that

$$g(x) = g'(x) = \ldots = g^{(n-1)}(x) = 0 \quad \text{if} \quad x = -1 \quad \text{or} \quad x = 1. \tag{4.6}$$

Now suppose m is an integer satisfying the condition $0 \leqslant m \leqslant n-1$. Integrating by parts and applying (4.6), we have

$$\begin{aligned} 2^n n! \int_{-1}^{1} x^m P_n(x)dx &= \int_{-1}^{1} x^m g^{(n)}(x)dx \\ &= [x^m g^{(n-1)}(x)]_{-1}^{1} \\ &\qquad - m \int_{-1}^{1} x^{m-1} g^{(n-1)}(x)dx \\ &= -m \int_{-1}^{1} x^{m-1} g^{(n-1)}(x)dx. \end{aligned}$$

Now repeatedly integrate by parts to obtain

$$\begin{aligned} 2^n n! \int_{-1}^{1} x^m P_n(x)dx &= (-1)^m m! \int_{-1}^{1} x^0 g^{(n-m)}(x)dx \\ &= (-1)^m m! [g^{(n-m-1)}(1) \\ &\qquad - g^{(n-m-1)}(-1)] \\ &= 0. \end{aligned}$$

To prove part (iii) we again use (4.6). By Rolle's theorem there exists a number ξ_{11} in $(-1,1)$ such that $g'(\xi_{11}) = 0$. Now g' is zero at -1, ξ_{11}, and 1. Again by Rolle's theorem there exist numbers

$\xi_{21} \in (-1, \xi_{11})$ and $\xi_{22} \in (\xi_{11}, 1)$ such that

$$g''(\xi_{21}) = g''(\xi_{22}) = 0.$$

Then g'' is zero at the four points: -1, ξ_{21}, ξ_{22}, and 1. Continuing in this fashion, we infer that $g^{(n)}$ has n distinct zeros in $(-1,1)$. It follows from (4.4) that P_n also has n distinct zeros in $(-1,1)$ and the proof is complete.

COROLLARY 4.5. *Let $x_1, x_2, \ldots, x_n$ be the zeros of the Legendre polynomial P_n, and let $w_1, w_2, \ldots, w_n$ be given by (4.2). If f' is continuous on $[-1,1]$, and $f^{(2n)}$ exists on $(-1,1)$, then*

$$\int_{-1}^{1} f(x)dx - \sum_{j=1}^{n} w_j f(x_j) = \frac{2^{2n+1}(n!)^4}{(2n+1)[(2n)!]^3} f^{(2n)}(\xi),$$

where $\xi \in (-1,1)$.

Proof. Let $\pi(x) = (x - x_1)(x - x_2)\ldots(x - x_n)$. Now $\pi(x) = P_n(x)/a_n$, where a_n is the leading coefficient of P_n. Differentiating (4.5) we see that the leading coefficient of $g^{(n)}$ is $(-1)^n(2n)!/n!$. Thus the leading coefficient of P_n is

$$\frac{(-1)^n(2n)!}{2^n(n!)^2}.$$

From the proof of Corollary (4.2) we have

$$\beta_n = \int_{-1}^{1} \frac{\pi^2(x)dx}{(2n)!} = \frac{2^{2n}(n!)^4}{[(2n)!]^3} \int_{-1}^{1} P_n^2(x)dx.$$

The proof is completed by showing (see Problem 4.2) that

$$\int_{-1}^{1} P_n^2(x)dx = \frac{2}{2n+1}.$$

Tables of the zeros of the Legendre polynomials and corresponding weights can be found in the NBS Handbook [1] or Stroud and Secrest [17]. For convenience in working problems, we list below the values for n = 2, 3, and 4.

n	x_1	x_2	w_1	w_2
2	−0.577 350	$\lvert x_1 \rvert$	1.000 000	w_1
3	−0.774 597	0	0.555 556	0.888 889
4	−0.861 136	−0.339 981	0.347 855	0.652 145

For $n = 3$: $x_3 = |x_1|$ and $w_3 = w_1$
For $n = 4$: $x_3 = |x_2|$, $x_4 = |x_1|$, $w_3 = w_2$, and $w_4 = w_1$

TABLE 4.1. Abscissas and Weights for Gaussian Quadrature.

EXAMPLE 4.1. Let us approximate $\int_{-1}^{1}\sqrt{x+1.5}\,dx$. Using Table 4.1 with $n = 2$, we get

$$\int_{-1}^{1}\sqrt{x+1.5}\,dx \doteq \sqrt{0.922\,650} + \sqrt{2.077\,350} = 2.401\,848.$$

The Newton-Cotes rule involving two points ($n = 1$) gives

$$\int_{-1}^{1}\sqrt{x+1.5}\,dx \doteq 2(\tfrac{1}{2}\sqrt{0.5} + \tfrac{1}{2}\sqrt{2.5}) = 2.288\,246.$$

These can be compared with the "exact" value of

$$\int_{-1}^{1}\sqrt{x+1.5}\,dx = \tfrac{2}{3}\,[(2.5)^{3/2} - (0.5)^{3/2}] = 2.399\,529.$$

For $n = 3$, the Gauss rule gives

$$\int_{-1}^{1}\sqrt{x+1.5}\,dx \doteq 0.555\,556(\sqrt{0.725\,403} + \sqrt{2.274\,597}) + 0.888\,889\sqrt{1.5} = 2.399\,709.$$

This can be compared with Simpson's rule:

$$\int_{-1}^{1}\sqrt{x+1.5}\,dx \doteq \tfrac{1}{3}(\sqrt{0.5} + 4\sqrt{1.5} + \sqrt{2.5}) = 2.395\,742.$$

An error bound can be obtained from Corollary 4.5. For example, if $n = 3$,

$$|E| = \frac{2^7 6^4}{7(720)^3} f^{(6)}(\xi) = \frac{2^7 6^4 945(\xi + 1.5)^{-11/2}}{7(720)^3 64}$$

$$\leqslant \frac{2^7 6^4 (945) 2^{11/2}}{7(720)^3 64} \doteq 0.0424.$$

The corresponding bound for Simpson's rule is

$$|E| \leqslant \frac{1^5 15(\xi + 1.5)^{-7/2}}{90(16)} \leqslant \frac{2^{7/2}}{96} = \frac{\sqrt{2}}{12} = 0.118.$$

Gaussian integration formulas can also be obtained for weighted integrals of the form

$$\int_{-1}^{1} w(x)f(x)dx,$$

where $w(x) \geqslant 0$ for $x \in [-1,1]$. Here one obtains formulas of the form

$$\int_{-1}^{1} w(x)f(x)dx \doteq \sum_{j=1}^{n} w_j f(x_j),$$

which are exact for polynomials in $\mathcal{P}_{2n-1}$. (Note that w_j is not $w(x_j)$ in this formula.) The theory is based on polynomials which are orthogonal to $\varphi_0, \varphi_1, \ldots, \varphi_{n-1}$ with respect to the inner product

$$f \cdot g = \int_{-1}^{1} w(x)f(x)g(x)dx.$$

For details see Davis and Rabinowitz [6] or Stroud and Secrest [17].

PROBLEMS

4.1. Approximate $\int_{-1}^{1} (x + 2)^{-1} dx$ using a Gauss rule with $n = 3$ and then $n = 4$. Compare your results with the Newton-Cotes rule which uses the same number of points. Find a bound on the error for $n = 4$.

4.2. Show that $\int_{-1}^{1} P_n^2(x)dx = 2/(2n + 1)$ if P_n is the nth-degree Legendre polynomial. Hint: Show that

$$\begin{aligned}\int_{-1}^{1} g^{(n)}(x)g^{(n)}(x)dx &= (-1)^n \int_{-1}^{1} g(x)g^{(2n)}(x)dx \\ &= (2n)! \int_{-1}^{1} (1-x)^n(1+x)^n dx \\ &= \frac{(n!)^2 2^{2n+1}}{2n+1}.\end{aligned}$$

See Courant and Hilbert [5].

5. INTEGRATION OF OSCILLATORY FUNCTIONS

The Newton-Cotes and Gauss rules are not well-suited for integrands which are highly oscillatory.

EXAMPLE 5.1. Let us apply the composite Simpson's rule with $h = 3\pi/20$ to $\int_0^{3\pi/2} \cos 15x\, dx$. This yields the approximation

$$\begin{aligned}\int_0^{3\pi/2} \cos 15x\, dx &= \frac{\pi}{20} [\cos 0 + 4(\cos \tfrac{9}{4}\pi + \cos \tfrac{27}{4}\pi + \cos \tfrac{45}{4}\pi \\ &\qquad + \cos \tfrac{63}{4}\pi + \cos \tfrac{81}{4}\pi) + 2(\cos \tfrac{9}{2}\pi + \cos 9\pi \\ &\qquad + \cos \tfrac{27}{2}\pi + \cos 18\pi) + \cos \tfrac{45}{2}\pi] \\ &= \frac{\pi}{20} [1 + 2.82844] = 0.60137.\end{aligned}$$

This is to be compared with the exact value

$$\int_0^{3\pi/2} \cos 15x\, dx = \frac{1}{15} \sin 15x \Big|_0^{3\pi/2} = \frac{1}{15} \doteq 0.06667.$$

From (3.8) we obtain the error bound

$$|E| = \frac{5}{90} \left(\frac{3\pi}{20}\right)^5 (15)^4 \,|\cos 15\xi| \leqslant \frac{1}{270} \left(\frac{9\pi}{4}\right)^5 \leqslant 65.4.$$

Thus there is no conflict; the approximate value is certainly within 65.4 units of the exact value. However, some of the practical readers would probably consider a relative error of 802% to be excessive.

There is a technique due to Filon [9] which is designed for integrals of the form

$$\int_a^b f(x) \cos kx \, dx \qquad \text{or} \qquad \int_a^b f(x) \sin kx \, dx.$$

The Filon method is quite similar to the composite Simpson's rule. First, subdivide the interval $[a,b]$ into an even number $(2n)$ of subintervals of length h. Let $x_j = a + jh$, and let p_j be the polynomial which interpolates f at the points x_{2j-2}, x_{2j-1}, and x_{2j}. Then

$$\int_a^b f(x) \sin kx \, dx = \sum_{j=1}^{n} \int_{x_{2j-2}}^{x_{2j}} f(x) \sin kx \, dx$$

$$\doteq \sum_{j=1}^{n} \int_{x_{2j-2}}^{x_{2j}} p_j(x) \sin kx \, dx = \sum_{j=0}^{2n} w_j f(x_j),$$

where the w_j's are constants independent of f. We will not go through the somewhat tedious computation required to evaluate the w_j's. The results are summarized in the following formulas:

$$\int_a^b f(x) \sin kx \, dx = h[\alpha(f(a) \cos ka - f(b) \cos kb) + \beta S_2 + \gamma S_1] + E_1, \tag{5.1}$$

$$\int_a^b f(x) \cos kx \, dx = h[\alpha(f(b) \sin kb - f(a) \sin ka) + \beta C_2 + \gamma C_1] + E_2, \tag{5.2}$$

where

$$\alpha = \frac{1}{\phi} + \frac{\cos\varphi \sin\varphi}{\varphi^2} - \frac{2\sin^2\varphi}{\varphi^3},$$

$$\beta = 2\left[\frac{1+\cos^2\varphi}{\varphi^2} - \frac{2\sin\varphi\cos\varphi}{\varphi^3}\right],$$

$$\gamma = 4\left[\frac{\sin\varphi}{\varphi^3} - \frac{\cos\varphi}{\varphi^2}\right],$$

$$\varphi = kh,$$

$$S_2 = \sum_{j=1}^{n-1} f(x_{2j}) \sin kx_{2j} + \tfrac{1}{2}[f(a) \sin ka + f(b) \sin kb],$$

$$S_1 = \sum_{j=1}^{n} f(x_{2j-1}) \sin kx_{2j-1},$$

$$C_2 = \sum_{j=1}^{n-1} f(x_{2j}) \cos kx_{2j} + \tfrac{1}{2}[f(a) \cos ka + f(b) \cos kb],$$

$$C_1 = \sum_{j=1}^{n} f(x_{2j-1}) \cos kx_{2j-1},$$

$$E_1 = \frac{(b-a)h^3}{12}\left(1 - \frac{1}{16\cos(\frac{kh}{4})}\right) \sin\frac{kh}{2}\, f^{(4)}(\xi) \qquad a < \xi < b.$$

E_2 is given by the same expression as E_1, but the value of ξ would not in general be the same for both.

Davis and Rabinowitz [6] recommend that h be chosen sufficiently small so that $\varphi < 1$. If φ is small, then α, β, and γ should be computed from their Taylor series expansions. Filon [9] has tabulated α, β, and γ for values of φ between 0 and $\pi/4$.

EXAMPLE 5.2. Let us compare Simpson's and Filon's rules on $\int_0^1 e^x \cos (5/2)\pi x \, dx$ with $h = 0.1$. The quantities C_1 and C_2 occur in both formulas.

$$\begin{aligned} C_1 &= e^{0.1} \cos \frac{\pi}{4} + e^{0.3} \cos \frac{3\pi}{4} + e^{0.5} \cos \frac{5\pi}{4} \\ &\quad + e^{0.7} \cos \frac{7\pi}{4} + e^{0.9} \cos \frac{9\pi}{4} \\ &= 0.707\,107(e^{0.1} - e^{0.3} - e^{0.5} + e^{0.7} + e^{0.9}) \\ &= 1.824\,298. \end{aligned}$$

$$\begin{aligned} C_2 &= \tfrac{1}{2}[1 + e \cos \tfrac{5}{2}\pi] + e^{0.2} \cos \frac{\pi}{2} + e^{0.4} \cos \pi \\ &\quad + e^{0.6} \cos \frac{3\pi}{2} + e^{0.8} \cos 2\pi \\ &= 1.233\,716. \end{aligned}$$

Now $\varphi = kh = (5/2)\pi(.1) = \pi/4$. Filon gives the following values of α, β, and γ for $\varphi = \pi/4$.

$$\begin{aligned} \alpha &= 0.019\,711, \\ \beta &= 0.735\,220, \\ \gamma &= 1.252\,878. \end{aligned}$$

Thus Filon's rule gives

$$\begin{aligned} \int_0^1 e^x \cos \tfrac{5}{2}\pi x \, dx &\doteq 0.1[0.019\,711e + 0.735\,220(1.233\,716) \\ &\quad + 1.252\,878(1.824\,298)] \\ &= 0.324\,626. \end{aligned}$$

Simpson's rule gives

$$\begin{aligned} \int_0^1 e^x \cos \tfrac{5}{2}\pi x \, dx &\doteq \frac{h}{3}[2C_2 + 4C_1] \\ &= \frac{0.1}{3}[2(1.233\,716) + 4(1.824\,298)] \\ &= 0.325\,487. \end{aligned}$$

The exact value is

$$\int_0^1 e^x \cos \tfrac{5}{2}\pi x \, dx = \left[\frac{e^x (\cos \frac{5}{2}\pi x + \frac{5}{2}\pi \sin \frac{5}{2}\pi x)}{1 + (\frac{5}{2}\pi)^2} \right]_0^1$$

$$= \frac{\frac{5}{2}\pi e - 1}{1 + (\frac{5}{2}\pi)^2} \doteq 0.324\,628.$$

A bound on the error in Filon's rule is

$$|E| \leqslant \frac{10^{-3}}{12} \left(1 - \frac{1}{16 \cos \frac{\pi}{16}} \right) e \sin \frac{\pi}{8}$$

$$\leqslant 8.12 \cdot 10^{-5}.$$

Several other methods have been devised for integrals of this type. See, for example, Goldberg and Varga [10], or Stetter [16].

PROBLEMS

5.1. Sketch a graph of the curve $y = \cos 15x$ and show by means of this graph why Simpson's rule provides a poor approximation in Example 5.1.

5.2. Use Filon's method with $h = 1/12$ to approximate

$$\int_0^{1/2} e^x \sin 6\pi x \, dx.$$

Compute a bound on the error and compare with the exact value.

6. NUMERICAL DIFFERENTIATION

The basic idea in numerical differentiation is the same as that used for integration; namely, construct an interpolation polynomial p for the function f and use $p'(x)$ to approximate $f'(x)$. Various formulas can be developed which depend on the choice of the interpolation points. Many of the formulas are simpler if the point where the derivative is to be evaluated is chosen as an interpolation point.

Let us begin by deriving several formulas based on a second-degree interpolation polynomial. Suppose that an approximate derivative for f is to be calculated at the point x_1. Let h be a positive number, $x_0 = x_1 - h$, and $x_2 = x_1 + h$. Now the polynomial $p \in \mathcal{P}_2$ which interpolates f at x_0, x_1, and x_2 is given by

$$p(x) = \frac{f(x_0)(x-x_1)(x-x_2)}{2h^2} - \frac{f(x_1)(x-x_0)(x-x_2)}{h^2} + \frac{f(x_2)(x-x_0)(x-x_1)}{2h^2}.$$

Let $z = (x - x_1)/h$ to obtain

$$p(x) = \frac{f(x_0)(z^2 - z)}{2} - f(x_1)(z^2 - 1) + \frac{f(x_2)(z^2 + z)}{2}.$$

Now differentiate and evaluate at $z = 0$. This gives

$$p'(x_1) = [f(x_0)(z - \tfrac{1}{2}) - 2zf(x_1) + f(x_2)(z + \tfrac{1}{2})] \frac{dz}{dx} = \frac{f(x_2) - f(x_0)}{2h}.$$

Thus we have the numerical differentiation formula

$$f'(x) \doteq \frac{f(x+h) - f(x-h)}{2h}. \tag{6.1}$$

In a similar manner, one can compute p'' to obtain

$$f''(x) \doteq \frac{f(x+h) - 2f(x) + f(x-h)}{h^2} = \frac{\Delta^2 f(x-h)}{h^2}, \tag{6.2}$$

where Δ is the forward difference operator defined in Section 5 of Chapter 5. Similar differentiation formulas can be derived using higher-degree interpolation polynomials. Such formulas are tabulated in the NBS Handbook [1].

Before deriving an error formula, let us prove a theorem about the error term in Lagrange interpolation.

THEOREM 6.1. *Suppose f'' is continuous on $[a,b]$, $x_0, x_1, \ldots, x_n$ are distinct points in $[a,b]$, and $p \in \mathcal{P}_n$ is the Lagrange interpolation polynomial for f at these points. Let $\pi(x) = (x - x_0)(x - x_1)\ldots(x - x_n)$ and*

$$g(x) = \begin{cases} \dfrac{f(x) - p(x)}{\pi(x)} & \text{if } x \neq x_j, \\[2ex] \dfrac{f'(x) - p'(x)}{\pi'(x)} & \text{if } x = x_j \text{ for some } j. \end{cases} \tag{6.3}$$

Then g' exists on $[a,b]$.

Proof. Let $\varphi(x) = f(x) - p(x)$, and for a fixed value of j, define

$$\psi(x) = \begin{cases} \dfrac{\varphi(x)}{x - x_j} & \text{if } x \neq x_j, \\[2ex] \varphi'(x) & \text{if } x = x_j. \end{cases}$$

We wish to show that $\psi'(x_j)$ also exists. Using Taylor's theorem and the fact that $\varphi(x_j) = 0$, we have

$$\varphi(x) = (x - x_j)\varphi'(x_j) + \frac{(x - x_j)^2}{2}\varphi''(\xi),$$

where ξ lies between x and x_j. Hence

$$\psi(x) - \psi(x_j) = \frac{\varphi(x)}{x - x_j} - \varphi'(x_j) = \frac{(x - x_j)}{2}\varphi''(\xi).$$

It then follows that

$$\lim_{x \to x_j} \frac{\psi(x) - \psi(x_j)}{x - x_j} = \lim_{x \to x_j} \frac{\varphi''(\xi)}{2} = \frac{\varphi''(x_j)}{2}.$$

Hence $\psi'(x_j)$ exists.

Next, let

$$R(x) = \prod_{\substack{k=0 \\ k \neq j}}^{n} (x - x_k)$$

and note that $g(x) = \psi(x)/R(x)$ in the vicinity of x_j. But R is a polynomial and $R(x_j) \neq 0$. Hence $g'(x_j)$ exists. Since j was fixed but arbitrary, we conclude that $g'(x_j)$ exists for all j, $j = 0, 1, 2, \ldots, n$. There was never any doubt that $g'(x)$ exists if $x \neq x_j$; so the proof is complete.

THEOREM 6.2. *Suppose $n \geqslant 1$ and $f^{(n+1)}$ is continuous on $[a,b]$. Let p and π be as in Theorem 6.1. If x_j is an interpolation point, then*

$$f'(x_j) - p'(x_j) = \frac{f^{(n+1)}(\xi)}{(n+1)!} \prod_{\substack{k=0 \\ k \neq j}}^{n} (x_j - x_k),$$

where $\xi \in [a,b]$.

Proof. Let g be defined by (6.3) and note that

$$f(x) - p(x) = g(x)\pi(x) \tag{6.4}$$

for all $x \in [a,b]$. Differentiate (6.4) and use the fact that $\pi(x_j) = 0$ to obtain

$$f'(x_j) - p'(x_j) = g(x_j)\pi'(x_j). \tag{6.5}$$

Now if $x \neq x_j$, then

$$g(x) = \frac{f(x) - p(x)}{\pi(x)} = \frac{f^{(n+1)}(\xi(x))}{(n+1)!}.$$

Let m and M be the minimum and maximum values of $f^{(n+1)}$ over $[a,b]$. Then for $x \neq x_j$ we have

$$m \leqslant (n+1)!g(x) \leqslant M.$$

But g is continuous at x_j. Hence

$$m \leqslant \lim_{x \to x_j} (n+1)!g(x) = (n+1)!g(x_j) \leqslant M.$$

Thus

$$g(x_j) = \frac{f^{(n+1)}(\xi)}{(n+1)!} \tag{6.6}$$

for some $\xi \in [a,b]$. The result now follows from (6.5) and (6.6) after noting that

$$\pi'(x_j) = \prod_{\substack{k=0 \\ k \neq j}}^{n} (x_j - x_k).$$

EXAMPLE 6.1. Let $f(x) = \sqrt{x}$. Using (6.1) we have

$$f'(2) \doteq \frac{\sqrt{2+h} - \sqrt{2-h}}{2h}.$$

The table below gives the approximate value of $f'(2)$ for several different values of h. These were calculated from this formula using fixed-point arithmetic with numbers truncated to four decimal places. The exact value is $1/(2\sqrt{2}) \doteq 0.353\,553$.

h	$\sqrt{2+h}$	$\sqrt{2-h}$	*Approx. Value* $f'(2)$	*Approx. Error*
1	1.7320	1.0000	0.3660	−0.012447
0.5	1.5811	1.2247	0.3564	−0.002847
0.1	1.4491	1.3784	0.3535	0.000053
0.05	1.4317	1.3964	0.3530	0.000553
0.01	1.4177	1.4106	0.3550	−0.001447
0.005	1.4159	1.4124	0.3500	0.003553
0.001	1.4145	1.4138	0.3500	0.003553
0.0005	1.4143	1.4140	0.3000	0.053553
0.0001	1.4142	1.4141	0.5000	−0.146447

TABLE 6.1. Approximate Values of $f'(2)$ as a Function h.

The thing to notice here is that although theoretically

$$\lim_{h \to 0} \frac{f(x+h) - f(x-h)}{2h} = f'(x),$$

in practice round-off error is important and the smallest value of h does not necessarily give the best approximation. A bound on the error, obtained from Theorem 6.2, is

$$|E| = \left| \frac{f^{(3)}(\xi)}{3!} \prod_{\substack{k=0 \\ k \neq 1}}^{2} (x_1 - x_k) \right| = \frac{3h^2}{6 \cdot 8\xi^{5/2}} .$$

For example, if $h = 0.1$, this gives

$$|E| \leqslant \frac{10^{-2}}{16(1.9)^{5/2}} \doteq 1.26 \cdot 10^{-4}.$$

Actually, we also have

$$|E| \geqslant \frac{10^{-2}}{10(2.1)^{5/2}} \doteq 9.78 \cdot 10^{-5}.$$

This is inconsistent with the observed error of $5.3 \cdot 10^{-5}$, but the discrepancy is caused by round-off. A more accurate computation with $h = 0.1$ gives

$$f'(2) \doteq 0.353\ 664.$$

The error is now about $-1.11 \cdot 10^{-4}$, which is consistent with the upper and lower bounds computed for $|E|$.

EXAMPLE 6.2. Let $f(x) = e^x$. Using a four-place exponential table and (6.1) with $h = 0.01$, we get

$$f'(2) \doteq \frac{7.4633 - 7.3155}{0.02} = 7.39.$$

The NBS Handbook [1] gives the five-point formula

$$f'(x) \doteq \frac{8[f(x+h) - f(x-h)] - [f(x+2h) - f(x-2h)]}{12h} . \tag{6.7}$$

Again using a four-place exponential table with $h = 0.01$, we get

$$f'(2) \doteq \frac{8(0.1478) - 0.2956}{0.12} = 7.39.$$

Note that our result only has three significant digits even though the exponential table had five significant digits. (The actual value of $f'(2)$ is $e^2 = 7.38905. \ldots$) This illustrates an inherent difficulty in numerical differentiation. By the very nature of differentiation we are forced to subtract quantities which are close to each other. This frequently leads to a loss of significant digits. With very accurate table entries, the error in approximating $f'(2)$ by (6.7) would be about $2.4 \cdot 10^{-9}$.

An error formula which is applicable to higher derivatives is given in Isaacson and Keller [13].

PROBLEMS

6.1. Let $f(x) = ln\ x$. Approximate $f'(3)$ using:
(a) Formula (6.1) with $h = 0.1$.
(b) Formula (6.7) with $h = 0.1$.
Find upper and lower bounds on the magnitude of the error in each case and compare with the observed error. If the magnitude of the observed error in (b) is less than the lower bound or greater than the upper bound, dig out a table which gives $ln\ x$ to many significant digits and see if the new observed error lies between the upper and lower bounds.

6.2. Let $f(x) = \cosh(x)$ (the hyperbolic cosine). Approximate $f''(1.5)$ and $f'''(1.5)$ using formulas obtained from the NBS Handbook [1]. Compare your results with the exact values.

6.3. Suppose the values of $f(x + h)$ and $f(x - h)$ in formula (6.1) are obtained from a table in which the entries have been rounded to m decimal digits. Find an upper bound on the error produced by the table rounding.

7. NUMERICAL SOLUTION OF DIFFERENTIAL EQUATIONS

In this section we will consider the problem of finding a function y which satisfies the equations

$$y'(x) = f(x,y(x)), \qquad y(x_0) = y_0.$$

This type of problem is referred to as an *initial-value problem*. The fact that y depends on x is usually suppressed and the equations are written in the form

$$(7.1) \qquad y' = f(x,y), \qquad y(x_0) = y_0.$$

We will assume that f is a continuous function of two variables which satisfies the inequality (6.5) in Chapter 4 on a region containing the point (x_0,y_0). It was shown in Chapter 4 that under this assumption the initial-value problem will have a unique solution in the vicinity of the point x_0.

The Euler-Cauchy Method. Let us first describe the Euler-Cauchy method for obtaining an approximate solution to (7.1). This method, although not the most practical, illustrates in a simple way the basic idea involved in a number of more complicated techniques. Let h denote a positive number, $x_1 = x_0 + h$, $x_2 = x_0 + 2h$, and so forth. The number h will be called the *step size*. Let y be the exact solution to (7.1). The basic steps in the Euler-Cauchy method are as follows:

(1) Find the first-degree Taylor polynomial for y about the point x_0.

(2) Use this polynomial to obtain a value y_1 which is an approximation to $y(x_1)$.

(3) Use the value y_1 to find an *approximate* first-degree Taylor polynomial for y about x_1.

(4) Use this new polynomial to obtain a value y_2 which is an approximation to $y(x_2)$.

(5) Continue in this fashion to obtain values $y_3, y_4, \ldots$ which are approximations to $y(x_3)$, $y(x_4)$, and so on.

Since y satisfies the equations (7.1), we must have $y(x_0) = y_0$ and $y'(x_0) = f(x_0,y_0)$. Hence the first-degree Taylor polynomial for y about x_0 is given by

$$p_1(x) = y_0 + (x - x_0)f(x_0,y_0).$$

Then the approximation to $y(x_1)$ is given by

$$y_1 = p_1(x_1) = y_0 + (x_1 - x_0)f(x_0,y_0) = y_0 + hf(x_0,y_0).$$

Next, let p_2 be defined by

$$p_2(x) = y_1 + (x - x_1)f(x_1,y_1).$$

Then $y_2 = p_2(x_2) = y_1 + hf(x_1,y_1)$ is an approximation to $y(x_2)$. In general, we have the recurrence formula

$$y_{n+1} = y_n + hf(x_n,y_n). \tag{7.2}$$

Of course, when applying the Euler-Cauchy method, we simply use (7.2) and don't actually compute p_1, p_2, etc.

It should be noted that errors can accumulate rapidly in the Euler-Cauchy method. Roughly speaking, the value of y_1 is in error because we neglected all terms in the Taylor series after the first-degree term. Then p_2 is not exactly the right Taylor polynomial for y about x_1. Thus y_2 is doubly in error; first, because p_2 was not exactly the right polynomial, and second, because terms after the first degree in the Taylor series were neglected. When these errors are all in the same direction, then the error in y_n will grow relatively fast as n increases.

EXAMPLE 7.1. Consider the initial-value problem

$$y' = 2xy, \qquad y(0) = 1.$$

Let us apply the Euler-Cauchy method with $h = 0.1$. From (7.2) we have

$$y_1 = y_0 + hf(x_0,y_0) = 1 + (0.1)(2)(0)(1) = 1,$$

$$y_2 = y_1 + hf(x_1,y_1) = 1 + (0.1)(2)(0.1)(1) = 1.02,$$

$$y_3 = y_2 + hf(x_2,y_2) = 1.02 + (0.1)(2)(0.2)(1.02) = 1.0608.$$

Table 7.1 gives to four decimals (truncated) the values of $y_0, y_1, \ldots, y_{10}$ along with values of the actual solution, e^{x^2}. Note the increase in the size of the error as n increases.

n	x_n	y_n	e^{x^2}	*Error*
0	0.0	1.0000	1.0000	0.0000
1	0.1	1.0000	1.0101	0.0101
2	0.2	1.0200	1.0408	0.0208
3	0.3	1.0608	1.0942	0.0334
4	0.4	1.1244	1.1735	0.0491
5	0.5	1.2143	1.2840	0.0697
6	0.6	1.3357	1.4333	0.0976
7	0.7	1.4959	1.6323	0.1364
8	0.8	1.7053	1.8965	0.1912
9	0.9	1.9781	2.2479	0.2698
10	1.0	2.3341	2.7183	0.3842

TABLE 7.1. The Euler-Cauchy Method for $y' = 2xy$ with $h = 0.1$.

It can be shown under fairly general hypotheses that the error in the Euler-Cauchy method tends to zero as h tends to zero [3, p. 59]. Thus it is only a matter of taking a sufficiently small step size in order to obtain any desired accuracy. However, round-off error can be troublesome. In the following example, the best the Euler-Cauchy method can do is to provide two-place accuracy, even though the computations are carried out to four decimal places.

EXAMPLE 7.2. Consider the initial-value problem

$$y' = y, \qquad y(0) = 1.$$

Table 7.2 gives the approximate value of $y(1)$ for various values of h obtained by the Euler-Cauchy method with fixed-point arithmetic truncated to four decimals. The correct value is $e \doteq 2.7183$. We will see that better accuracy can be obtained using more refined methods, such as the improved Euler-Cauchy or the Runge-Kutta method.

The Improved Euler-Cauchy Method. The improved Euler-Cauchy method is a first step in the development of a more accurate scheme for solving first-order differential equations. It is described by the following formulas:

$$\begin{aligned} y^*_{n+1} &= y_n + hf(x_n, y_n), \\ y_{n+1} &= y_n + \frac{h}{2}[f(x_n, y_n) + f(x_{n+1}, y^*_{n+1})]. \end{aligned} \tag{7.3}$$

h	*Approximate Value* $y(1)$	*Error*
1	2.0000	0.7183
0.5	2.2500	0.4683
0.1	2.5934	0.1249
0.05	2.6518	0.0665
0.01	2.6975	0.0208
0.005	2.6955	0.0228
0.001	2.6329	0.0854
0.0005	2.5518	0.1665
0.0001	2.0000	0.7183

TABLE 7.2. The Euler-Cauchy Method for $y' = y$.

The basic idea is still similar to the original Euler-Cauchy method. However, after obtaining an approximation y^*_{n+1} for the value of $y(x_{n+1})$, we refine this estimate by using the average of $f(x_n,y_n)$ and $f(x_{n+1},y^*_{n+1})$ for the slope in computing y_{n+1}.

EXAMPLE 7.3. Let us consider again the initial-value problem of Example 7.2,

$$y' = y, \qquad y(0) = 1.$$

With $h = 0.1$, we obtain from (7.3)

$$y^*_1 = y_0 + hf(x_0,y_0) = 1 + (0.1)(1) = 1.1,$$

$$y_1 = y_0 + \frac{h}{2}[f(x_0,y_0) + f(x_1,y^*_1)] = 1 + 0.05[1 + 1.1] = 1.105,$$

$$y^*_2 = 1.105 + (0.1)(1.105) = 1.2155,$$

$$y_2 = 1.105 + 0.05[1.105 + 1.2155] = 1.221\,025.$$

Table 7.3 gives the error in $y(1)$ for various values of h computed with fixed-point arithmetic truncated to four decimal places. It should

	Error in $y(1)$	
h	*Euler-Cauchy*	*Improved Euler-Cauchy*
1	0.7183	0.2183
0.5	0.4683	0.0777
0.1	0.1249	0.0048
0.05	0.0665	0.0051
0.01	0.0208	0.0208
0.005	0.0228	0.0228
0.001	0.0854	0.0854

TABLE 7.3. Comparison of Euler-Cauchy and Improved Euler-Cauchy Methods.

be noted that each step in the improved method is about twice as much work as a step in the Euler-Cauchy method. In this example, the improved method had a better accuracy with $h = 0.1$ than the Euler-Cauchy did with $h = 0.01$. The better result required about one-fifth as much work.

PROBLEMS

7.1. Use the Euler-Cauchy method with $h = 0.2$ to approximate the solution to the initial-value problem

$$y' = 3x + y, \qquad y(0) = 2.$$

Compare your value for $y(1)$ with the exact value.

7.2. Compute the last entry in Table 7.2 by hand. Hint: The problem requires only a finite amount of time.

7.3. Work Problem 7.1 using the improved Euler-Cauchy method and compare with the previous results.

7.4. Write flow diagrams for the Euler-Cauchy and improved Euler-Cauchy methods.

7.5. Write a FORTRAN program for the Euler-Cauchy method.

7.6. Write a FORTRAN program for the improved Euler-Cauchy method.

7.7. Consider the initial-value problem

$$y' = y, \qquad y(0) = 1.$$

Derive a formula for y_n as a function of h and n using:
(a) The Euler-Cauchy method.
(b) The improved Euler-Cauchy method.

7.8. Suppose h tends to zero through the values $1, 1/2, 1/3, \ldots, 1/k, \ldots$. Use the result of Problem 7.7(a) to show that

$$\lim_{h \to 0} y_k = e = y(1).$$

8. THE RUNGE-KUTTA FORMULA

The Runge-Kutta method is one of the most popular methods for solving differential equations. Actually, there are many formulas of the Runge-Kutta type, one of which is the improved Euler-Cauchy formula. The particular technique which we will discuss is more accurately referred to as the *classical fourth-order Runge-Kutta method.*

Let us first explain why the classical Runge-Kutta method is considered to be better than either the Euler-Cauchy or improved Euler-Cauchy method. Let x be a fixed number and let $E_1(h)$, $E_2(h)$, and $E_3(h)$ be the errors in approximating $y(x)$ by the Euler-Cauchy, improved Euler-Cauchy, and classical Runge-Kutta methods, respectively, with step size h. Then it turns out that

$$|E_1(h)| \leqslant M_1 h,$$
$$|E_2(h)| \leqslant M_2 h^2,$$
$$|E_3(h)| \leqslant M_3 h^4,$$

where M_1, M_2, and M_3 are constants which depend on x and f but not on h. The error in the Runge-Kutta method tends to zero like h^4, whereas the errors in the two Euler-Cauchy methods only tend to zero like h^2 and h. From this, one can imagine that the convergence in the Runge-Kutta method would normally be much more rapid than in the Euler-Cauchy methods.

There are systematic ways, based on Taylor series, for deriving Runge-Kutta formulas. (See Problem 8.4 for a derivation of the improved Euler-Cauchy method by the Runge-Kutta technique.) These derivations, which can be found in Scheid [14] or Collatz [3], are quite laborious; so instead of presenting them, we have made up a little story which will hopefully make it seem plausible that the Runge-Kutta formula is an improvement over the Euler-Cauchy methods. Consider the usual problem:

$$y' = f(x,y), \qquad y(x_0) = y_0.$$

It was shown in Section 6 of Chapter 4 that these two equations are equivalent to the integral equation

$$y(x) = y_0 + \int_{x_0}^{x} f(t,y(t))\,dt. \tag{8.1}$$

Applying Simpson's rule at x_0, $x_0 + (h/2)$, and $x_1 = x_0 + h$, we have

$$y(x_1) \doteq y_0 + \frac{h}{6}\left[f(x_0,y_0) + 4f(x_0 + \tfrac{h}{2}, y(x_0 + \tfrac{h}{2})) + f(x_1, y(x_1))\right]. \tag{8.2}$$

The problem in applying (8.2) is that we don't know the values of $f(x_0 + h/2, y(x_0 + h/2))$ and $f(x_1, y(x_1))$. In the Runge-Kutta method these are estimated as follows:

$$\begin{aligned} f(x_0 + \tfrac{h}{2}, y(x_0 + \tfrac{h}{2})) &\doteq \frac{m_2 + m_3}{2}, \\ f(x_1,y(x_1)) &\doteq m_4, \end{aligned} \tag{8.3}$$

where

$$\begin{aligned} m_1 &= f(x_0,y_0), \\ m_2 &= f(x_0 + \tfrac{h}{2}, y_0 + \frac{m_1 h}{2}), \\ m_3 &= f(x_0 + \tfrac{h}{2}, y_0 + \frac{m_2 h}{2}), \\ m_4 &= f(x_0 + h, y_0 + m_3 h). \end{aligned} \tag{8.4}$$

Substitution of the values from (8.3) and (8.4) on the right side of (8.2) yields the equation

$$y(x_1) \doteq y_0 + \frac{h}{6}[m_1 + 2m_2 + 2m_3 + m_4].$$

In general, the Runge-Kutta process is described by the recurrence formula

$$y_{n+1} = y_n + \frac{h}{6}[m_1 + 2m_2 + 2m_3 + m_4], \tag{8.5}$$

where

$$m_1 = f(x_n, y_n),$$

$$m_2 = f\left(x_n + \frac{h}{2}, y_n + \frac{m_1 h}{2}\right),$$

$$m_3 = f\left(x_n + \frac{h}{2}, y_n + \frac{m_2 h}{2}\right),$$

$$m_4 = f(x_n + h, y_n + m_3 h).$$

A simple practical error bound for the Runge-Kutta method does not appear to have been discovered (see Collatz [3, p. 51]). A rule of thumb can be derived on the assumption that the error is given by

$$E(h) = Mh^4, \tag{8.6}$$

where M is a constant independent of h. (What is really true is that $|E(h)| \leqslant Mh^4$, where M is a constant.) Under this assumption, we have

$$E(2h) - E(h) = 16Mh^4 - Mh^4 = 15Mh^4. \tag{8.7}$$

Let y_h and y_{2h} be the approximate values of $y(x)$ corresponding to a step length of h and $2h$, respectively. Then from (8.7) we have

$$y(x) - y_{2h} - [y(x) - y_h] = 15Mh^4.$$

Simplifying, and using (8.6) we have

$$E(h) \doteq \frac{y_h - y_{2h}}{15}. \tag{8.8}$$

The approximation symbol was used in (8.8) to emphasize the fact that (8.6) is probably at best an approximation. In addition to using (8.8) as an error estimate, one can also use it as a correction for the value of y_h; that is,

$$y(x) = y(x) - y_h + y_h = E(h) + y_h \doteq y_h + \frac{y_h - y_{2h}}{15}.$$

EXAMPLE 8.1. Consider again the equation used in Example 7.1:

$$y' = 2xy, \qquad y(0) = 1.$$

For the first step in the Runge-Kutta method with $h = 0.2$, we have

$$m_1 = f(x_0, y_0) = 2(0)1 = 0,$$

$$m_2 = f(x_0 + \tfrac{h}{2}, y_0 + \tfrac{m_1 h}{2}) = 2(0.1)1 = 0.2,$$

$$m_3 = f(x_0 + \tfrac{h}{2}, y_0 + \tfrac{m_2 h}{2}) = 2(0.1)(1.02) = 0.204,$$

$$m_4 = f(x_0 + h, y_0 + m_3 h) = 2(0.2)(1.0408) = 0.41632.$$

Then

$$\begin{aligned} y_1 &= y_0 + \tfrac{h}{6}(m_1 + 2m_2 + 2m_3 + m_4) \\ &= 1 + \tfrac{0.2}{6}(0 + 0.4 + 0.408 + 0.41632) \\ &= 1.0408\,1067. \end{aligned}$$

To estimate the error using (8.8), Collatz [3] recommends the simultaneous calculation of the approximate values for $y(x)$ using step sizes of $2h$ and h. Taking $h = 0.1$, we have

$$m_1 = 0,$$

$$m_2 = 2(0.05)1 = 0.1,$$

$$m_3 = 2(0.05)(1.005) = 0.1005,$$

$$m_4 = 2(0.1)(1.01005) = 0.202\,010.$$

Thus

$$\begin{aligned} y_1 &= 1 + \tfrac{0.1}{6}(0 + 0.2 + 0.2010 + 0.202\,010) \\ &= 1.0100\,5016. \end{aligned}$$

Continuing with $h = 0.1$, we have

$$m_1 = 2(0.1)(1.0100\,5016) = 0.2020\,1003,$$

$$m_2 = 2(0.15)(1.0201\,5066) = 0.3060\,4519,$$

$$m_3 = 2(0.15)(1.0253\,5241) = 0.3076\,0572,$$

$$m_4 = 2(0.2)(1.0408\,1073) = 0.4163\,2429,$$

$$y_2 = y_1 + \tfrac{h}{6}(m_1 + 2m_2 + 2m_3 + m_4) = 1.0408\,1076.$$

The rule-of-thumb error estimate (8.8) now gives

$$E(0.1) \doteq \frac{1.0408\,1076 - 1.0408\,1067}{15} = \frac{9 \cdot 10^{-8}}{15} = 6 \cdot 10^{-9}.$$

The actual error is $e^{0.04} - 1.0408\,1076 \doteq 1.42 \cdot 10^{-8}$. In this calculation the round-off error (fixed-point arithmetic truncated to eight decimals) would be expected to be on the order of 10^{-8}; so it may well be the dominant factor in the error of $1.42 \cdot 10^{-8}$.

In order to compare the three methods which we have studied, let us return to the equation

$$y' = y, \qquad y(0) = 1.$$

Table 8.1 gives the error in $y(1)$ as a function of the step size for these three methods. The results were carried out with fixed-point arithmetic truncated to four decimals.

		Errors in $y(1)$	
h	*Runge-Kutta*	*Imp. Euler-Cauchy*	*Euler-Cauchy*
1	0.0101	0.2183	0.7183
0.5	0.0011	0.0777	0.4683
0.1	0.0022	0.0048	0.1249
0.05	0.0051	0.0051	0.0665
0.01	0.0208	0.0208	0.0208
0.005	0.0228	0.0228	0.0228
0.001	0.0854	0.0854	0.0854

TABLE 8.1. Comparison of Errors for Runge-Kutta, Improved Euler-Cauchy, and Euler-Cauchy Methods.

PROBLEMS

8.1. Consider the initial-value problem

$$y' = \frac{y}{x} + x^2, \qquad y(1) = 2.$$

Approximate $y(1.2)$ using the Runge-Kutta method with $h = 0.1$. As a check on the accuracy, also calculate $y(1.2)$ with $h = 0.2$. Estimate the error with the rule-of-thumb error estimate and compare with the actual error.

8.2. Consider the initial-value problem

$$y' = y, \qquad y(0) = 1.$$

For this problem we can write the recurrence formula

$$y_{n+1} = y_n \varphi(h).$$

Find $\varphi(h)$ for the Euler-Cauchy, improved Euler-Cauchy, and Runge-Kutta methods. Use $\varphi(h)$ and properties of the true solution to derive a general formula for the errors. What is the error in $y(1)$ as a function of n if $h = 1/n$? To what extent is the rule-of-thumb error estimate valid for this example?

8.3. Estimate the relative amounts of work involved in one step of the Euler-Cauchy, improved Euler-Cauchy, and Runge-Kutta methods. Check your estimates against a careful calculation of the time for arithmetic operations in Problem 8.1 if addition and subtraction take 5μ sec each and multiplications and divisions take 10μ sec each.

8.4. Consider the usual problem

$$y' = f(x,y), \qquad y(x_0) = y_0.$$

Define ϕ and z by the equations

$$\phi(h) = y_0 + h(rf(x_0, y_0) + sf(x_0 + h, y_0 + th)),$$
$$z(h) = y(x_0 + h).$$

We wish to choose the constants r, s, and t so that ϕ and z will have the same second-degree Taylor polynomials about zero, and then use $\phi(h)$ to approximate $z(h) = y(x_0 + h)$. Show that this requirement leads to the improved Euler-Cauchy formula.

9. PREDICTOR-CORRECTOR METHODS

The basic idea of a predictor-corrector formula has already been illustrated in our discussion of the Runge-Kutta method. Let

$$y(x) = y_0 + \int_{x_0}^{x} f(t,y(t))dt \tag{9.1}$$

be the integral equation equivalent to the initial-value problem $y' = f(x,y)$, $y(x_0) = y_0$. To start a predictor-corrector process, we obtain values of $y_1, y_2, \ldots, y_k$ which approximate $y(x_1), y(x_2), \ldots, y(x_k)$. These values, called *starting values*, are obtained by some other method such as the Runge-Kutta. From (9.1) we obtain

$$y(x_{k+1}) = y(x_m) + \int_{x_m}^{x_{k+1}} f(t,y(t))dt. \tag{9.2}$$

Now a predictor method is obtained by approximating the integral in (9.2) with a quadrature rule which involves some or all of the points $x_0, x_1, \ldots, x_k$. In addition, the value of $y(x_j)$ is approximated by y_j, $j = 0, 1, \ldots, k$. This results in an approximation for $y(x_{k+1})$, called a *predicted value*, given by an equation of the form

$$y_{k+1}^* = y_m + h[\alpha_0 f(x_0,y_0) + \alpha_1 f(x_1,y_1) + \ldots + \alpha_k f(x_k,y_k)].$$

Of course, the α_j's depend on the value of m and the particular quadrature rule used to approximate the integral in (9.2). A (hopefully) better approximation to $y(x_{k+1})$, called a *corrected value*, is then obtained by approximating the integral in (9.2) by a quadrature rule involving x_{k+1} as well as some or all of the points $x_0, x_1, \ldots, x_k$. This results in an equation of the form

$$y_{k+1} = y_\mu + h[\beta_0 f(x_0,y_0) + \ldots + \beta_k f(x_k,y_k) + \beta_{k+1} f(x_{k+1},y_{k+1}^*)].$$

In general, the predictor-corrector scheme is described by the recurrence formulas:

$$y^*_{n+1} = y_{n+m-k} + h[\alpha_0 f(x_{n-k}, y_{n-k}) + \ldots + \alpha_k f(x_n, y_n)], \tag{9.3}$$

$$\begin{aligned} y_{n+1} = y_{n+\mu-k} &+ h[\beta_0 f(x_{n-k}, y_{n-k}) + \ldots + \beta_k f(x_n, y_n) \\ &+ \beta_{k+1} f(x_{n+1}, y^*_{n+1})]. \end{aligned} \tag{9.4}$$

Equation (9.3) is called the *predictor* and (9.4) is called the *corrector*.

The principal advantages of the predictor-corrector schemes are that they are faster than the Runge-Kutta and they provide an error estimate with less work than the Runge-Kutta. Their speed is accounted for by the fact that they require only two evaluations of f per step whereas the Runge-Kutta requires four. (We assume that the values of $f(x_{n-k}, y_{n-k}), \ldots, f(x_{n-1}, y_{n-1})$ will be stored.) The main disadvantages are that the predictor-corrector procedures are more complicated to program and are sometimes unstable (see Conte [4], Henrici [11], and Section 11).

EXAMPLE 9.1. *Milne's three-point method* uses an open-type quadrature formula over the interval $[x_{n-3}, x_{n+1}]$ for a predictor. By an open-type formula we mean that the end points x_{n-3} and x_{n+1} are not used as interpolation points. Using a slight modification of formula (3.3), we obtain

$$\alpha_1 = \int_0^4 \frac{(z-2)(z-3)}{2}\, dz = \frac{8}{3},$$

$$\alpha_2 = \int_0^4 \frac{(z-1)(z-3)}{-1}\, dz = \frac{-4}{3},$$

$$\alpha_3 = \int_0^4 \frac{(z-1)(z-2)}{2}\, dz = \frac{8}{3}.$$

The predictor is obtained by substituting these values along with $\alpha_0 = 0$, $m = 0$, and $k = 3$ in (9.3). Simpson's rule is used as a corrector with $\mu = 2$, $\beta_0 = \beta_1 = 0$; so Milne's method is described by the recurrence formulas:

$$y^*_{n+1} = y_{n-3} + \frac{4h}{3}[2f(x_{n-2}, y_{n-2}) - f(x_{n-1}, y_{n-1}) + 2f(x_n, y_n)],$$

$$y_{n+1} = y_{n-1} + \frac{h}{3}[f(x_{n-1}, y_{n-1}) + 4f(x_n, y_n) + f(x_{n+1}, y^*_{n+1})].$$

An estimate of the truncation error incurred in each step is given by

$$E_n \doteq \frac{-1}{29}(y_{n+1} - y^*_{n+1}). \tag{9.5}$$

(See Problem 9.2.) As in the Runge-Kutta method, E_n can be added to y_{n+1} to correct the corrected value. This term is sometimes referred to as the *mop-up term*.

Another popular predictor-corrector scheme is the *Adams-Moulton method*. The formulas for this method are:

$$y_{n+1}^* = y_n + \frac{h}{24}[-9f_{n-3} + 37f_{n-2} - 59f_{n-1} + 55f_n],$$

$$y_{n+1} = y_n + \frac{h}{24}[f_{n-2} - 5f_{n-1} + 19f_n + 9f_{n+1}],$$

where $f_{n+1} = f(x_{n+1}, y_{n+1}^*)$ and $f_{n-j} = f(x_{n-j}, y_{n-j})$, $j = 0, 1, 2, 3$. The estimated error for each step is given by

$$E_n \doteq \frac{-1}{14}(y_{n+1} - y_{n+1}^*).$$

The Adams-Moulton method has somewhat more desirable stability properties than the Milne method (see Conte [4]).

PROBLEMS

9.1. Show that the improved Euler-Cauchy method is of the predictor-corrector type.

9.2. Show how formula (9.5) is derived. Hint: Look up the error expressions for the open-quadrature rule and Simpson's rule.

9.3. Derive the formulas for the Adams-Moulton method. Hint: The predictor approximates

$$\int_{x_n}^{x_{n+1}} f(t,y(t))dt$$

by interpolating at x_{n-3}, x_{n-2}, x_{n-1}, and x_n.

9.4. Apply the Adams-Moulton method to the equation

$$y' = 2xy, \qquad y(0) = 1.$$

Use Table 7.1 for starting values and compare your results with the rest of that table. (Actually, it would be better to use the Runge-Kutta method to obtain the starting values.)

10. SYSTEMS AND HIGHER-ORDER EQUATIONS

A common way to handle a higher-order differential equation is to replace it by an equivalent system of first-order equations. To see how this is done, consider, for example, the initial-value problem

$$y''' = f(x,y,y',y''), \qquad (10.1)$$
$$y(x_0) = y_0,\ y'(x_0) = u_0,\ y''(x_0) = v_0.$$

Now let $u = y'$ and $v = y''$. The initial-value problem (10.1) is then equivalent to the system of first-order equations

$$
\begin{aligned}
y' &= u, \\
u' &= v, \\
v' &= f(x,y,u,v),
\end{aligned}
\tag{10.2}
$$

with initial conditions $u(x_0) = u_0$, $v(x_0) = v_0$ and $y(x_0) = y_0$. The idea is similar for equations of order higher than three; however, the notation needed to discuss the problem in general tends to be complicated. Instead of considering only the problem (10.2), we will consider a general system of three equations given by

$$
\begin{aligned}
u' &= F(x,u,v,y), \qquad u(x_0) = u_0; \\
v' &= G(x,u,v,y), \qquad v(x_0) = v_0; \\
y' &= H(x,u,v,y), \qquad y(x_0) = y_0.
\end{aligned}
\tag{10.3}
$$

Such systems arise in various physical problems as well as coming from higher-order equations such as (10.1).

The general idea in treating the system (10.3) is to step each solution along as in the one-dimensional case. For example, the Euler-Cauchy method is carried out by the three recurrence formulas

$$
\begin{aligned}
u_{n+1} &= u_n + hF(x_n,u_n,v_n,y_n), \\
v_{n+1} &= v_n + hG(x_n,u_n,v_n,y_n), \\
y_{n+1} &= y_n + hH(x_n,u_n,v_n,y_n),
\end{aligned}
\tag{10.4}
$$

where, as usual, h is the step size. The Runge-Kutta method is given by the recurrence formulas

$$
\begin{aligned}
u_{n+1} &= u_n + \frac{h}{6}(k_1 + 2k_2 + 2k_3 + k_4), \\
v_{n+1} &= v_n + \frac{h}{6}(l_1 + 2l_2 + 2l_3 + l_4), \\
y_{n+1} &= y_n + \frac{h}{6}(m_1 + 2m_2 + 2m_3 + m_4),
\end{aligned}
\tag{10.5}
$$

where

$$
\begin{aligned}
k_1 &= F(x_n,u_n,v_n,y_n), \\
k_2 &= F(x_n + \tfrac{h}{2}, u_n + \frac{k_1h}{2}, v_n + \frac{l_1h}{2}, y_n + \frac{m_1h}{2}), \\
k_3 &= F(x_n + \tfrac{h}{2}, u_n + \frac{k_2h}{2}, v_n + \frac{l_2h}{2}, y_n + \frac{m_2h}{2}), \\
k_4 &= F(x_n + h, u_n + k_3h, v_n + l_3h, y_n + m_3h);
\end{aligned}
\tag{10.6}
$$

h is the step size, and the l_i's and m_i's are defined by formulas similar to (10.6) involving G and H, respectively. The error estimate (8.8) can be used to obtain a rough idea of the error in u, v, and y.

EXAMPLE 10.1. Consider the initial-value problem

$$y'' + 4y' - 5y = 5x - 4,$$

$$y(0) = 1, \qquad y'(0) = 0.$$

Let $u = y'$. Then this equation is equivalent to the system

$$y' = u, \qquad y(0) = 1;$$

$$u' = 5(x + y) - 4(u + 1), \qquad u(0) = 0.$$

Applying the Runge-Kutta formulas with $h = 0.5$, we have

$$k_1 = 5(x_0 + y_0) - 4(u_0 + 1) = 1,$$

$$m_1 = u_0 = 0,$$

$$k_2 = 5(x_0 + \frac{h}{2} + y_0 + \frac{m_1 h}{2}) - 4(u_0 + \frac{k_1 h}{2} + 1) = 1.25,$$

$$m_2 = u_0 + \frac{k_1 h}{2} = 0.25,$$

$$k_3 = 5(x_0 + \frac{h}{2} + y_0 + \frac{m_2 h}{2}) - 4(u_0 + \frac{k_2 h}{2} + 1) = 1.3125,$$

$$m_3 = u_0 + \frac{k_2 h}{2} = 0.3125,$$

$$k_4 = 5(x_0 + h + y_0 + m_3 h) - 4(u_0 + k_3 h + 1) = 1.65625,$$

$$m_4 = u_0 + k_3 h = 0.65625,$$

$$u(0.5) \doteq u_1 = u_0 + \frac{h}{6}(k_1 + 2k_2 + 2k_3 + k_4) = 0.6484\,3750,$$

$$y(0.5) \doteq y_1 = y_0 + \frac{h}{6}(m_1 + 2m_2 + 2m_3 + m_4) = 1.1484\,3750.$$

Using $h = 0.25$ and two steps, we get

$$y(0.25) \doteq y_1 = 1.0340\,1692,$$

$$y(0.5) \doteq y_2 = 1.1486\,9945.$$

The error estimate (8.8) gives

$$E \doteq \frac{1.1486\,9945 - 1.1484\,3750}{15} \doteq 1.746 \cdot 10^{-5}.$$

The actual solution is given by $y = e^x - x$; hence

$$y(0.5) = e^{0.5} - 0.5 \doteq 1.1487\,2127$$

and

$$E = e^{0.5} - 0.5 - 1.1486\,9945 \doteq 2.182 \cdot 10^{-5}.$$

It should be noted that there are some direct methods for handling higher-order equations—that is, methods which do not require the replacement of the higher-order equation by a system of first-order equations. A discussion of such methods is given in Henrici [11, Chap. 6].

PROBLEMS

10.1. Consider the initial-value problem

$$y'' + 4y = 0, \qquad y(0) = 1, \qquad y'(0) = 0.$$

Approximate $y(0.4)$ using the Runge-Kutta method with $h = 0.4$ and $h = 0.2$. Use (8.8) to estimate the error and compare with the actual error.

10.2. Use slide rule accuracy and the Euler-Cauchy method with $h = 0.1$ to approximate $y(1)$ in the initial-value problems:
(a) $y'' = y, \quad y(0) = 1, \quad y'(0) = 1$
(b) $y'' = 11y' + 180y, \quad y(0) = 45, \quad y'(0) = -405$
Compare your result with the actual value of $y(1)$. In part (b) the relative error might be quite high. Can you explain the difficulty?

11. STABILITY PROBLEMS

Speaking very loosely, a numerical process is called unstable if it goes wild after a while. Let us first illustrate an unstable process by a numerical example.

EXAMPLE 11.1. Consider the initial-value problem

$$(11.1) \qquad y' = 4(x - y) + 1, \qquad y(0) = 1.$$

In view of formula (6.1), it would appear reasonable to find an approximate solution to (11.1) by solving the difference equation

$$\frac{y_{n+1} - y_{n-1}}{2h} = 4(x_n - y_n) + 1, \qquad y_0 = 1,$$

where h, as usual, is the step size. For computational purposes, let us write this difference equation in the form

$$(11.2) \qquad y_{n+1} = y_{n-1} + 2h[4(x_n - y_n) + 1].$$

In order to get the solution of (11.2) started, we need a value for y_1, as well as y_0. Using the improved Euler-Cauchy method with $h = 0.1$, we obtain $y_1 = 0.78$. Then from ((11.2) we have

$$\begin{aligned} y_2 &= y_0 + 2h[4(x_1 - y_1) + 1] \\ &= 1 + 0.2[4(0.1 - 0.78) + 1] \\ &= 0.6560 \end{aligned}$$

$$\begin{aligned} y_3 &= y_1 + 2h[4(x_2 - y_2) + 1] \\ &= 0.78 + 0.2[4(0.2 - 0.656) + 1] \\ &= 0.6152 \end{aligned}$$

The values of $y_4, \ldots, y_{22}$ were calculated in this fashion using fixed-point arithmetic truncated to four decimals. The value at every other point is given in Table 11.1 along with the corresponding value for the Euler-Cauchy method. The actual solution, $y = e^{-4x} + x$, is tabulated in the next column, while the last column shows the values computed by means of (11.2) with $h = 0.05$.

		Formula (11.2) $h = 0.1$	Euler-Cauchy $h = 0.1$		Formula (11.2) $h = 0.05$
n	x_n	y_n	y_n	$e^{-4x_n} + x_n$	y_{2n}
0	0.0	1.0000	1.0000	1.0000	1.0000
2	0.2	0.6560	0.5600	0.6493	0.6516
4	0.4	0.6039	0.5296	0.6019	0.6036
6	0.6	0.6823	0.6466	0.6907	0.6911
8	0.8	0.8134	0.8167	0.8408	0.8394
10	1.0	0.9531	1.0060	1.0183	1.0139
12	1.2	1.0628	1.2021	1.2082	1.1978
14	1.4	1.0848	1.4007	1.4037	1.3802
16	1.6	0.9052	1.6002	1.6017	1.5496
18	1.8	0.2809	1.8000	1.8007	1.6854
20	2.0	−1.3155	2.0000	2.0003	1.7451
22	2.2	−5.0339	2.2000	2.2002	1.6354

TABLE 11.1. Example of an Unstable Process.

In Table 11.1 note that the values of y_n computed by (11.2) with $h = 0.1$ bear no resemblance to the actual solution for $x_n \geqslant 1.6$. With the step size halved (last column), the resemblance is better but not very good. This is one example of an unstable numerical process.

In order to explain the cause of the instability, we will need to examine the nature of the solution to a difference equation such as (11.2). The theory of such difference equations was outlined in Example 5.1 of Chapter 2. Let us rewrite (11.2) in the form

(11.3) $$y_{n+1} + 8hy_n - y_{n-1} = 2h(4x_n + 1).$$

The characteristic polynomial for (11.3) is given by

$$p(z) = z^2 + 8hz - 1.$$

Thus every solution of (11.3) is of the form

(11.4) $$y_n = c_1 z_1^n + c_2 z_2^n + w_n,$$

where z_1 and z_2 are the zeros of p, c_1 and c_2 are constants, and w_n is a particular solution to (11.3). There are systematic ways to compute a particular solution; but following our custom, we will pull it out of a hat. It is easy to check that

$$w_n = x_n = nh$$

satisfies the difference equation (11.3) and is thus a particular solution. Substituting this in (11.4) and computing the zeros of p, we have

$$y_n = c_1 z_1^n + c_2 z_2^n + nh, \tag{11.5}$$

where

$$\begin{aligned} z_1 &= -4h + \sqrt{16h^2 + 1}\,, \\ z_2 &= -4h - \sqrt{16h^2 + 1}\,. \end{aligned} \tag{11.6}$$

The error in the approximation to the solution of (11.1) is then given by

$$E_n = y_n - e^{-4nh} - nh = c_1 z_1^n + c_2 z_2^n - e^{-4nh}.$$

Next, it can be shown (Problem 11.3) that $|z_1| < 1$ and $|z_2| > 1$. Hence for a fixed h,

$$\lim_{n\to\infty} |z_1|^n = 0 \qquad \text{and} \qquad \lim_{n\to\infty} |z_2|^n = \infty.$$

Thus, unless $c_2 = 0$, we will have

$$\lim_{n\to\infty} |E_n| = \infty.$$

For approximation purposes, it would appear desirable to have $c_2 = 0$. But a small round-off error, or a small inaccuracy in the initial values will cause the solution (11.5) to pick up a small non-zero value for c_2. Then this error will grow exponentially since c_2 is multiplied by z_2^n and $|z_2| > 1$. This is the cause of the instability observed in Table 11.1.

In order to discuss the stability problem in general, let us consider the class of approximation procedures known as *multistep methods.* A technique for approximating the solution to the initial-value problem

$$y' = f(x,y), \qquad y(x_0) = y_0 \tag{11.7}$$

is called a multistep method if the differential equation in (11.7) is replaced by a difference equation of the form

$$\begin{aligned} &\alpha_k y_{n+k} + \alpha_{k-1} y_{n+k-1} + \ldots + \alpha_0 y_n \\ &\quad = h[\beta_k f(x_{n+k}, y_{n+k}) + \ldots + \beta_0 f(x_n, y_n)]. \end{aligned} \tag{11.8}$$

This class includes the Euler-Cauchy and predictor-corrector methods if one recognizes the fact that a predictor formula helps to provide an approximate solution to a corrector equation of the form (11.8). The classical fourth-order Runge-Kutta procedure is not strictly a multistep method, but much of the analysis presented here can be readily carried out for that technique.

The general solution to a difference equation of the form (11.8) can be written in the form

$$(11.9) \qquad y_n = c_1 u_{1n} + c_2 u_{2n} + \ldots + c_m u_{mn},$$

where $c_1, \ldots, c_m$, and $u_{1n}, \ldots, u_{mn}$ depend on h. If x is fixed and $h = (x - x_0)/n$, then normally one term in (11.9), say $c_1 u_{1n}$, converges as $n \to \infty$ to the true solution $y(x)$ of the initial-value problem (11.7). The other terms in (11.9) are extraneous and the ideal situation would be to have $c_2 = c_3 = \ldots = c_m = 0$. However, round-off errors and errors in the initial values $y_0, y_1, \ldots, y_{k-1}$ usually cause some of these constants to take on non-zero values. Then if one of the extraneous solutions grows large with increasing n, it may become an important source of error. We will not give a precise mathematical description of the term "stability," but the general idea is that a procedure is considered to be stable if there is no tendency for the extraneous solutions in (11.9) to cause significant errors in the approximate solution of the initial-value problem.

Stability analysis for a multistep method is difficult because it is hard to obtain general expressions for the solutions $u_{1n}, u_{2n}, \ldots, u_{mn}$ in (11.9). The usual approach is to study the behavior of these solutions for certain simple choices of f such as $f(x,y) = 0$, $f(x,y) = 1$, and $f(x,y) = \lambda y$. By insisting that the method exhibit stable behavior for these choices of f, we will derive several conditions which are considered to be necessary for stability.

Let us first consider the choice $f(x,y) = 0$ and $y_0 = 0$. In this case the solution to (11.8) is given by

$$(11.10) \qquad y_n = c_1 z_1^n + c_2 z_2^n + \ldots + c_k z_k^n,$$

where $z_1, \ldots, z_k$ are the zeros of the polynomial

$$\rho(z) = \alpha_k z^k + \alpha_{k-1} z^{k-1} + \ldots + \alpha_0.$$

If a root, say z_1, were repeated j times, we would have instead

$$(11.11) \qquad y_n = z_1^n (c_1 + nc_2 + \ldots + n^{j-1} c_j) + c_{j+1} z_{j+1}^n + \ldots + c_k z_k^n.$$

For this problem all of the solutions are extraneous. Thus, in order to have all the solutions remain bounded as $n \to \infty$, we require that:

(i) No zero of ρ should have a magnitude greater than 1, and
(ii) No zero of ρ having magnitude 1 should be repeated.

These requirements are often referred to as the *root condition.* A multistep method which violates the root condition is said to be strongly unstable—the growth of round-off error would tend to be a serious problem when such a method is applied to any initial-value problem.

Next, consider the initial-value problems

$$y' = 0, \qquad y_0 = 1;$$

$$y' = 1, \qquad y_0 = 0.$$

It would seem reasonable to require that a multistep method provide exact solutions to these problems when the starting values $y_0, \ldots, y_{k-1}$ are exact. Letting $n=0$, $f=0$, and $y_0=y_1 = \ldots = y_{k-1} = 1$ in (11.8), we obtain the equation

$$1 = y_k = -(\alpha_{k-1} + \alpha_{k-2} + \ldots + \alpha_0)/\alpha_k$$

or

$$\alpha_k + \alpha_{k-1} + \ldots + \alpha_0 = 0.$$

This can be expressed more compactly as

(iii) $\quad \rho(1) = 0.$

The second initial-value problem has the solution

$$y(x_k) = x_k - x_0 = kh.$$

Thus the equation

$$\alpha_k(kh) + \alpha_{k-1}(k-1)h + \ldots + \alpha_0 \cdot 0 = h[\beta_k + \ldots + \beta_0]$$

should be satisfied. After dividing by h, we can express this more compactly as

(iv) $\quad \rho'(1) = \sigma(1),$

where

$$\sigma(z) = \beta_k z^k + \beta_{k-1} z^{k-1} + \ldots + \beta_0.$$

Conditions (iii) and (iv) are referred to as the *consistency conditions.* They are not strictly stability conditions, but they are usually discussed in this context because they provide further information on the polynomials ρ and σ which are important in stability considerations.

The other test problem which is often used to get an indication of stability is

$$y' = \lambda y, \qquad y(0) = 1,$$

where λ is a constant. Again, the general solution of the difference equation (11.8) is of the form (11.10) or (11.11), where $z_1, \ldots, z_k$ are now the zeros of the polynomial

$$\phi(z) = (\alpha_k - \lambda h \beta_k) z^k + \ldots + (\alpha_0 - \lambda h \beta_0).$$

Usually one of the terms in (11.10), say $c_1 z_1^n$, will converge to the true solution $e^{\lambda x}$ as $n \to \infty$ if $h = x/n$. Then one would expect stability to be a problem if and only if one of the extraneous zeros $z_2, \ldots, z_k$ violates the root condition. This type of analysis will be illustrated in the following example.

EXAMPLE 11.2. Let us examine the method

$$y_{n+1} - y_{n-1} = 2hf(x_n, y_n),$$

which was used in Example 11.1, from the point of view of our general discussion of stability. To make the notation consistent with (11.8), let us shift the index and write the above equation in the equivalent form

$$(11.12) \qquad y_{n+2} - y_n = 2hf(x_{n+1}, y_{n+1}).$$

The polynomials ρ and σ for this method are given by

$$\rho(z) = z^2 - 1,$$

$$\sigma(z) = 2z.$$

It is easy to check that the root condition and the consistency conditions are satisfied. Let us examine the behavior of the approximate solutions when this method is applied to the problem

$$y' = \lambda y, \qquad y(0) = 1.$$

The polynomial ϕ is given by

$$\phi(z) = z^2 - 2\lambda h z - 1.$$

Thus the general solution to the difference equation (11.12) with $f(x,y) = \lambda y$ is given by

$$y_n = c_1 z_1^n + c_2 z_2^n,$$

where

$$z_1 = \lambda h + \sqrt{1 + \lambda^2 h^2},$$

$$z_2 = \lambda h - \sqrt{1 + \lambda^2 h^2}.$$

With $h = x/n$, one can use L'Hospital's rule to show that

$$\lim_{n\to\infty} z_1^n = e^{\lambda x}, \qquad \lim_{n\to\infty} z_2^n = e^{-\lambda x}.$$

For large n, the solution is then approximately

$$y_n \doteq c_1 e^{\lambda x} + c_2 e^{-\lambda x}.$$

The second term is extraneous since the desired solution is $e^{\lambda x}$. If $\lambda > 0$, the extraneous solution will tend to zero as x increases and thus there is no stability problem. On the other hand, if $\lambda < 0$, then the extraneous solution will increase with x while the desired solution decreases. The extraneous solution would tend to mask the true solution, and one would expect a stability problem. More generally, one would expect stability when $\partial f/\partial y > 0$ and instability when $\partial f/\partial y < 0$. This type of behavior is referred to as weak instability.

PROBLEMS

11.1. Write out a difference equation for the Euler-Cauchy method applied to (11.1). Solve this difference equation and show that for a fixed h, $0 < h < 0.5$,

$$\lim_{n\to\infty} (y_n - e^{-4x_n} - x_n) = 0.$$

11.2. Consider the difference equation

$$y_{n+1} + ay_n + by_{n-1} = 0.$$

What conditions on a and b would insure that

$$\lim_{n\to\infty} y_n = 0$$

for any solution of the difference equation?

11.3. Suppose $h > 0$ and that z_1 and z_2 are given by (11.6). Show that:
(a) $0 < z_1 < 1$
(b) $z_2 < -1$

11.4. Investigate the stability and consistency of:
(a) The improved Euler-Cauchy method.
(b) The Milne three-point predictor-corrector method.
(c) The Adams-Moulton method.
Hint: When studying (c) under the test equation $y' = \lambda y$, either use numerical techniques or study the behavior as $h \to 0$.

11.5. Examine the behavior of the Runge-Kutta method on the various test problems described in this section.

12. EIGENVALUE PROBLEMS

Let L be a linear differential operator—that is, an operator defined by an equation of the form

$$L(y) = \sum_{j=0}^{m} a_j y^{(j)},$$

where the a_j's are functions of one variable and $y^{(j)}$ represents the jth derivative of y. In this section we will consider the boundary value problem

$$L(y) + \lambda \rho y = 0, \qquad y(a) = 0, \qquad y(b) = 0, \tag{12.1}$$

where λ is a constant and ρ is a function of one variable which does not vanish on $[a,b]$. Now it is easy to see that the equation

$$y(x) = 0 \qquad \text{for all} \quad x \in [a,b]$$

defines a solution to (12.1). This solution, which is called the *trivial solution*, usually has little physical significance. Instead, we seek a solution of (12.1) other than the trivial solution. However, it turns out that (12.1) has a non-trivial solution only for special values of λ. A number λ for which (12.1) has a non-trivial solution is called an *eigenvalue*, and a corresponding non-trivial solution is called an *eigen-function of the boundary value problem.*

EXAMPLE 12.1. The following problem arises in the study of vibrating strings:

$$y'' + \lambda y = 0, \qquad y(0) = 0 = y(\pi). \tag{12.2}$$

Now if $\lambda > 0$, the general solution of this differential equation is given by

$$y(x) = c_1 \sin \sqrt{\lambda}\, x + c_2 \cos \sqrt{\lambda}\, x.$$

The condition $y(0) = 0$ requires that $c_2 = 0$. Then the condition $y(\pi) = 0$ requires that $\sqrt{\lambda}$ be an integer. Thus the positive eigenvalues are given by

$$\lambda_1 = 1, \;\; \lambda_2 = 4, \;\; \ldots, \;\; \lambda_n = n^2, \;\; \ldots. \tag{12.3}$$

If $\lambda < 0$, the solution to the differential equation is given by

$$y(x) = c_1 e^{\sqrt{-\lambda x}} + c_2 e^{-\sqrt{-\lambda x}}.$$

The two boundary conditions then require that

$$c_1 + c_2 = 0,$$

$$c_1 e^{\sqrt{-\lambda}\pi} + c_2 e^{-\sqrt{-\lambda}\pi} = 0.$$

The determinant of this system is $e^{-\sqrt{-\lambda}\pi} - e^{\sqrt{-\lambda}\pi} \neq 0$; hence the boundary condition can only be satisfied by the trivial solution. Similarly, only the trivial solution satisfies (12.2) when $\lambda = 0$. Thus all of the eigenvalues are given by (12.3). Corresponding eigenfunctions are given by the equations

$$y_n(x) = \sin nx, \qquad n = 1, 2, \ldots.$$

Let us now describe a method for obtaining approximate eigenvalues and functions for (12.1). As usual, we will subdivide the interval $[a,b]$ using the points

$$x_0 = a, \quad x_1 = a + h, \quad \ldots, \quad x_n = a + nh = b,$$

where n is a positive integer and $h = (b-a)/n$. Now let y_n be an approximate value for $y(x_n)$. We will replace all of the derivatives in (12.1) by approximate derivatives of the form

$$y^{(j)}(x_i) \doteq \sum_{k=0}^{n} c_{ijk} y_k .$$

If we insist that (12.1) be satisfied in terms of the approximate derivatives at the points $x_1, \ldots, x_{n-1}$, we are led to the system of equations

$$\text{(12.4)} \qquad \sum_{j=0}^{m} a_j(x_i) \sum_{k=0}^{n} c_{ijk} y_k + \lambda \rho(x_i) y_i = 0, \qquad i = 1, 2, \ldots, n-1.$$

We will also insist that $y_0 = 0 = y_n$ so that the boundary conditions will be satisfied. Dividing (12.4) by $\rho(x_i)$ and rearranging the sums, we obtain

$$\text{(12.5)} \qquad \sum_{k=0}^{n} y_k \sum_{j=0}^{m} \frac{a_j(x_i) c_{ijk}}{\rho(x_i)} + \lambda y_i = 0, \qquad i = 1, 2, \ldots, n-1.$$

Now let

$$b_{ik} = \sum_{j=0}^{m} \frac{a_j(x_i) c_{ijk}}{\rho(x_i)}$$

Then after noting that $y_0 = 0 = y_n$, (12.5) reduces to

$$\sum_{k=1}^{n-1} b_{ik} y_k + \lambda y_i = 0, \qquad i = 1, 2, \ldots, n-1.$$

These equations can be written in matrix form as

$$\text{(12.6)} \qquad (B + \lambda I) y = 0.$$

Now (12.6) is the usual matrix eigenvalue problem (see Chapter 3, Section 7) except for the sign on λ. Thus the approximate eigenvalues of (12.1) are the negatives of the eigenvalues of the matrix B.

EXAMPLE 12.2. Let us find approximate eigenvalues for the equation

$$y'' + \lambda x y = 0, \qquad y(0) = 0 = y(1).$$

We will subdivide the interval $[0,1]$ with the points 0, 1/4, 1/2, 3/4, 1, and replace $y''(x_i)$ by

$$\frac{y_{i+1} - 2y_i + y_{i-1}}{h^2} = 16y_{i+1} - 32y_i + 16y_{i-1}.$$

This leads to the system of equations

$$16y_2 - 32y_1 + \tfrac{1}{4}\lambda y_1 = 0,$$

$$16y_3 - 32y_2 + 16y_1 + \tfrac{1}{2}\lambda y_2 = 0,$$

$$-32y_3 + 16y_2 + \tfrac{3}{4}\lambda y_3 = 0.$$

After simplifying, we can write this in matrix form as

$$(B + \lambda I)y = 0,$$

where

$$B = \begin{bmatrix} -128 & 64 & 0 \\ 32 & -64 & 32 \\ 0 & \frac{64}{3} & \frac{-128}{3} \end{bmatrix}.$$

The approximate eigenvalues are the zeros of the polynomial p defined by

$$p(\lambda) = |B + \lambda I| = \tfrac{1}{3}(3\lambda^3 - 704\lambda^2 + 40{,}960\lambda - 524{,}288).$$

By trial and error we find that p has a positive zero between 17 and 18. Using four iterations of Newton's method with $x_0 = 18$, we obtain

$$x_1 = 18 - p(18)/p'(18) = 17.8709\,2597,$$

$$x_2 = 17.8714\,0990,$$

$$x_3 = 17.8714\,0991,$$

$$x_4 = 17.8714\,0991.$$

The other two approximate eigenvalues can now be obtained by dividing $\lambda - 17.8714\,0991$ into $p(\lambda)$ and finding the zeros of the resulting quadratic. The approximate values for the corresponding eigenfunctions can then be obtained by solving the equations

$$(B + \lambda_i I)y = \emptyset, \qquad i = 1, 2, 3.$$

PROBLEMS

12.1. Find approximate eigenvalues and eigenfunctions for the boundary value problem

$$y'' + \lambda y = 0, \qquad y(0) = 0 = y(0.5).$$

Let $h = 1/6$ and replace $y''(x_i)$ by $(y_{i+1} - 2y_i + y_{i-1})/h^2$. Compare your results with the actual eigenvalues and eigenfunctions.

12.2. In Example 12.2 approximate y by its tenth-degree Taylor expansion about zero, i.e., let $y = \Sigma_{j=0}^{10} c_j x^j$, and use the differential equation and the condition $y(0) = 0$ to compute the c_j's in terms of c_1. The condition $y(1) = 0$ leads to a cubic equation in λ. Show that this equation has a solution between 18 and 19 and calculate this solution correct to five decimals.

REFERENCES

1. M. Abramowitz and I. A Stegun, (Eds.), *Handbook of Mathematical Functions*, Applied Math. Series 55, National Bureau of Standards, 1964.
2. L. Brand, *Advanced Calculus*, Wiley, 1955.
3. L. Collatz, *The Numerical Treatment of Differential Equations*, Springer-Verlag, 1960.
4. S. D. Conte, *Elementary Numerical Analysis*, McGraw-Hill, 1965.
5. R. Courant and D. Hilbert, *Methods of Mathematical Physics*, Vol. I, Interscience, 1953.
6. P. J. Davis and P. Rabinowitz, *Numerical Integration*, Blaisdell, 1967.
7. A. Erdélyi, *Asymptotic Expansions*, Dover, 1956.
8. W. Feller, *An Introduction to Probability Theory and Its Applications*, Vol. I, (Second Edition), Wiley, 1957.
9. L. N. G. Filon, "On a quadrature formula for trigonometric integrals," *Proc. Roy. Soc. Edinburgh Sect. A*, 49 (1928–29), 38–47.
10. R. R. Goldberg and R. S. Varga, "Moebius inversion of Fourier transformations," *Duke Math. J.*, 23 (1956), 553–559.
11. P. Henrici, *Discrete Variable Methods in Ordinary Differential Equations*, Wiley, 1962.
12. P. Henrici, *Elements of Numerical Analysis*, Wiley, 1964.
13. E. Isaacson and H. B. Keller, *Analysis of Numerical Methods*, Wiley, 1966.
14. F. Scheid, *Theory and Problems of Numerical Analysis*, McGraw-Hill, 1968.
15. S. M. Selby, (Ed.), *C.R.C. Standard Mathematical Tables*, 19th Ed., Chemical Rubber Publishing Co., Cleveland, 1971.
16. H. J. Stetter, "Numerical approximation of Fourier transforms," *Numerische Mathematik*, 8 (1966), 235–249.
17. A. H. Stroud and D. Secrest, *Gaussian Quadrature Formulas*, Prentice-Hall, 1966.
18. G. Szegö, *Orthogonal Polynomials*, Amer. Math. Soc. Colloquium Pub., 23, 1959.

CHAPTER 7

Best Approximations

1. THE GENERAL APPROXIMATION PROBLEM

Let f be an element and S a subset of a normed linear space X. A basic problem of approximation theory is to find an element of S which is as close as possible to f; that is, we seek an element s^* of S such that

$$\|f - s^*\| \leqslant \|f - s\|$$

for all $s \in S$. In this case s^* is said to be a best approximation to f from S relative to the given norm. Three fundamental questions immediately arise in connection with this type of problem:

1. Does such a best approximation exist?
2. If a best approximation exists, is it unique?
3. What properties distinguish the best approximation from other approximations?

These questions will be studied in later sections in connection with specific approximation problems.

EXAMPLE 1.1. Let $X = C[a,b]$, $\|f\| = \max_{x \in [a,b]} |f(x)|$, and $S = \mathcal{P}_n$, the set of all polynomials of degree at most n. The approximation problem here is to find the polynomial in $\mathcal{P}_n$ which yields the smallest possible maximum error as an approximation to f on the interval $[a,b]$. This problem, which is known as the *Chebyshev approximation problem*, will be studied in Section 4.

EXAMPLE 1.2. With X and $\|f\|$ as in Example 1.1, let $S = \mathcal{R}(m,n)$, the set of all rational functions $r = p/q$, such that $p \in \mathcal{P}_m$ and $q \in \mathcal{P}_n$. The object here will be to find the rational function $r \in \mathcal{R}(m,n)$ which yields the smallest maximum error when used as an approximation to f on $[a,b]$. This type of approximation, which is known as *rational Chebyshev approximation*, will be studied in Section 6.

An approximation problem is said to be linear iff $S = \text{span}\{\varphi_1, \varphi_2, \ldots, \varphi_n\}$ for some finite set $\{\varphi_1, \varphi_2, \ldots, \varphi_n\}$ of elements from X. The problem in Example 1.1 is linear, while that in Example 1.2 is non-linear. In general, the theory of non-linear approximations is more complicated than the linear theory.

2. FOURIER SERIES

Let us define an inner product and norm on $C[-\pi,\pi]$ by the equations

$$f \cdot g = \int_{-\pi}^{\pi} f(x)g(x)dx, \tag{2.1}$$

$$\|f\| = \left[\int_{-\pi}^{\pi} f^2(x)dx\right]^{1/2}. \tag{2.2}$$

Let $\mathfrak{T}_n$ be the set of all trigonometric polynomials of degree at most n (see Section 9 in Chapter 5). In this section we will study the problem of finding the best approximation from $\mathfrak{T}_n$ to a given function f relative to the norm (2.2). Equivalently, we seek a member T^* of $\mathfrak{T}_n$ such that the integral

$$\int_{-\pi}^{\pi} [f(x) - T^*(x)]^2 dx$$

is a minimum.

This problem can be readily solved by applying the theory of least squares approximation developed in Chapter 2. To see this, let

$$\begin{aligned} \varphi_0(x) &= \frac{1}{\sqrt{2\pi}}, \\ \varphi_1(x) &= \frac{\cos x}{\sqrt{\pi}}, \\ \varphi_2(x) &= \frac{\sin x}{\sqrt{\pi}}, \\ &\vdots \\ \varphi_{2n-1}(x) &= \frac{\cos nx}{\sqrt{\pi}}, \\ \varphi_{2n}(x) &= \frac{\sin nx}{\sqrt{\pi}}. \end{aligned} \tag{2.3}$$

It is left as an exercise (Problems 2.2 and 2.3) to prove the following: If k, m, and n are positive integers, then

$$\int_{-\pi}^{\pi} \sin kx \cos mx \, dx = 0, \tag{2.4}$$

$$\int_{-\pi}^{\pi} \sin kx \sin mx \, dx = \begin{cases} \pi & \text{if } k = m, \\ 0 & \text{if } k \neq m, \end{cases} \tag{2.5}$$

$$\int_{-\pi}^{\pi} \cos kx \cos mx \, dx = \begin{cases} \pi & \text{if } k = m, \\ 0 & \text{if } k \neq m, \end{cases} \tag{2.6}$$

$$\text{span } \{\varphi_0, \varphi_1, \ldots, \varphi_{2n}\} = \mathfrak{J}_n. \tag{2.7}$$

It follows that $\{\varphi_0, \varphi_1, \ldots, \varphi_{2n}\}$ is an orthonormal set. Using (2.7) above and Theorem 8.1 of Chapter 2, we see that the best approximation to f from $\mathfrak{J}_n$ is given by

$$T^* = \sum_{j=0}^{2n} \alpha_j \varphi_j,$$

where

$$\alpha_j = f \cdot \varphi_j, \qquad j = 0, 1, 2, \ldots, 2n.$$

Replacing the inner products by integrals and the φ_j's by sines and cosines, we obtain the more customary expression

$$T^*(x) = \frac{a_0}{2} + \sum_{j=1}^{n} (a_j \cos jx + b_j \sin jx),$$

where

$$\begin{aligned} a_j &= \frac{1}{\pi} \int_{-\pi}^{\pi} f(x) \cos jx \, dx, \qquad j = 0, 1, \ldots, n, \\ b_j &= \frac{1}{\pi} \int_{-\pi}^{\pi} f(x) \sin jx \, dx, \qquad j = 1, 2, \ldots, n. \end{aligned} \tag{2.8}$$

The numbers a_j and b_j in (2.8) are referred to as the *Fourier coefficients of f*. Note that in this case there is a reasonably simple expression relating the best approximation T^* to f. We will see later that approximation problems are not always so easy to solve.

EXAMPLE 2.1. Let $f(x) = e^x$ and $n = 2$. From (2.8) we obtain

$$a_0 = \frac{1}{\pi} \int_{-\pi}^{\pi} e^x \, dx = \frac{2}{\pi} \sinh \pi,$$

$$a_1 = \frac{1}{\pi} \int_{-\pi}^{\pi} e^x \cos x \, dx = \frac{-1}{\pi} \sinh \pi,$$

$$a_2 = \frac{1}{\pi} \int_{-\pi}^{\pi} e^x \cos 2x \, dx = \frac{2}{5\pi} \sinh \pi,$$

$$b_1 = \frac{1}{\pi} \int_{-\pi}^{\pi} e^x \sin x \, dx = \frac{1}{\pi} \sinh \pi,$$

$$b_2 = \frac{1}{\pi} \int_{-\pi}^{\pi} e^x \sin 2x \, dx = \frac{-4}{5\pi} \sinh \pi.$$

Thus the best approximation to f from $\mathfrak{I}_2$ is given by

$$T^*(x) = \frac{2 \sinh \pi}{\pi} \left[\frac{1}{2} - \frac{1}{2} \cos x + \frac{1}{2} \sin x + \frac{1}{5} \cos 2x - \frac{2}{5} \sin 2x\right].$$

The distance between f and T^*, $\|f - T^*\|$, can be computed by means of Corollary 8.2 in Chapter 2 (see Problem 2.4).

The series

$$\frac{a_0}{2} + \sum_{j=1}^{\infty} (a_j \cos jx + b_j \sin jx),$$

with a_j and b_j given by (2.8), is called the *Fourier series of f.* The theory of Fourier series has played an extremely important role in the development of mathematics, and the literature on this subject is huge. A thorough study of this theory would be quite time-consuming; so we prefer to just state several theorems which bear on the approximation problems being considered in this chapter. First, if $f \in C[-\pi,\pi]$, the approximation relative to the norm (2.2) can be made as accurate as we please by taking n sufficiently large.

THEOREM 2.1. *Let a_j and b_j be given by* (2.8), *and let*

$$S_n(x) = T^*(x) = \frac{a_0}{2} + \sum_{j=1}^{n} (a_j \cos jx + b_j \sin jx). \tag{2.9}$$

Then

$$\lim_{n \to \infty} \|f - S_n\| = 0. \tag{2.10}$$

A proof of this theorem can be found in Tolstov [8]. When the norm (2.2) is used and (2.10) holds, we say that S_n converges to f in the mean.

In Section 4 we will be concerned with approximations relative to the uniform norm. Further conditions on f are required in order to guarantee convergence of the Fourier series in this norm.

THEOREM 2.2. *Let $\| \ \|_U$ be the uniform norm on $[-\pi,\pi]$ and let S_n be given by* (2.9). *If $f \in C[-\pi,\pi]$, $f(\pi) = f(-\pi)$, and f' is bounded on* $[-\pi,\pi]$, then

$$\lim_{n \to \infty} \|f - S_n\|_U = 0. \tag{2.11}$$

A proof of this theorem can be found in [9]. When equation (2.11) holds, we say that S_n converges to f uniformly on $[-\pi,\pi]$.

Students of higher analysis should note that the requirement that f' be bounded can be replaced by the less restrictive requirement that f be of bounded variation on $[-\pi,\pi]$.

PROBLEMS

2.1. The function f is said to be even on $[-\pi,\pi]$ iff $f(-x) = f(x)$ for all $x \in [-\pi,\pi]$, and it is said to be odd iff $f(-x) = -f(x)$ for all $x \in [-\pi,\pi]$.

(a) Show that if f is even, then the sine coefficients in the Fourier series are all zero.

(b) Show that if f is odd, then the cosine coefficients in the Fourier series are all zero.

2.2. Prove formulas (2.4), (2.5), and (2.6).

2.3. Prove (2.7).

2.4. Let $f \in C[-\pi,\pi]$ and let S_n be given by (2.9). Show that

$$\|f - S_n\|^2 = \|f\|^2 - \pi\left[\frac{a_0^2}{2} + \sum_{j=1}^{n} (a_j^2 + b_j^2)\right].$$

Hint: Use Corollary 8.2 in Chapter 2.

2.5. Use Theorem 2.1 and Problem 2.4 to show that

(a) $$\|f\|^2 = \pi\left[\frac{a_0^2}{2} + \sum_{j=1}^{\infty} (a_j^2 + b_j^2)\right],$$

(b) $$\|f - S_n\|^2 = \pi \sum_{j=n+1}^{\infty} (a_j^2 + b_j^2),$$

(c) $$f \cdot g = \pi\left[\frac{1}{2}a_0c_0 + \sum_{j=1}^{\infty} (a_jc_j + b_jd_j)\right],$$

where a_j, b_j are the Fourier coefficients of f and c_j, d_j are the Fourier coefficients of g.

2.6. Suppose the Fourier coefficients of f satisfy the condition $a_j^2 + b_j^2 \leqslant 2^{-j}$. Let S_n be given by (2.9). Find a bound on $\|f - S_n\|$, where the norm is that defined by (2.2).

3. CHEBYSHEV SERIES

In this section we will discuss the expansion of a function in terms of Chebyshev polynomials. These expansions, which are closely related to Fourier series, are important because they are relatively easy to construct and frequently provide an approximation which is almost as good as the best approximation relation to the uniform norm.

Let us define an inner product and norm on $C[-1,1]$ by the equations

(3.1) $$f \cdot g = \int_{-1}^{1} \frac{f(x)g(x)dx}{\sqrt{1-x^2}},$$

(3.2) $$\|f\| = \left[\int_{-1}^{1} \frac{f^2(x)dx}{\sqrt{1-x^2}}\right]^{1/2}.$$

For a given function $f \in C[-1,1]$, we wish to find the best polynomial approximation from $\mathcal{P}_n$ when the norm (3.2) is used to measure the quality of the approximation. This problem can be solved by means of the least squares theory. Since the development is similar to that of the preceding section, we will just summarize the results and leave the details for an exercise (Problem 3.1). Let $\theta(x) = \text{arc} \cos x$ and let $T_j(x) = \cos j\theta(x)$ be the jth-degree Chebyshev polynomial of the first kind. If $f \in C[-1,1]$, then the best approximation to f from $\mathcal{P}_n$ relative to the norm (3.2) is given by the equations

(3.3) $$s_n = \frac{a_0 T_0}{2} + \sum_{j=1}^{n} a_j T_j,$$

where s_n is the best approximation and

(3.4) $$a_j = \frac{2}{\pi}(f \cdot T_j), \qquad j = 0, 1, \ldots, n.$$

The series $a_0 T_0/2 + \Sigma_{j=1}^{\infty} a_j T_j$, with the a_j's given by (3.4), is called the *Chebyshev* or *Fourier-Chebyshev series for f*. There is a close connection between the theory of Chebyshev series and that of Fourier series. To see this, let $x = \cos\theta$, $0 \leqslant \theta \leqslant \pi$, and note that

$$\sqrt{1-x^2} = \sqrt{1-\cos^2\theta} = \sin\theta,$$

$$T_j(x) = T_j(\cos\theta) = \cos j\theta.$$

Now use (3.4) and make the change of variable $x = \cos\theta$ to obtain

(3.5) $$a_j = \frac{2}{\pi}\int_{-1}^{1} \frac{f(x)T_j(x)}{\sqrt{1-x^2}} = \frac{2}{\pi}\int_0^{\pi} f(\cos\theta)\cos j\theta \, d\theta.$$

Let $g(\theta) = f(\cos\theta)$ and note that g is an even function. This implies (see Problem 2.1) that the sine coefficients in its Fourier series are all zero. From (3.5) we see that the Chebyshev coefficients of f are the same as the Fourier cosine coefficients of g. Furthermore, if $x = \cos\theta$, then

(3.6) $$f(x) - \frac{a_0 T_0(x)}{2} - \sum_{j=1}^{n} a_j T_j(x) = g(\theta) - \frac{a_0}{2} - \sum_{j=1}^{n} a_j \cos j\theta.$$

Thus the convergence of the Chebyshev series to f on $[-1,1]$ is equivalent to the convergence of the Fourier series to g on $[-\pi,\pi]$. This relationship can be used to establish several convergence theorems for the Chebyshev series.

THEOREM 3.1. *Let $f \in C[-1,1]$ and let s_n be the best approximation to f from $\mathcal{P}_n$ relative to the norm (3.2). Then*

$$\lim_{n\to\infty} \|f - s_n\| = 0.$$

Proof. Let $\|\ \|$ and $\|\ \|_0$ be the norms defined by (3.2) and (2.2), respectively. Let $S_n(\theta) = a_0/2 + \sum_{j=1}^{n} a_j \cos j\theta$. Making the change of variable $x = \cos\theta$ and using (3.6), we obtain

$$\|f - s_n\|^2 = \int_{-1}^{1} \frac{[f(x) - s_n(x)]^2 dx}{\sqrt{1-x^2}} = \int_0^\pi [g(\theta) - S_n(\theta)]^2 d\theta$$
$$= \tfrac{1}{2}\|g - S_n\|_0^2.$$

The result now follows from Theorem 2.1.

THEOREM 3.2. *If f is continuous and f' is bounded on $[-1,1]$, then*

$$\lim_{n\to\infty} \|f - s_n\|_U = 0,$$

where $\|\ \|_U$ represents the uniform norm on $[-1,1]$.

Proof. Let $g(\theta) = f(\cos\theta)$ and note that g' is bounded on $[-\pi,\pi]$. Using the same notation as in Theorem 3.1, we have

$$\max_{x\in[-1,1]} |f(x) - s_n(x)| = \max_{\theta\in[-\pi,\pi]} |g(\theta) - S_n(\theta)|.$$

Thus $\|f - s_n\|_U = \|g - S_n\|_U^*$, where $\|\ \|_U^*$ represents the uniform norm on $[-\pi,\pi]$. The result now follows from Theorem 2.2.

The truncated Chebyshev series s_n is important not only because it gives the best approximation from $\mathcal{P}_n$ in the norm (3.2), but also because it often gives nearly the best approximation in the uniform norm. As we will see in the next two sections, the coefficients of the best polynomial approximation in the uniform norm are harder to compute than those of the truncated Chebyshev series. Inequalities which compare the effectiveness of these two types of approximations are given in Cheney [2] and Rivlin [7].

In this section the approximation interval has always been $[-1,1]$ rather than an arbitrary interval $[a,b]$. However, it is easy to convert the problem of approximating a function g on $[a,b]$ into one involving

[−1,1]. To see this, let

$$f(z) = g\left(a + \frac{(b-a)(z+1)}{2}\right)$$

for each $z \in [-1,1]$. Now approximate f by its truncated Chebyshev series s_n and use

$$s_n\left(-1 + \frac{2(x-a)}{b-a}\right)$$

as an approximation to $g(x)$.

PROBLEMS

3.1. Let $f \in C[-1,1]$ and let s_n and a_j be given by (3.3) and (3.4). Prove that s_n is the best approximation to f from $\mathcal{P}_n$ relative to the norm (3.2). You may want to look at Example 8.2 in Chapter 2 and the corresponding problems.

3.2. Show that $T_j(\cos\theta) = \cos j\theta$. Is this true for all θ or only for $\theta \in [0,\pi]$?

3.3. Prove formula (3.5).

3.4. Let $f(x) = x^5$. Compute the coefficients of the best approximations to f from $\mathcal{P}_2$ and $\mathcal{P}_3$ relative to the norm (3.2). Sketch the graph of the error curves.

3.5. Prove or disprove: If f is an even function, then a_j (given by (3.5)) is zero if j is odd.

3.6. Let $\{f_n\}$ be a sequence of functions from $C[-1,1]$. Consider the following statements:

A: $\lim_{n\to\infty} f_n(x) = g(x)$ for every $x \in [-1,1]$.

B: The sequence $\{f_n\}$ converges to g in the uniform norm on $[-1,1]$.

C: The sequence $\{f_n\}$ converges to g in the norm (3.2).

Prove or disprove:

(a) A implies B
(b) B implies A
(c) A implies C
(d) C implies A
(e) B implies C
(f) C implies B

4. CHEBYSHEV APPROXIMATION

Let $\mathcal{P}_n$ be the collection of all polynomials whose degree is at most n, and let f be a continuous function on the interval $[a,b]$. The polynomial p is said to be the best approximation to f from $\mathcal{P}_n$ in the sense of Chebyshev iff $p \in \mathcal{P}_n$ and

$$\max_{x \in [a,b]} |f(x) - p(x)| \leqslant \max_{x \in [a,b]} |f(x) - q(x)|$$

for all $q \in \mathcal{P}_n$. In other words, p is a best approximation to f from $\mathcal{P}_n$ relative to the uniform (or Chebyshev) norm which is defined by

$$\|f\| = \max_{x \in [a,b]} |f(x)|. \tag{4.1}$$

In this section we will discuss some of the theoretical aspects of Chebyshev approximation. Let us first prove that a best approximation exists whenever f is continuous.

THEOREM 4.1. *If f is continuous on $[a,b]$, then a best approximation to f from $\mathcal{P}_n$ exists.*

Proof. (The reader not familiar with the notion of compactness may wish to skip over this proof.) We will first show that our attention can be restricted to a certain bounded set of coefficients. Let $E = \inf_{p \in \mathcal{P}_n} \|f - p\|$. Note that the zero polynomial belongs to $\mathcal{P}_n$; hence $E \leqslant \|f\|$. Let

$$S = \{p \in \mathcal{P}_n\colon\ \|f - p\| \leqslant \|f\|\}.$$

Then

$$\inf_{p \in S} \|f - p\| = \inf_{p \in \mathcal{P}_n} \|f - p\| = E, \tag{4.2}$$

since $E \leqslant \|f\|$. Now if $p \in S$, then

$$\|p\| = \|p - f + f\| \leqslant \|p - f\| + \|f\| \leqslant 2\|f\|.$$

Now let $x_0 = a$, $x_1 = a + h$, . . . , $x_n = a + nh = b$, where $h = (b - a)/n$. Using the Lagrange interpolation formula, we have

$$p(x) = \sum_{j=0}^{n} p(x_j) l_j(x),$$

where

$$l_j(x) = \prod_{\substack{k=0 \\ k \neq j}}^{n} (x - x_k)/(x_j - x_k).$$

Let $c = (c_0, c_1, \ldots, c_n)$ and $\alpha_j = (\alpha_{j0}, \alpha_{j1}, \ldots, \alpha_{jn})$ be the coefficient vectors associated with p and l_j, respectively. Now for $i = 0, 1, 2, \ldots, n$,

$$c_i = \sum_{j=0}^{n} p(x_j) \alpha_{ji}.$$

Let

$$M = \max_{0 \leqslant i \leqslant n} \sum_{j=0}^{n} |\alpha_{ji}|.$$

Then if $p \in S$,

$$|c_i| \leqslant \sum_{j=0}^{n} |p(x_j)|\,|\alpha_{ji}| \leqslant 2\,\|f\| \sum_{j=0}^{n} |\alpha_{ji}| \leqslant 2M\,\|f\|.$$

If we let $\|c\| = \sqrt{c_0^2 + c_1^2 + \ldots + c_n^2}$ be the usual Euclidean norm, then $p \in S$ implies that

$$\|c\| \leqslant 2M\,\|f\|\sqrt{n+1}.$$

Let $U = \{c\colon \|c\| \leqslant 2M\,\|f\|\sqrt{n+1}\}$, and let T be the set of polynomials in $\mathcal{P}_n$ whose coefficient vectors are in U. Since $S \subset T \subset \mathcal{P}_n$ and (4.2) holds, we have

$$E \leqslant \inf_{p \in T} \|f-p\| \leqslant \inf_{p \in S} \|f-p\| = E.$$

Thus we have

$$(4.3) \qquad E = \inf_{p \in T} \|f-p\|.$$

For each $c \in U$ let $\varphi(c) = \|f-p\|$, where p is the polynomial with coefficient vector c. It follows from (4.3) that

$$E = \inf_{c \in U} \varphi(c).$$

But φ is continuous on the compact set U (see Problem 4.1). Hence there is a vector $c^* \in U$ such that

$$\varphi(c^*) = \inf_{c \in U} \varphi(c) = E.$$

Then the polynomial p^* whose coefficient vector is c^* is a best approximation to f.

A fundamental fact in the Chebyshev theory is that the error curve associated with the best polynomial has a certain oscillatory character. Before discussing this theorem we need to explain the notion of an *oscillation set*.

DEFINITION 4.1. *Let g be continuous on $[a,b]$. The set of points $\{x_0, x_1, \ldots, x_k\}$ with $a \leqslant x_0 < x_1 < \ldots < x_k \leqslant b$ is called an oscillation set for g iff*

$$(4.4) \qquad g(x_j) = (-1)^j \sigma\,\|g\|, \qquad j = 0, 1, \ldots, k,$$

where σ is either 1 or -1.

EXAMPLE 4.1. Let $g(x) = \cos 5\pi x$ on the interval $[0,1]$. Then $\|g\| = 1$, and if j is an integer between 0 and 5,

$$g(\tfrac{j}{5}) = \cos j\pi = (-1)^j = (-1)^j \|g\|.$$

Thus (4.4) is satisfied with $\sigma = 1$, and the points $\{0, 1/5, 2/5, 3/5, 4/5, 1\}$ form an oscillation set for g on the interval $[0,1]$.

EXAMPLE 4.2. Let $f(x) = x \cos 5x$ on the interval $[0,1]$. Then $\|f\| = 1$, but the points $\{0, 1/5, 2/5, 3/5, 4/5, 5/5\}$ are no longer an oscillation set even though f alternates sign at these points. This is because $|f(j/5)| = |j \cos j\pi|/5 \neq \|f\|$ when $j = 0, 1, \ldots, 4$.

The following theorem, known as the *Chebyshev equioscillation theorem*, is one of the most important in the Chebyshev theory.

THEOREM 4.2. *Let $f \in C[a,b]$ and $p \in \mathcal{P}_n$. Then p is a best Chebyshev approximation to f from $\mathcal{P}_n$ if and only if the error curve $f - p$ has an oscillation set containing at least $n+2$ points.*

Proof. We will only prove the "if" part of this theorem. A proof of the converse can be found in Davis [4]. Suppose that $f - p$ has an oscillation set containing the $n + 2$ points $x_0 < x_1 < \ldots < x_{n+1}$, and that $q \in \mathcal{P}_n$ provides a better approximation to f. Then we have

$$\|f - q\| < \|f - p\| \tag{4.5}$$

and

$$f(x_j) - p(x_j) = (-1)^j \sigma \|f - p\|. \tag{4.6}$$

We will assume that $\sigma = 1$; a similar argument applies if $\sigma = -1$. From (4.5) and (4.6) it follows that

$$f(x_0) - q(x_0) \leqslant \|f - q\| < \|f - p\| = f(x_0) - p(x_0).$$

Hence $p(x_0) - q(x_0) < 0$. Similarly,

$$f(x_1) - q(x_1) \geqslant - \|f - q\| > - \|f - p\| = f(x_1) - p(x_1).$$

Thus $p(x_1) - q(x_1) > 0$. Continuing in this fashion, one infers that $p - q$ alternates sign at the $n + 2$ points $x_0, x_1, \ldots, x_{n+1}$. It follows that $p - q$ has at least $n + 1$ zeros on $[a,b]$. But $p - q \in \mathcal{P}_n$; so $p - q$ can have $n + 1$ zeros only if it is the zero polynomial. This implies that $p - q$, which contradicts (4.5). The contradiction establishes the fact that p is a best approximation to f from $\mathcal{P}_n$.

COROLLARY 4.3. *If $f \in C[a,b]$, then the best Chebyshev approximation to f from $\mathcal{P}_n$ is unique.*

Proof. Let p_1 and p_2 be best approximations to f from $\mathcal{P}_n$. Now define $p_0 = (p_1 + p_2)/2$ and let $E = \|f - p_1\| = \|f - p_2\|$. Then

$$E \leqslant \|f - p_0\| = \left\| \frac{f - p_1}{2} + \frac{(f - p_2)}{2} \right\|$$

$$\leqslant \frac{\|f - p_1\|}{2} + \frac{\|f - p_2\|}{2} = E.$$

Thus p_0 is also a best approximation to f. Let $x_0 < x_1 < \ldots < x_{n+1}$ be points satisfying the equations

$$f(x_j) - p_0(x_j) = (-1)^j \sigma E, \qquad j = 0, 1, \ldots, n+1.$$

Then for $j = 0, 1, \ldots, n+1$,

$$E = |f(x_j) - p_0(x_j)| = \left| \frac{f(x_j) - p_1(x_j)}{2} + \frac{f(x_j) - p_2(x_j)}{2} \right|.$$

But $|f(x_j) - p_1(x_j)|$ and $|f(x_j) - p_2(x_j)|$ are both $\leqslant E$. It follows that $f(x_j) - p_1(x_j) = f(x_j) - p_2(x_j) = (-1)^j \sigma E$. Thus $p_1(x_j) = p_2(x_j)$, $j = 0, 1, \ldots, n+1$. This implies that $p_1 - p_2$ has $n + 2$ zeros and hence $p_1 = p_2$.

COROLLARY 4.4. *If $f \in C[a,b]$, then its best polynomial is a Lagrange interpolation polynomial for f at $n + 1$ points in $[a,b]$.*

Proof. Let p be the best polynomial. From Theorem 4.2 $f - p$ is either identically zero or it alternates sign at $n + 2$ points in $[a,b]$. In either case, it follows that $f - p$ has at least $n + 1$ zeros on $[a,b]$.

EXAMPLE 4.3. Let

$$[a,b] = [-1,1],$$

$$f(x) = 4x^3 + 2x^2 + x + 1,$$

$$p(x) = 2x^2 + 4x + 1,$$

$$F(x) = f(x) - p(x) = 4x^3 - 3x.$$

Now $F'(x) = 12x^2 - 3 = 3(4x^2 - 1)$. Thus F' is zero at $\pm$ 0.5 and

$$\|F\| = \max[|F(-1)|, |F(-0.5)|, |F(0.5)|, |F(1)|] = 1.$$

Then note that -1, -0.5, 0.5, and 1 form an oscillation set of four points. Thus p is the best polynomial for f from $\mathcal{P}_2$. (For the moment don't worry about how we came up with p—see Problems 4.4 and 4.7.)

The fact that the best polynomial is an interpolation polynomial does not help us compute the best polynomial because in general we don't know the values of the interpolation points. Explicit formulas for the best polynomial or its error curve have been discovered for only a few functions [1, 6]. In general, it is necessary to compute the best polynomial by a successive approximation technique such as the algorithm discussed in the next section.

PROBLEMS

4.1. Let $\varphi(c) = \|f - p\|$, where p is the polynomial in $\mathcal{P}_n$ whose coefficient vector is c. Show that φ is continuous.

4.2. Let $f(x) = x^2$. Find the best polynomial approximation to f (in the Chebyshev sense) from
(a) $\mathcal{P}_0$ (b) $\mathcal{P}_1$

4.3. Let $f(x) = 8x^4$ and $p(x) = 8x^2 - 1$. Show that p is the best approximation to f from $\mathcal{P}_3$ on the interval $[-1,1]$.

4.4. Let $B_n(f)$ represent the best approximation to f from $\mathcal{P}_n$. Show that if $q \in \mathcal{P}_n$, then

$$B_n(f + q) = B_n(f) + B_n(q). \tag{4.7}$$

Give an example to show that equation (4.7) does not always hold when $q \notin \mathcal{P}_n$. Then show that for any real number α,

$$B_n(\alpha f) = \alpha B_n(f).$$

4.5. Let $f \in C[-b,b]$, where $b > 0$. Show that if f is an odd (even) function, then its best polynomial is also odd (even).

4.6. Find a formula for the best approximation to a function f from $\mathcal{P}_0$.

4.7. Let $f(x) = x^{n+1}$ on the interval $[-1,1]$. Show that the best approximation to f from $\mathcal{P}_n$ is given by $p = f - 2^{-n}T_{n+1}$, where T_{n+1} is the $(n+1)$st degree Chebyshev polynomial (see Chapter 5, Section 4).

4.8. Let $f \in C[a,b]$ and suppose $f^{(n+1)} > 0$ on (a,b). Show that the oscillation set for the error curve of the best approximation is unique. Furthermore, a and b are in this oscillation set.

5. THE REMEZ ALGORITHM

The Remez algorithm is a successive approximation technique for computing the coefficients of the polynomial in $\mathcal{P}_n$ which provides the best Chebyshev approximation to a given continuous function f. The general approach is to find the best approximation to f on a set of $n+2$ points, and then vary the set of $n+2$ points in such a manner that the successive best approximations on the finite sets approach the best approximation over the whole interval. Before describing the algorithm, let us establish some results concerning the best approximation on a set of $n+2$ points.

THEOREM 5.1. *Let $x_0 < x_1 < \ldots < x_{n+1}$ be $n+2$ distinct numbers. Then the system of equations*

$$(5.1) \qquad f(x_k) - \sum_{j=0}^{n} a_j x_k^j = (-1)^k \lambda, \qquad k = 0, 1, \ldots, n+1$$

has a unique solution in $a_0, a_1, \ldots, a_n, \lambda$.

Proof. The determinant of the system (5.1) is given by

$$(5.2) \qquad D = \begin{vmatrix} 1 & x_0 & x_0^2 & \cdots & x_0^n & 1 \\ 1 & x_1 & x_1^2 & \cdots & x_1^n & -1 \\ \vdots & & & & & \\ 1 & x_{n+1} & x_{n+1}^2 & \cdots & x_{n+1}^n & (-1)^{n+1} \end{vmatrix}.$$

Let $v(x_0, x_1, \ldots, x_n)$ denote the Vandermonde determinant obtained by crossing out the last row and column in (5.2). It was shown in Section 2 of Chapter 5 that

$$(5.3) \qquad v(x_0, x_1, \ldots, x_n) = \prod_{i=1}^{n} \prod_{k=0}^{i-1} (x_i - x_k).$$

It follows from (5.3) that $v(x_0, x_1, \ldots, x_n) > 0$ if $x_0 < x_1 < \ldots < x_n$. Now expanding the determinant in (5.2) about the last column, we obtain

$$(5.4) \qquad \begin{aligned} D = (-1)^{n+1}[v(x_1, x_2, \ldots, x_{n+1}) + v(x_0, x_2, \ldots, x_{n+1}) \\ + \ldots + v(x_0, x_1, \ldots, x_n)]. \end{aligned}$$

But each of the Vandermonde determinants in (5.4) is positive. It follows that $D \neq 0$ and hence the system (5.1) has a unique solution.

As another preliminary result, we will prove a lemma concerning functions which weakly alternate sign at a number of points. In this lemma, each zero of f which coincides with a zero of f' will be counted twice.

LEMMA 5.2. *Suppose* $f \in C[a,b]$, f' *exists on* (a,b), *and* $a_1 < a_2 < \ldots < a_m$ *are points of* $[a,b]$. *If* $(-1)^k f(a_k) \geqslant 0$, $1 \leqslant k \leqslant m$, *then* f *has at least* $m-1$ *zeros on* $[a_1,a_m]$.

Proof. Suppose the contrary. Let j be the first index such that there are fewer than $j-1$ zeros on $[a_1,a_j]$. Note that $j > 2$, $f(a_j) \neq 0$, and $f(a_{j-1}) = 0$. Let l be the largest integer $< j-1$ such that $f(a_l) \neq 0$. Now there are at least $l-1$ zeros on $[a_1,a_l]$, and exactly $j-1-l$ zeros on $(a_l,a_{j-1}]$, namely the simple zeros $a_{l+1}, a_{l+2}, \ldots, a_{j-1}$. Since f' exists on (a_l,a_j), we can factor out the zeros in this interval to obtain

(5.5) $$f(x) = (x - a_{l+1}) \ldots (x - a_{j-1}) g(x),$$

where g is a continuous function which does not vanish on $[a_l, a_j]$. The hypotheses imply that

$$\operatorname{sgn} f(a_j) = (-1)^j,$$

where sgn denotes the signum function defined by

$$\operatorname{sgn} z = \begin{cases} 1 & \text{if } z > 0, \\ 0 & \text{if } z = 0, \\ -1 & \text{if } z < 0. \end{cases}$$

But (5.5) implies that

$$\begin{aligned} \operatorname{sgn} f(a_j) &= \operatorname{sgn} g(a_j) = \operatorname{sgn} g(a_l) \\ &= (-1)^{j-1-l} \operatorname{sgn} f(a_l) = (-1)^{j-1}. \end{aligned}$$

This contradiction establishes the lemma.

For a given set $T = \{x_0, x_1, \ldots, x_{n+1}\}$ of $n + 2$ distinct numbers, let us define a norm by the equation

$$\|f\|_T = \max_{x \in T} |f(x)|.$$

We say that $p \in \mathcal{P}_n$ is a best approximation to f on the set T iff

$$\|f - p\|_T \leqslant \|f - q\|_T$$

for all $q \in \mathcal{P}_n$. The next theorem shows that the best polynomial for f on T can be found by solving the system (5.1).

THEOREM 5.3. *Suppose $T = \{x_0, x_1, \ldots, x_{n+1}\}$ is a set of $n + 2$ points indexed so that $x_0 < x_1 < \ldots < x_{n+1}$, and f is a function defined on T. Let $a_0, a_1, \ldots, a_n, \lambda$ be the unique solution to (5.1) and define p by $p(x) = \sum_{j=0}^{n} a_j x^j$. Then p is the unique best approximation to f on the set T.*

Proof. Suppose $q \in \mathcal{P}_n$ provides at least as good an approximation to f as p; that is,

$$\|f - q\|_T \leqslant \|f - p\|_T.$$

Clearly, $\|f - p\|_T = |\lambda|$. Suppose, for definiteness, that $\lambda \geqslant 0$. Then we have $\lambda = \|f - p\|$, and from (5.1)

$$f(x_j) - p(x_j) = (-1)^j \lambda.$$

Then

$$f(x_0) - q(x_0) \leqslant \|f - q\|_T \leqslant \|f - p\|_T = f(x_0) - p(x_0).$$

Hence $p(x_0) - q(x_0) \leqslant 0$. Similarly,

$$f(x_1) - q(x_1) \geqslant -\|f-q\|_T \geqslant -\|f-p\|_T = f(x_1) - p(x_1).$$

Thus $p(x_1) - q(x_1) \geqslant 0$. Continuing in this fashion, we see that

$$(-1)^j[q(x_j) - p(x_j)] \geqslant 0, \qquad j = 0, 1, \ldots, n+1.$$

Then Lemma 5.2 implies that $q - p$ has at least $n + 1$ zeros on $[x_0, x_{n+1}]$. But p and $q \in \mathcal{P}_n$; hence $p = q$ and the proof is complete.

We will also need a result known as *Vallee Poussin's theorem.*

THEOREM 5.4. *Let f be continuous on $[a,b]$, and let $T = \{x_0, \ldots, x_{n+1}\}$ be a set of $n + 2$ distinct points in $[a,b]$ indexed so that $x_0 < x_1 < \ldots < x_{n+1}$. Suppose $p \in \mathcal{P}_n$ and $f - p$ alternates sign at the points $x_0, x_1, \ldots, x_{n+1}$. Then*

$$E_n(f) \geqslant E_{n,T}(f) \geqslant \min_{0 \leqslant j \leqslant n+1} |f(x_j) - p(x_j)|,$$

where

$$E_n(f) = \inf_{q \in \mathcal{P}_n} \|f - q\|$$

and

$$E_{n,T}(f) = \inf_{q \in \mathcal{P}_n} \|f - q\|_T.$$

The proof of Theorem 5.4 is left as an exercise (Problem 5.1).

The Remez algorithm proceeds as follows: Choose a set $T_0 = \{x_0^0, x_1^0, \ldots, x_{n+1}^0\}$ of $n + 2$ points in $[a,b]$. Solve the system of equations

$$f(x_k^0) - \sum_{j=0}^{n} a_j^0 (x_k^0)^j = (-1)^k \lambda_0, \qquad k = 0, 1, \ldots, n+1$$

for $a_0^0, a_1^0, \ldots, a_n^0$, and λ_0. (If $\lambda_0 = 0$, choose a new set of points for T_0.) Now proceed inductively. Having chosen a set of points $T_m = \{x_0^m, x_1^m, \ldots, x_{n+1}^m\}$, let p_m be the best approximation to f on T_m and let $\lambda_m = f(x_0^m) - p_m(x_0^m)$. Then choose a new set $T_{m+1} = \{x_0^{m+1}, x_1^{m+1}, \ldots, x_{n+1}^{m+1}\}$ so that the following conditions are satisfied:

(i) $x_0^{m+1} < x_1^{m+1} < \ldots < x_{n+1}^{m+1}$,

(ii) $|f(x_k^{m+1}) - p_m(x_k^{m+1})| = \|f - p_m\|$ for some k,

(iii) $f - p_m$ alternates sign at the points of T_{m+1}, and

$$|f(x_k^{m+1}) - p_m(x_k^{m+1})| \geqslant |\lambda_m|, \qquad k = 0, 1, \ldots, n+1.$$

We prefer to pick points corresponding to the relative maxima and minima of $f - p_m$, but this is not essential. If $|\lambda_m| = \|f - p_m\|$, then p_m is the best approximation and the process is terminated.

It is shown in Cheney [2] that the sequence of polynomials produced by the Remez algorithm converges to the best approximation. A stopping rule for the process can be based on the following inequality.

THEOREM 5.5. *Let λ_m and p_m, $m = 0, 1, 2, \ldots$, be determined by the Remez algorithm, and let*

$$E_n(f) = \inf_{p \in \mathcal{P}_n} \|f - p\|.$$

Then

$$|\lambda_m| \leqslant E_n(f) \leqslant \|f - p_m\|. \tag{5.6}$$

The proof of this inequality is saved for Problem 5.2.

From (5.6) we see that

$$0 \leqslant \|f - p_m\| - E_n(f) \leqslant \|f - p_m\| - |\lambda_m|. \tag{5.7}$$

A reasonable stopping rule is to terminate the process when

$$\|f - p_m\| - |\lambda_m| < \epsilon, \tag{5.8}$$

where ϵ is a preassigned tolerance. Then (5.7) shows that the norm of the error curve cannot be improved by more than ϵ, even if the best polynomial is used.

Ralston and Wilf [5] have published flow diagrams for the Remez algorithm. An ALGOL 60 program can be found in Cody, Fraser, and Hart [3]. Although these flow diagrams and programs are designed for rational functions, they can be adapted to polynomial approximation by requiring that the denominator of the rational function have degree zero.

PROBLEMS

5.1. Prove Theorem 5.4.

5.2. Prove Theorem 5.5.

5.3. Draw a flow chart of the major steps in the Remez algorithm. Assume that subroutines are available for solving linear systems of equations and finding zeros of functions.

6. RATIONAL APPROXIMATION

Let $f \in C[a,b]$ and let $\mathcal{R}(m,n)$ be the set of all rational functions $r = p/q$ such that $p \in \mathcal{P}_m$ and $q \in \mathcal{P}_n$. The problem here is to find the best approximation to f from $\mathcal{R}(m,n)$ relative to the uniform norm on $[a,b]$. This will be called the *rational Chebyshev approximation problem.*

The basic theory of rational approximation is similar to the polynomial theory. However, the rational theory is more difficult, largely because the problem is non-linear (see Problem 6.1). As a consequence, we will primarily state results in this section, leaving the easy proofs for exercises and the hard proofs for more advanced courses. Let us first state an existence theorem.

THEOREM 6.1. *Let $f \in C[a,b]$. Then the rational Chebyshev problem has a solution.*

A proof of this theorem can be found in Cheney [2].

An important result, known as Vallee Poussin's theorem, gives a means for obtaining a lower bound on the best approximation.

THEOREM 6.2. *Suppose $f \in C[a,b]$, $\rho = \inf_{s \in \mathcal{R}(m,n)} \|f - s\|$, and $r = p/q \in \mathcal{R}(m,n)$, where p has degree $m - \mu$, $0 \leqslant \mu \leqslant m$, q has degree $n - \nu$, $0 \leqslant \nu \leqslant n$, and q has no zeros on $[a,b]$. Let $d = \min(\mu,\nu)$ and $N = m + n + 2 - d$. If $f - r$ alternates sign at N consecutive points of $[a,b]$, say $x_1 < x_2 < \ldots < x_N$, then*

$$\rho \geqslant \min_{1 \leqslant j \leqslant N} |f(x_j) - r(x_j)|.$$

Furthermore, if f alternates sign at the $m + 2$ consecutive points $x_1 < x_2 < \ldots < x_{m+2}$, then

$$\rho \geqslant \min_{1 \leqslant j \leqslant m+2} |f(x_j)|.$$

The proof of this theorem is left for an exercise (Problem 6.2).

As in the polynomial case, the best approximation is characterized by the fact that the error curve must have a certain oscillatory property.

THEOREM 6.3. *Let $f \in C[a,b]$.*

(i) *The zero function $\emptyset$ is a best approximation to f from $\mathcal{R}(m,n)$ if and only if f has an oscillation set of at least $m + 2$ points on $[a,b]$.*

(ii) *Let* $r = p/q \in \mathcal{R}(m,n)$, *where* p *has degree* $m - \mu$, $0 \leqslant \mu \leqslant m$, q *has degree* $n - \nu$, $0 \leqslant \nu \leqslant n$, p *and* q *have no common factors, and* q *has no zeros on* $[a,b]$. *Let* $d = \min(\mu,\nu)$. *Then* r *is a best approximation to* f *from* $\mathcal{R}(m,n)$ *if and only if* $f - r$ *has an oscillation set of at least* $m + n + 2 - d$ *points on* $[a,b]$.

A proof of this theorem can be found in Achieser [1]. The "if" part follows readily from Vallee Poussin's theorem (see Problem 6.4), while the "only if" part is somewhat more difficult.

COROLLARY 6.4. *Let* $f \in C[a,b]$. *The best approximation to* f *from* $\mathcal{R}(m,n)$ *is unique.*

Proof. Let $r_1 = p_1/q_1$ and $r_2 = p_2/q_2$ be best approximations. The case where r_1 or $r_2 = \emptyset$ is left as an exercise (Problem 6.5). We will use the following notation:

$$\mu_j = m - \deg(p_j), \qquad \nu_j = n - \deg(q_j), \qquad j = 1, 2,$$

$$d_j = \min(\mu_j,\nu_j), \qquad j = 1, 2,$$

$$N = n + m + 2 - d_1,$$

$$\Delta = r_1 - r_2,$$

$$\varphi = p_1 q_2 - p_2 q_1,$$

$$\rho = \|f - r_1\| = \|f - r_2\|,$$

$$x_1, \ldots, x_N \qquad \text{oscillation points of } f - r_1.$$

Now $f(x_1) - r_1(x_1)$ is either ρ or $-\rho$; suppose for definiteness that it is ρ. Note that $\Delta = r_1 - f + f - r_2$. Then

$$\Delta(x_1) = -\rho + f(x_1) - r_2(x_1) \leqslant -\rho + \rho = 0,$$

$$\Delta(x_2) = \rho + f(x_2) - r_2(x_2) \geqslant \rho - \rho = 0.$$

Continuing in this fashion, we see that

$$(6.1) \qquad (-1)^j \Delta(x_j) \geqslant 0, \qquad j = 1, 2, \ldots, N.$$

Note that $\Delta = \varphi/(q_1 q_2)$. But we may assume that $q_1 q_2$ has no zeros on $[a,b]$ and hence has constant sign there. From (6.1) we have either

$$(-1)^j \varphi(x_j) \geqslant 0, \qquad j = 1, 2, \ldots, N$$

or

$$(-1)^j \varphi(x_j) \leqslant 0, \qquad j = 1, 2, \ldots, N.$$

It follows from Lemma 5.2 that φ has at least $N - 1$ zeros on $[a,b]$.

But

$$\begin{aligned}\deg(\varphi) &= \max(\deg(p_1q_2), \deg(p_2q_1)) \\ &\leqslant \max(m - \mu_1 + n, m + n - \nu_1) \\ &= m + n - d_1 = N - 2.\end{aligned}$$

Thus $\varphi = \emptyset$, $p_1q_2 = p_2q_1$, and $r_1 = r_2$.

The Remez algorithm can be modified in order to compute best rational approximations. Flow charts for this algorithm can be found in Ralston and Wilf [5], and an ALGOL 60 program has been published by Cody, Fraser, and Hart [3]. Some other techniques are described in Cheney [2].

PROBLEMS

6.1. Show that $\mathcal{R}(m,n)$ is not the span of any set.

6.2. Prove Theorem 6.2.

6.3. Prove part (i) of Theorem 6.3.

6.4. Prove the "if" part of part (ii) of Theorem 6.3. Is the assumption that p and q have no common factors needed here?

6.5. Prove Corollary 6.4 for the case where r_1 or $r_2 = \emptyset$.

REFERENCES

1. N. I. Achieser, *Theory of Approximation*, Ungar, 1956.
2. E. W. Cheney, *Introduction to Approximation Theory*, McGraw-Hill, 1966.
3. W. J. Cody, W. Fraser, and J. F. Hart, "Rational Chebyshev approximation using linear equations," *Numerische Mathematik*, 12 (1968), 242–251.
4. P. J. Davis, *Interpolation and Approximation*, Blaisdell, 1963.
5. A. Ralston and H. S. Wilf, *Mathematical Methods for Digital Computers*, Vol. II, Wiley, 1967.
6. T. J. Rivlin, "Polynomials of best uniform approximation to certain rational functions," *Numerische Mathematik*, 4 (1962), 345–349.
7. T. J. Rivlin, *An Introduction to the Approximation of Functions*, Blaisdell, 1969.
8. G. P. Tolstov, *Fourier Series*, Prentice-Hall, 1962.
9. A. Zygmund, *Trigonometrical Series*, Dover, 1955.

Answers to Selected Problems

CHAPTER 1

3.1. 1.5125

3.2. (a) $(\epsilon_1 - \epsilon_2 b \cot \phi) \csc \phi$
(b) Absolute error: better to measure c
Relative error: can be quite larger in either case

4.1. (a) Holds, (b) Does not hold, (c) Does not hold

4.5. (a) 0.02, relative error 33%
(b) 0.0300, relative error 0.073%

CHAPTER 2

1.2. (a) (6, 8, 8, 15), (b) (2, 6, 2, 3), (c) (32, 56, 40, 72),
(d) (40, 45, 55, 105)

1.4. Yes, every real number is a complex number.

1.5. (a) $x = (-1, -2, -3, -6, -2)$

2.1. (i) (a) 0, (c) 25 (e) z^2, (f) $(x + z)^2$, (g) $x^2 + z^2$
(ii) Yes
(iii) No
(iv) No, because the answer to (iii) is no.

2.2. (i) (a) (10, 5, 21)
(c) $5z + (0, 0, 1)$
(d) $5(x + z) + (0, 0, 1)$
(e) $5(x + z) + (0, 0, 2)$
(ii) No

2.3. (a), (b), (d), and (f) are linear.

2.4. (a), (c), (d), (e), (f), (g)

3.1. 3

4.1. (1, 0, 0, 0), (0, 1, 0, 0), (0, 0, 1, 0) Independent
(1, 1, 1, 1), (2, 0, 1, 3), (3, 1, 2, 4) Dependent

5.4. (a) $x_n = c_1 + c_2 \cdot 4^n - 2^{n-1}$
(b) $x_n = c_1 i^n + c_2(-i)^n + \frac{5}{2} n - \frac{5}{2}$

6.1. $\|f - g\| = \frac{1}{4}$, $\|f - h\| = \frac{1}{\sqrt{2}}$; thus g is better.

6.3. $\sqrt{\frac{e^2}{2} - 2e + \frac{11}{6}}$

7.1. (a) $\frac{1}{2}$, (b) $\sqrt{\frac{8}{15}}$

8.1. $a_o = \frac{1}{2\pi}\int_{-\pi}^{\pi} f(t)dt, \quad a_j = \frac{1}{\pi}\int_{-\pi}^{\pi} f(t)\cos jt\,dt,$

$b_j = \frac{1}{\pi}\int_{-\pi}^{\pi} f(t)\sin jt \sin jt\,dt$

8.3. $\alpha_o = \frac{\pi^2}{3}\sqrt{2\pi}, \quad \alpha_1 = \alpha_{-1} = -2\sqrt{2\pi}$

Norm: $\sqrt{\frac{8}{45}\pi^5 - 16\pi}$

8.5. Bound: $\frac{5}{16}\pi$

$\alpha_o = 0, \quad \alpha_1 = \frac{3}{4}, \quad \alpha_2 = 0$

8.7. $\|T_o\| = \sqrt{\pi}, \quad \|T_j\| = \sqrt{\frac{\pi}{2}}$ for $j > 0$

CHAPTER 3

1.1. $A' = \begin{bmatrix} 1 & 4-i & 4 \\ 2+i & 2i & 5i \\ 3 & 6 & 6+3i \end{bmatrix} \quad \bar{A} = \begin{bmatrix} 1 & 2-i & 3 \\ 4+i & -2i & 6 \\ 4 & -5i & 6-3i \end{bmatrix}$

$A^* = \begin{bmatrix} 1 & 4+i & 4 \\ 2-i & -2i & -5i \\ 3 & 6 & 6-3i \end{bmatrix}$

2.1. (a) $\begin{bmatrix} 2 & 4 & 3 \\ 4 & 7 & 11 \\ 1 & 0 & 2 \end{bmatrix}$ (c) $\begin{bmatrix} -2 & 1 & 2 \\ 2 & -2 & 3 \\ 1 & 0 & -2 \end{bmatrix}$

(e) $\begin{bmatrix} 8 & 4 & 12 \\ 22 & 8 & 30 \\ 1 & 0 & 1 \end{bmatrix}$

2.4. (a) $\begin{bmatrix} 21 \\ 5 \\ 12 \end{bmatrix}$ (c) $\begin{bmatrix} x_1 + 2x_3 + z_1 + 2z_3 \\ x_2 + z_2 \\ 2x_1 + x_3 + 2z_1 + z_3 \end{bmatrix}$

2.9. (a) $X = A^{-1}(C - B)$
(c) Solution not guaranteed by the assumptions.
$X = (B^2 - A)^{-1} C$, assuming $B^2 - A$ has an inverse.

3.1. 10

3.3. (a) 0.000590 sec, (c) 0.016415 sec

5.1. (a) (1) (0.00, 1.00), (2) (1.00, 1.00)
(b) (1) (0.0, 1.0, 0.0) (2) (1.0, 1.0, −2.0)

6.1. Divisions: $\frac{n}{2}(n^3 - 3n^2 + 5n - 1)$

Multiplications:

$$\frac{n^2[(n-1)(n-2)(2n-3) + 6(n-2)] + n(n-1)(2n-1) + 6(n-1)}{6}$$

Additions: $\frac{n^2(n-1)(n-2)(2n-3) + n(n-1)(2n-1)}{6}$

These results assume that the determinants were computed by the triangular method. Cramer's rule requires roughly $n^5/3$ multiplications as compared to roughly n^3 for the Jordan or Gauss method.

7.3. $p(\lambda) = -\lambda^3 + 9\lambda^2 - 21\lambda + 13$

8.2. (a) $\phi = 22^\circ 30'$

$$A_1 = \begin{bmatrix} 5.41420 & 0.00001 & 0.09239 \\ 0.00001 & 2.58578 & -0.03827 \\ 0.09239 & -0.03827 & 1 \end{bmatrix}$$

(c) $|\lambda_1 - 5.41420| \leq 0.09240$
$|\lambda_2 - 2.58578| \leq 0.03828$
$|\lambda_3 - 1| \leq 0.13066$

(d) $\phi = 1^\circ 12'$

(e) $$A_2 = \begin{matrix} 5.41612 & -0.00079 & -0.00010 \\ -0.00079 & 2.58578 & -0.03826 \\ -0.00010 & -0.03826 & 0.99807 \end{matrix}$$

Eigenvalues to eight decimals computed from the characteristic polynomial of A: 5.4161 4658, 2.5867 0765, 0.9971 4576.

9.1.

$$P = \begin{bmatrix} 1 & 0 & 0 & 0 \\ 0 & 1 & 0 & 0 \\ 0 & 0 & -\frac{1}{\sqrt{5}} & -\frac{2}{\sqrt{5}} \\ 0 & 0 & -\frac{2}{\sqrt{5}} & \frac{1}{\sqrt{5}} \end{bmatrix} \qquad P'AP = \begin{bmatrix} 3 & 5 & 0 & 0 \\ 5 & 6 & -\sqrt{5} & 0 \\ 0 & -\sqrt{5} & 7.2 & 1.4 \\ 0 & 0 & 1.4 & 1.8 \end{bmatrix}$$

9.4. $p(\lambda) = \lambda^4 - 16\lambda^3 + 30\lambda^2 + 144\lambda - 111$

9.6. Eigenvalues in $(-3, -2)$, $(0, 1)$, $(4, 5)$, and $(12, 13)$. Largest eigenvalue to nearest tenth is 12.8.

10.2. (a) Maximum profit of \$1800 per week with 300 colonial and 200 Danish chairs.
(b) Maximum profit of \$2800 per week with 700 colonial, no Danish, no new carpenters, and 7.5 new lathes. This solution, which is physically impossible to implement, illustrates the need for integer programming—a technique for finding integer solutions to linear programs.

10.3. (a) No solution
(b) $T(1, 0) = T(\frac{3}{2}, \frac{1}{2}) = 1$ is the minimum value.

CHAPTER 4

1.1. $x_5 = 0.984\ 375$

2.2. (a) 0.1001 0030, (c) Divergent, (d) 0.99, (e) 0.999

3.1. (a) 2.205 569

3.2. 1.746

3.7. (b) −3.63119, 2.12910, 7.50209

4.1. Intersection point correct to six decimals: (0.916875,1.681321)

5.1. $(u_0, v_0) = (4, \frac{25}{3})$ or $(4, 8)$
$(u_1, v_1) = (4.03, 8.24)$
$(u_2, v_2) = (4.03408, 8.23972)$

6.1. (a) $y_3(x) = 1 + 2x^2 + 2x^4 + \frac{4}{3}x^6$

$$|y(0.2) - y_3(0.2)| \leq \frac{1.6e^{0.16}(0.8)^3(0.2)^4}{4!}$$

$$R = \{(x, y)\colon -0.2 \leq x \leq 0.2, 0 \leq y \leq 2\}$$

(b) $y_4(x) = \frac{x^2}{2} + \frac{x^3}{6} + \frac{x^4}{24} + \frac{x^5}{120}$

$|y(0.2) - y_4(0.2)| \leq \frac{2e^{0.2}(0.2)^5}{5!}$

$R = \{(x, y): \ |x| \leq 1, \quad |y| \leq 1\}$

7.1. (a) $y_2(x) = \cos x + \lambda\,(\frac{\pi}{4}\sin x - \frac{1}{2}\cos x) + \frac{\lambda^2}{4}(1 - \frac{\pi^2}{4})\cos x$

(b) $|\lambda| < \frac{2}{\pi}$

(c) $y(\frac{\pi}{4}) \doteq y_2(\frac{\pi}{4}) = 0.724\ 693$

Bound: $\frac{(0.05\pi)^3}{1-0.05\pi}$

(d) $\frac{2}{\pi}$

(e) $\lambda = \pm i\sqrt{\frac{\pi^2}{16} - \frac{1}{4}} = \pm\ 0.60568i;\ \frac{2}{\pi} = 0.63662$

8.1. (a)

$$X_2 = \begin{bmatrix} .250525 & -.0050 & -.00832 \\ -.0050 & .2001 & .00016 \\ -.00832 & .00016 & .166912 \end{bmatrix}$$

(b) $\|E\| \leq 0.0785$

$\|B\| = 0.25$

$\|A^{-1} - X_2\| \leq \frac{(0.25)(0.0785)^3}{0.9215} \leq 1.32 \cdot 10^{-4}$

CHAPTER 5

1.1. (a) $B^{-1} = \begin{bmatrix} 0 & 1 & 0 \\ -\frac{1}{2} & 0 & \frac{1}{2} \\ \frac{1}{2} & -1 & \frac{1}{2} \end{bmatrix}$

(b) $p_0(x) = -\frac{1}{2}x + \frac{1}{2}x^2$

$p_1(x) = 1 - x^2$

$p_2(x) = \frac{1}{2}x + \frac{1}{2}x^2$

(d) $p(x) = \frac{-5}{2}x^2 + \frac{x}{2} + 7$

2.3. $p(x) = \frac{(x-0.5)(x-1)}{0.5} - \frac{1.6487x(x-1)}{0.25} + \frac{2.7183x(x-0.5)}{0.5}$

2.7. (iii) $30\ \mu$ sec

3.1. (a) $3.4 \cdot 10^{-5}$

(b) $8.4 \cdot 10^{-5}$

(c) $1.34 \cdot 10^{-4}$
(This bound depends on the order in which the operations are performed.)

3.3. (a) 6, (b) $\frac{3}{32}$

3.4. 0.0425

4.1. (a) 2.03407, 2.29289, 2.74118
3.25882, 3.70711, 3.96593

(b) $\frac{1}{3 \cdot 2^{12}}$

(c) 11

5.3. (a) 190, (b) 141.3125, (c) 1, (d) -0.0390625

6.1. $e^{0.2} \doteq 1.2214\ 0000$, error bound $7.47 \cdot 10^{-6}$
(obtained by replacing e^{ξ} by 2.8)

6.3. (a) $p(x) = 3 + \frac{x}{6} - \frac{x^2}{216} + \frac{x^3}{3888}$

(b) $p(.01) \doteq 3.0166\ 2063$

(c) $|E| \leq \frac{5 \cdot 10^{-4}}{128 \cdot 2187} < 1.8 \cdot 10^{-9}$

(In (b) we rounded to eight decimals; thus round-off error dominates.)

7.1. $\frac{x^3 + 21x^2 + 30x}{9x^2 + 36x + 30}$

7.2. $p(0.01) = 1.1046\ 2210$ $r(0.01) = 1.1046\ 2211$
$p(0.02) = 1.2189\ 9360$ $r(0.02) = 1.2189\ 9392$
$p(0.05) = 1.6288\ 1250$ $r(0.05) = 1.6288\ 3435$

8.3. (a) $p(x) = 3084x^3 - 2071x^2 + 10x + 1$
(b) $p(0.02) = 0.396\ 272$
(c) 5.163

CHAPTER 6

1.1. Lagrange: $\int_{-0.1}^{0.1} f(x)dx \doteq \frac{0.1}{3}[f(-0.1) + 4f(0) + f(0.1)]$

Taylor: $\int_{-0.1}^{0.1} f(x)dx \doteq 0.2f(0) + \frac{0.001}{3} f''(0)$

Lagrange: $\int_{-0.1}^{0.1} e^x dx \doteq 0.2003$

Taylor: $\int_{-0.1}^{0.1} e^x dx \doteq 0.2003$

2.1. (b) 0.0999 99
Error bound: $4.63 \cdot 10^{-12}$

2.4. (a) $n = 1$, (c) $n = 99$,
(e) The first integer exceeding $\frac{1}{\epsilon} - 2$

3.2. 0.1061 6028
Error bound (from 3.8): $\frac{3^3}{4^8 \cdot 5} \doteq 8.24 \cdot 10^{-5}$

3.4. 1.0987 2534
Error bound (3.8): $1.04 \cdot 10^{-3}$
(3.7): $3.06 \cdot 10^{-4}$
Error estimate (3.9): $-1.286 \cdot 10^{-4}$

3.5. (a) $h = 0.1$, (b) Choose $h < 1.11 \sqrt[4]{\epsilon}$

4.1. Gauss: $n = 3$, 1.098 040; $n = 4$, 1.098 570
Newton-Cotes: $n = 2$, 1.111 111; $n = 3$, 1.104 762
Exact value: $\ln 3 \doteq 1.098\ 612$
Bound for $n = 4$: 0.0116

6.1. (a) 0.33345, $1.11 \cdot 10^{-4} < |E| < 1.37 \cdot 10^{-4}$
(b) 0.333 325 (5-place table)
$2.38 \cdot 10^{-7} < |E| < 4.65 \cdot 10^{-7}$
0.3333 3300 2805 (12-place table)

6.3. $\frac{5 \cdot 10^{-(m+1)}}{h}$

7.1. $y_5 = 6.4416$
$y(1) = 5e - 6 \doteq 7.5914$

7.3. $y_5 = 7.5135$

8.1. $h = 0.1,\ 2.6639\ 9932$
$h = 0.2,\ 2.6639\ 9081$
$E \doteq 5.67 \cdot 10^{-7}$ (rule of thumb)
$E = 6.8 \cdot 10^{-7}$ (actual error)

9.4. $y_4^* = 1.1372, y_4 = 1.1376$

10.1. $h = 0.4, y_1 = 0.6970\ 6667$
$h = 0.2, y_2 = 0.6967\ 8337$
Error estimate: $-1.888 \cdot 10^{-5}$
Actual error: $-7.666 \cdot 10^{-5}$

12.1. $\lambda_1 \doteq 36, \lambda_2 \doteq 108$
$\lambda_1 = (2\pi)^2 \doteq 39.5, \lambda_2 = (4\pi)^2 \doteq 157.9$
Approximate eigenfunction for $\lambda_1; y_0 = 0, y_1 = y_2, y_3 = 0$
(y_2 can be chosen arbitrarily).
Actual eigenfunction: $y = c \sin 2\pi x$
(c can be chosen arbitrarily).

12.2. Cubic equation: $1 - \dfrac{\lambda}{12} + \dfrac{\lambda^2}{42 \cdot 12} - \dfrac{\lambda^3}{90 \cdot 42 \cdot 12} = 0$
Solution: 18.43298

CHAPTER 7

2.6. $\sqrt{\dfrac{\pi}{2^n}}$

3.2. True for all θ

3.4. $s_2(x) = \dfrac{5}{8}x$

$s_3(x) = \dfrac{5}{4}x^3 - \dfrac{5}{16}x$

4.2. (a) $p(x) = \dfrac{1}{2}$ for all x

(b) $p(x) = \dfrac{1}{2}$ for all x

Symbols and Notations

Index

1 2 3 4 5 6 7 8 9 10 -KP- 80 79 78 77 76 75 74